Solid State Batteries: From Materials Research to Design and Applications

Solid State Batteries: From Materials Research to Design and Applications

Editors

Chunwen Sun
Siqi Shi
Yongjie Zhao

Basel • Beijing • Wuhan • Barcelona • Belgrade • Novi Sad • Cluj • Manchester

Editors

Chunwen Sun
China University of Mining &
Technology (Beijing)
Beijing
China

Siqi Shi
Shanghai University
Shanghai
China

Yongjie Zhao
Beijing Institute of
Technology
Beijing
China

Editorial Office
MDPI AG
Grosspeteranlage 5
4052 Basel, Switzerland

This is a reprint of articles from the Special Issue published online in the open access journal *Batteries* (ISSN 2313-0105) (available at: https://www.mdpi.com/journal/batteries/special_issues/012W4AR1M0).

For citation purposes, cite each article independently as indicated on the article page online and as indicated below:

Lastname, A.A.; Lastname, B.B. Article Title. *Journal Name* **Year**, *Volume Number*, Page Range.

ISBN 978-3-7258-2105-1 (Hbk)
ISBN 978-3-7258-2106-8 (PDF)
doi.org/10.3390/books978-3-7258-2106-8

© 2024 by the authors. Articles in this book are Open Access and distributed under the Creative Commons Attribution (CC BY) license. The book as a whole is distributed by MDPI under the terms and conditions of the Creative Commons Attribution-NonCommercial-NoDerivs (CC BY-NC-ND) license.

Contents

About the Editors vii

Junfeng Ma, Zhiyan Wang, Jinghua Wu, Zhi Gu, Xing Xin and Xiayin Yao
In Situ Solidified Gel Polymer Electrolytes for Stable Solid–State Lithium Batteries at High Temperatures
Reprinted from: *Batteries* 2023, *9*, 28, doi:10.3390/batteries9010028 1

Zelin Wang, Chunwen Sun, Liang Lu and Lifang Jiao
Recent Progress and Perspectives of Solid State Na-CO_2 Batteries
Reprinted from: *Batteries* 2023, *9*, 36, doi:10.3390/batteries9010036 12

Shuo Fu, Yulia Arinicheva, Claas Hüter, Martin Finsterbusch and Robert Spatschek
Grain Boundary Characterization and Potential Percolation of the Solid Electrolyte LLZO
Reprinted from: *Batteries* 2023, *9*, 222, doi:10.3390/batteries9040222 57

Yang Li, Zheng Sun, Haibo Jin and Yongjie Zhao
Engineered Grain Boundary Enables the Room Temperature Solid-State Sodium Metal Batteries
Reprinted from: *Batteries* 2023, *9*, 252, doi:10.3390/batteries9050252 71

Ruifan Lin, Yingmin Jin, Yumeng Li, Xuebai Zhang and Yueping Xiong
Recent Advances in Ionic Liquids—MOF Hybrid Electrolytes for Solid-State Electrolyte of Lithium Battery
Reprinted from: *Batteries* 2023, *9*, 314, doi:10.3390/batteries9060314 82

Yao Fu, Dangling Liu, Yongjiang Sun, Genfu Zhao and Hong Guo
Epoxy Resin-Reinforced F-Assisted $Na_3Zr_2Si_2PO_{12}$ Solid Electrolyte for Solid-State Sodium Metal Batteries
Reprinted from: *Batteries* 2023, *9*, 331, doi:10.3390/batteries9060331 103

Pengcheng Shi, Xu Wang, Xiaolong Cheng and Yu Jiang
Progress on Designing Artificial Solid Electrolyte Interphases for Dendrite-Free Sodium Metal Anodes
Reprinted from: *Batteries* 2023, *9*, 345, doi:10.3390/batteries9070345 113

Shujian Zhang, Hongmo Zhu, Lanfang Que, Xuning Leng, Lei Zhao and Zhenbo Wang
Effect of Carrier Film Phase Conversion Time on Polyacrylate Polymer Electrolyte Properties in All-Solid-State LIBs
Reprinted from: *Batteries* 2023, *9*, 471, doi:10.3390/batteries9090471 132

Tao Sun, Xiaopeng Cheng, Tianci Cao, Mingming Wang, Jiao Tian, Tengfei Yan, et al.
Optimizing Li Ion Transport in a Garnet-Type Solid Electrolyte via a Grain Boundary Design
Reprinted from: *Batteries* 2023, *9*, 526, doi:10.3390/batteries9110526 144

Mingyuan Chang, Mengli Yang, Wenrui Xie, Fuli Tian, Gaozhan Liu, Ping Cui, et al.
Micro-Sized MoS_6@15%$Li_7P_3S_{11}$ Composite Enables Stable All-Solid-State Battery with High Capacity
Reprinted from: *Batteries* 2023, *9*, 560, doi:10.3390/batteries9110560 156

About the Editors

Chunwen Sun

Prof. Chunwen Sun received his Ph.D. degree in Condensed Matter Physics from the Institute of Physics (IOP), Chinese Academy of Sciences (CAS) in 2006. He is a full professor and group leader of Energy Storage Materials and Devices at the school of chemical and environmental engineering at China University of Mining & Technology (Beijing). He is also the director of Joint Laboratory on Energy Materials between China and Spain, supported by the National Key R&D Program of China and the Ministry of Science and Innovation, Spain. His research interests include lithium-ion batteries, all-solid-state batteries, solid oxide fuel cells, solid oxide electrolyzers and electrocatalysis. He has published 180 peer-reviewed papers with a citation more than 17600 times (Google Scholar), edited 8 book chapters and filed 25 Chinese patents. He has also received a number of awards, including the Outstanding Overseas Talents by the Institute of Physics CAS (2011), the International Association of Advanced Materials (IAAM) Scientist Medal (2017), and the Chinese Most Cited Researchers for exceptional research performance in the field of Material Science and Chemistry, Elsevier (2018–2023), The Most Influential Paper Award of Chinese Physics Society in 2020.

Siqi Shi

Siqi Shi is a Professor and PhD supervision at School of Materials Science and Engineering, Shanghai University. Siqi Shi obtained his Ph.D. from Institute of Physics, Chinese Academy of Sciences in 2004. After that, he joined the National Institute of Advanced Industrial Science and Technology of Japan, University of Nebraska Lincoln, and Brown University as postdoctoral or visiting scholar. His current research focuses on the calculation and design of electrochemical energy storage materials, material databases, and machine learning, committed to promoting the research and development of artificial intelligence enabling materials. In 2001, he firstly applied the first principles calculation to investigate lithium-ion battery materials in China. He has published more than 180 research papers and a monograph called "Computation, Modeling and Simulation in Electrochemical Energy Storage". He also created an electrochemical energy storage material calculation and data platform with independent intellectual property rights. He is in charge of 12 projects supported by the National Natural Science Foundation of China or the National Key Research and Development Plan. Currently, he serves as a branch director of the Solid State Ionics of the Chinese Silicate Society, and a branch committee member of the Computational Materials Science, Chinese Society for Materials Research.

Yongjie Zhao

Dr Yongjie Zhao received his Ph.D. in Materials Science and Engineering from Tsinghua University in 2014. Currently, He is an Associate Professor at the Beijing Institute of Technology. His research activities focus on the research and development of inorganic materials in relation to energy storage/conversion.

Article

In Situ Solidified Gel Polymer Electrolytes for Stable Solid−State Lithium Batteries at High Temperatures

Junfeng Ma [1,2], Zhiyan Wang [2], Jinghua Wu [2], Zhi Gu [2], Xing Xin [1] and Xiayin Yao [2,3,*]

1. School of Material Science and Chemical Engineering, Ningbo University, Ningbo 315211, China
2. Ningbo Institute of Materials Technology and Engineering, Chinese Academy of Sciences, Ningbo 315201, China
3. Center of Materials Science and Optoelectronics Engineering, University of Chinese Academy of Sciences, Beijing 100049, China
* Correspondence: yaoxy@nimte.ac.cn

Abstract: Lithium metal batteries have attracted much attention due to their high energy density. However, the critical safety issues and chemical instability of conventional liquid electrolytes in lithium metal batteries significantly limit their practical application. Herein, we propose polyethylene (PE)−based gel polymer electrolytes by in situ polymerization, which comprise a PE skeleton, polyethylene glycol and lithium bis(trifluoromethylsulfonyl)imide as well as liquid carbonate electrolytes. The obtained PE−based gel polymer electrolyte exhibits good interfacial compatibility with electrodes, high ion conductivity, and wide electrochemical window at high temperatures. Moreover, the assembled LiFePO$_4$//Li solid−state batteries employing PE−based gel polymer electrolyte with 50% liquid carbonate electrolytes deliver good rate performance and excellent cyclic life at both 60 °C and 80 °C. In particular, they achieve high specific capacities of 158.5 mA h g^{-1} with a retention of 98.87% after 100 cycles under 80 °C at 0.5 C. The in situ solidified method for preparing PE−based gel polymer electrolytes proposes a feasible approach for the practical application of lithium metal batteries.

Keywords: in situ gel electrolyte; lithium metal battery; polymer electrolyte

Citation: Ma, J.; Wang, Z.; Wu, J.; Gu, Z.; Xin, X.; Yao, X. In Situ Solidified Gel Polymer Electrolytes for Stable Solid−State Lithium Batteries at High Temperatures. Batteries 2023, 9, 28. https://doi.org/10.3390/batteries9010028

Academic Editor: Claudio Gerbaldi

Received: 6 December 2022
Revised: 25 December 2022
Accepted: 28 December 2022
Published: 30 December 2022

Copyright: © 2022 by the authors. Licensee MDPI, Basel, Switzerland. This article is an open access article distributed under the terms and conditions of the Creative Commons Attribution (CC BY) license (https://creativecommons.org/licenses/by/4.0/).

1. Introduction

Over the past 30 years, rechargeable lithium−ion batteries have had tremendous success. With the demand for high energy density, lithium metal batteries have attracted widespread attention due to the high theoretical specific capacity of lithium as well as its lowest electrochemical potential [1–5]. However, organic liquid electrolyte−based lithium metal batteries usually exhibit dendritic structural evolution and continuous parasitic reactions at the electrode–electrolyte interface, leading to safety issues and low coulombic efficiency [6–10]. Significant efforts have been made to overcome these challenges, of which the exploration of safe and stable electrolytes compatible with lithium metal is particularly critical and indispensable [11–17]. Solid−state electrolytes with high stability have been considered as optional alternatives to conventional liquid electrolytes for achieving safer and more stable energy storage systems [18–20].

Among various solid electrolytes, gel polymer electrolytes with liquid organic electrolytes possess high ionic conductivity under room temperature and can form a flexible interface with the electrode [21–24]. Generally, the polymer films of gel polymer electrolytes are prepared by an ex situ method and then immersed in a liquid electrolyte for gelation. However, the gel polymer electrolyte only contacts with the top of the electrode and therefore the entire mass of active material cannot be directly utilized. Thus, additional liquid electrolyte is still required to ensure interfacial wettability when utilizing ex situ prepared gel polymer electrolytes [25,26].

In situ preparation of gel polymer electrolytes is considered a feasible method to address the aforementioned interfacial problems and active material utilization issues [27–32]. Gelation could be achieved in the presence of liquid electrolyte−containing monomers injected into the battery. This method takes advantage of low viscosity, ease of handling, and good wettability, which enable gel polymer electrolytes with enhanced interfacial contact with electrodes, creating well−connected pathways for ionic transport [27,29,31].

Herein, we present polyethylene (PE)−based gel polymer electrolytes composed of PE skeleton, polyethylene glycol (PEG), lithium bis(trifluoromethylsulfonyl)imide and liquid carbonate electrolyte (LCE) for high temperature solid−state lithium metal batteries. Poly(ethylene glycol) methyl ether acrylate (PEGMEA) is used as the monomers for in situ polymerization due to the formed PEG possessing good compatibility with lithium metal as well as high ionic conductivity [19], which enables the obtained gel polymer electrolyte to possess both good chemical stability and high ion conductivity at high temperatures. The PE−based gel polymer electrolyte with 50% liquid carbonate electrolytes (PE−50%LCE@PEG) shows a high ionic conductivity of 1.73×10^{-4} S cm^{-1} under 60 °C and good electrochemical stability with the lithium metal anode. The PE−50%LCE@PEG−based Li–Li symmetrical battery exhibits cycling stability for 1200 h at 0.1 mA cm^{-2} and 0.1 mAh cm^{-2} at 60 °C. Furthermore, the LiFePO$_4$/PE−50%LCE@PEG/Li cell exhibits excellent cyclic performance with no capacity decay over 150 cycles at 0.5 C under 60 °C. In addition, the cell shows a high initial reversible capacity of 160.3 mA h g^{-1} under a higher temperature of 80 °C at 0.5 C and delivers excellent cycling stability for 100 cycles. This in situ solidified PE−based gel polymer electrolyte exhibits promising potential application in lithium metal batteries.

2. Experimental Section

2.1. Synthesis of PE−Based Gel Polymer Electrolytes

Lithium bis(trifluoromethylsulfonyl)imide (LiTFSI, Sigma) and polyethylene glycol methyl ether acrylate (Sigma, 480 g mol^{-1}) monomer with dibenzoyl peroxide (Aladdin) initiator were mixed to form a homogeneous solution. The EO:Li ratio was controlled at 18:1 and the mass of initiator was 0.2% of PEGMEA. At the same time, a LCE solution of 1 M LiPF$_6$ in EC/DMC/EMC (1/1/1, $v/v/v$) was also prepared. The gel polymer electrolyte liquid precursor solutions were prepared by mixing different amounts of LCE (x = 40%, 50%, 60%) with the PEGMEA/LiTFSI solution, where x is the mass percent of PEGMEA. Then, the above liquid precursor solutions were incorporated into PE separator and then heating cured at 60 °C for 12 h, resulting in PE−x%LCE@PEG (x = 40, 50, and 60) gel polymer electrolytes.

2.2. Physical Characterization

Scanning electron microscopy (SEM, Hitachi S−4800) was used to observe the microstructure of the samples. A Nicolet 6700 spectrometer was used to collect Fourier transform infrared spectroscopy (FT−IR) spectra of the PEGMEA precursors and different electrolytes. Thermogravimetric analysis (TG209F1) was carried out under N$_2$ atmosphere with a heating rate of 10 °C min^{-1}.

2.3. Electrochemical Measurements

A Solartron 1470E electrochemical workstation was used to test electrochemical impedance spectroscopy (EIS), linear sweep voltammetry (LSV), and direct current polarization. EIS experiments were used to determine the ionic conductivity of the electrolytes using stainless steel (SS)/gel polymer electrolytes/SS symmetric cells and calculated based on Equation (1):

$$\sigma = L/(S \cdot R) \quad (1)$$

where L is the thickness, R represents the bulk resistance, and S represents the area.

The activation energy (E_a) of the electrolytes was obtained by the Arrhenius Equation (2):

$$\sigma(T) = A\exp\left(\frac{-E_a}{RT}\right) \quad (2)$$

where σ is the ionic conductivity, A is the frequency factor, R is the molar gas constant, and T is the absolute temperature. The electrochemical stability window of SS//Li cells was ascertained by linear scanning voltammetry (LSV) with a scan rate of 1 mV s^{-1} within 3~6 V. The Li$^+$ transference number (t_{Li^+}) of the polymeric solid electrolyte was determined by direct−current polarization on a Li//Li symmetric cell and calculated from the Bruce−Vincent−Evans Equation (3):

$$t_{Li^+} = \frac{I_{ss}(\Delta V - I_0 R_0)}{I_0\{\Delta V - I_{ss} R_{ss}\}} \quad (3)$$

where ΔV is the polarization voltage of 10 mV, I_0 is the initial current, I_{ss} is the steady−state current. R_0 and R_{ss} are the initial and steady−state interfacial resistance after the polarization process. Additionally, Li//Li symmetric batteries with various electrolytes were assembled to test the stability with lithium anode on commercial battery testing equipment (LAND Wuhan Electronics Co., Ltd., Wuhan, China).

2.4. Battery Fabrication and Evaluation

LiFePO$_4$, PVDF/LiClO$_4$ and Super p were mixed together with a mass ratio of 8:1:1 in N−methyl−2−pyrrolidone to prepare cathode slurry, which was coated on aluminum foil and vacuum dried at 60 °C for 24 h to obtain the cathode. The specific area capacity of the cathode was about 0.3 mAh cm^{-2}. The PE separator was sandwiched between the cathode and lithium metal. The gel polymer electrolyte liquid precursor solutions were injected into the cell and direct in situ solidified at 60 °C for 12 h to fully form PE−based gel polymer electrolytes in the battery. The charging–discharging of assembled LiFePO$_4$//Li cells were tested by the commercial battery testing system (Wuhan LAND Electronics Co., Ltd.) from 2.8–3.8 V at 30 °C, 60 °C and 80 °C, respectively.

3. Results and Discussion

The flowable precursor solutions containing PEGMEA, LiTFSI, different amounts of LCE (x = 40%, 50%, 60%) and initiator can be directly polymerized and transformed into solid−state LCE@PEG after thermally cured (Figure S1). Figure 1a displays the preparation schematic diagram of the PE−x%LCE@PEG (x = 40%, 50%, 60%) gel polymer electrolytes. A 16 μm−thick PE separator is selected as skeleton, which can be completely infiltrated by the transparent and flowable precursor solution. After in situ thermal curing, PE−x%LCE@PEG solid electrolytes formed. Figure 1b exhibits the FT−IR absorption spectra of PEGMEA, PEGMEA@LCE, PEG, PE−x%LCE@PEG (x = 40, 50, and 60) electrolytes. Clearly, the C=C stretching vibration of acrylate group in PEGMEA at 1600~1690 cm^{-1} disappears after in situ polymerization. Moreover, the C=O in the gel membrane guarantees a strong interaction with the liquid electrolyte. Figure 1c shows the TG curves of PEG, PE, PE@PEG, PE−50%LCE@PEG and LCE. The LCE evaporates rapidly at relatively low temperature and the weight loss reaches up to 41.88% when the temperature increases to 100 °C. In contrast, the weight loss of PE−50%LCE@PEG is nearly zero at 100 °C, indicating that the liquid electrolytes have been effectively immobilized in the polymerized PEG matrix and become thermally stable. Furthermore, PEG is thermally stable and its decomposition temperature can reach about 300 °C. By further incorporating with the PE separator, the PE−50%LCE@PEG gel polymer electrolyte exhibits good thermal stability at high temperature.

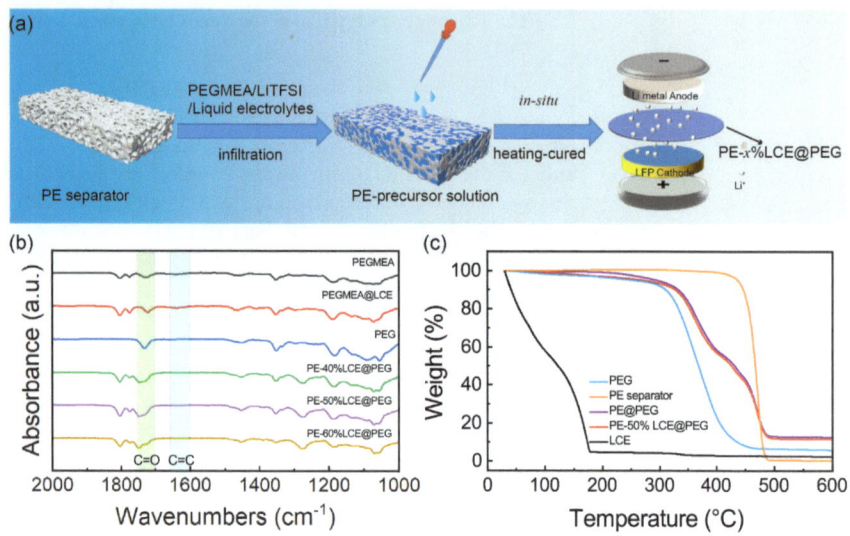

Figure 1. (**a**) Schematic diagram of the preparation process of PE−x%LCE@PEG via in situ polymerization. (**b**) FT−IR adsorption spectra of monomers and polymers. (**c**) TGA curves of monomers and polymers.

The morphology of PE and PE−x%LCE@PEG (x = 40, 50, and 60) gel polymer electrolytes were observed by SEM. As shown in Figure 2a, the surface of the pristine PE separator shows a porous structure, which can well accommodate the precursor solution before heating curing. After in situ polymerization, the pores in PE separator store LCE@PEG well to form a gel polymer electrolyte, and the surface of the electrolyte is uniform and flat (Figure 2b and Figure S2). The thickness of the PE−50% LCE@PEG electrolyte is about 18 µm with a dense structure (Figure 2c).

The ionic conductivity of gel polymer electrolytes is measured by EIS tests (Figure S3), showing 1.21×10^{-4}, 1.73×10^{-4}, and 2.11×10^{-4} S cm^{-1} for PE−x%LCE@PEG (x = 40, 50, and 60) at 60 °C, respectively. The activation energies of PE−x%LCE@PEG (x = 40, 50, and 60) are calculated to be 0.286, 0.281, and 0.22 eV, respectively (Figure 3a). Based on LSV testing (Figure 3b), no obvious oxidation peak for the PE−x%LCE@PEG was observed until 4.3 V vs. Li/Li$^+$, which is slightly higher than that of the liquid carbonate electrolyte at 4.1 V (Figure S4). The higher electrochemical window can be ascribed to the strong interaction between the C=O groups of PEGMEA and the anions in the electrolytic salt [19,33]. Moreover, the t_{Li^+} values for PE−x%LCE@PEG (x = 40, 50, and 60) gel polymer electrolytes are 0.415, 0.425, and 0.5, respectively (Figures 3c and S5), exceeding those of commercial liquid electrolytes (0.2–0.4) [34]. The high t_{Li^+} could be due to the C=O groups in the PE−x%LCE@PEG (x = 40, 50, and 60) gel polymer electrolytes limiting the movement of anions, which could further reduce the concentration polarization and improve the performance of the cells [35–37]. The electrochemical stability of gel polymer electrolytes with lithium metal anode is further investigated in Li/Li symmetric cells. The critical current densities of PE−x%LCE@PEG (x = 40, 50, and 60) are 1.1, 1.1, and 0.9 mA cm^{-2}, respectively (Figure 3d), indicating the PE−x%LCE@PEG (x = 40, 50, and 60) gel polymer electrolytes possess good ability to inhibit lithium dendrites [38]. In addition, the long−term cyclic performances of Li//Li symmetric cells employing PE−x%LCE@PEG (x = 40, 50, and 60) and LCE are shown Figure 3e and Figure S6. Clearly, PE−50%LCE@PEG gel polymer electrolyte displays the best cycling stability up to 1200 h at 0.1 mA cm^{-2} and 0.1 mAh cm^{-2}. The symmetrical cell with LCE suffers a short circuit after 875 h. The excellent interfacial stability is due to the confinement of polymer to the liquid phase, which reduces the interfacial reactions between lithium metal and reactive electrolyte components [39].

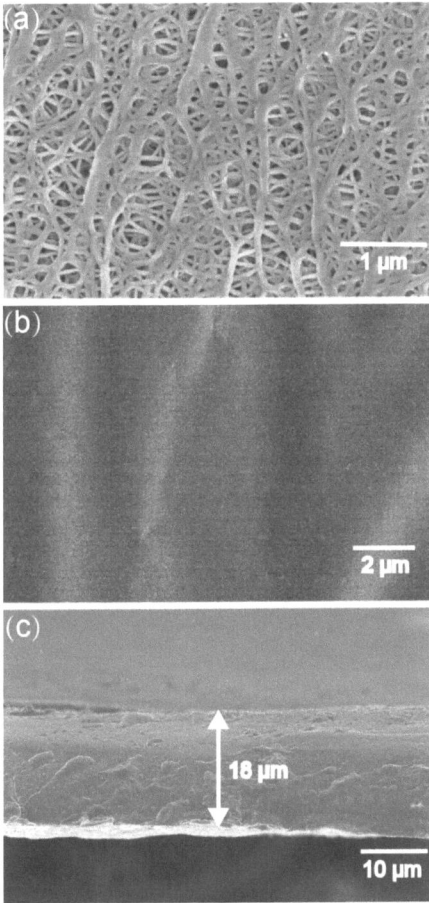

Figure 2. SEM images of (**a**) PE and (**b**) PE−50%LCE@PEG gel polymer electrolyte. (**c**) Cross−section SEM image of PE−50%LCE@PEG gel polymer electrolyte.

To demonstrate the feasibility of PE−x%LCE@PEG (x = 40, 50, and 60) gel polymer electrolytes in rechargeable lithium metal batteries, LiFePO$_4$//Li cells were assembled employing PE−x%LCE@PEG (x = 40, 50, and 60) and LCE, as shown in Figures 4a and S7. The PE−50%LCE@PEG cell shows excellent cycling stability over 150 cycles with no capacity decay at 0.5 C under 60 °C. However, the LCE−based cell shows dramatic decrease in capacity after 30 cycles under the same conditions. Furthermore, the rate capabilities of LiFePO$_4$/PE−50%LCE@PEG cells were assessed at various current densities from 2.8–3.8 V (Figure 4b). The reversible capacities for PE−50%LCE@PEG cells are 164.8, 165, 160, 157, 152, 142, 63.6 mA h g^{-1} at 0.1 C, 0.2 C, 0.5 C, 1 C, 2 C, 3 C, and 5 C, respectively. The corresponding charge–discharge curves are presented in Figure 4c. With the increase of current density, the discharge voltage plateau decreases slowly without serious polarization until 3 C.

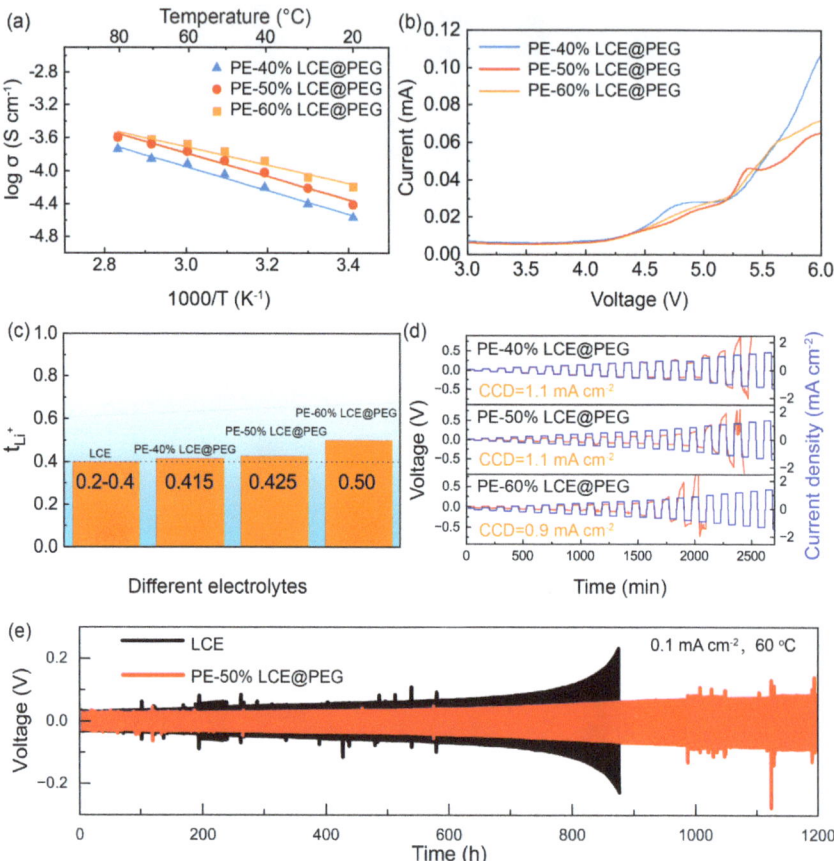

Figure 3. (**a**) Arrhenius plots of PE−x%LCE@PEG (x = 40, 50, and 60) gel polymer electrolytes. (**b**) Linear sweep voltammetry of PE−x%LCE@PEG (x = 40, 50, and 60) gel polymer electrolytes. (**c**) Li$^+$ transference number comparison of LCE and PE−x%LCE@PEG (x = 40, 50, and 60) gel polymer electrolytes. (**d**) Critical current density test of Li/PE−x%LCE@PEG (x = 40, 50, and 60)/Li cells at 60 °C. (**e**) Long−term cyclic performances of Li//Li symmetrical cells using the LCE and PE−50%LCE@PEG gel polymer electrolytes at 0.1 mA cm^{-2} and 0.1 mAh cm^{-2} under 60 °C.

In addition, the cyclic performances of LiFePO$_4$/PE−x%LCE@PEG/Li cells and LiFePO$_4$/LCE/Li cells at 0.5 C under 80 °C are also evaluated, as shown in Figures 4d and S8. For the commercial liquid electrolyte−based cell, it is difficult to operate at a high temperatures of 80 °C for a long cycle life [40]. Nevertheless, the PE−50%LCE@PEG based cell can stably cycle for 100 cycles at 0.5 C under 80 °C, exhibiting a reversible discharging specific capacity of 158.5 mA h g^{-1} at the 100th cycle with a capacity retention of 98.87%. Moreover, the LiFePO$_4$//Li battery with the PE−50%LCE@PEG electrolyte also possesses outstanding electrochemical performances under 0.5 C at 30 °C, showing a comparable performance with LCE−based cell (Figure S9). Furthermore, the lithium anode of the LiFePO$_4$/PE−50%LCE@PEG/Li cell after cycling shows a smooth and compact surface without cracks, and few lithium dendrites are observed (Figure S10). However, after cycling, the lithium anode of LiFePO$_4$/LCE/Li cell becomes rough and full of dendrites.

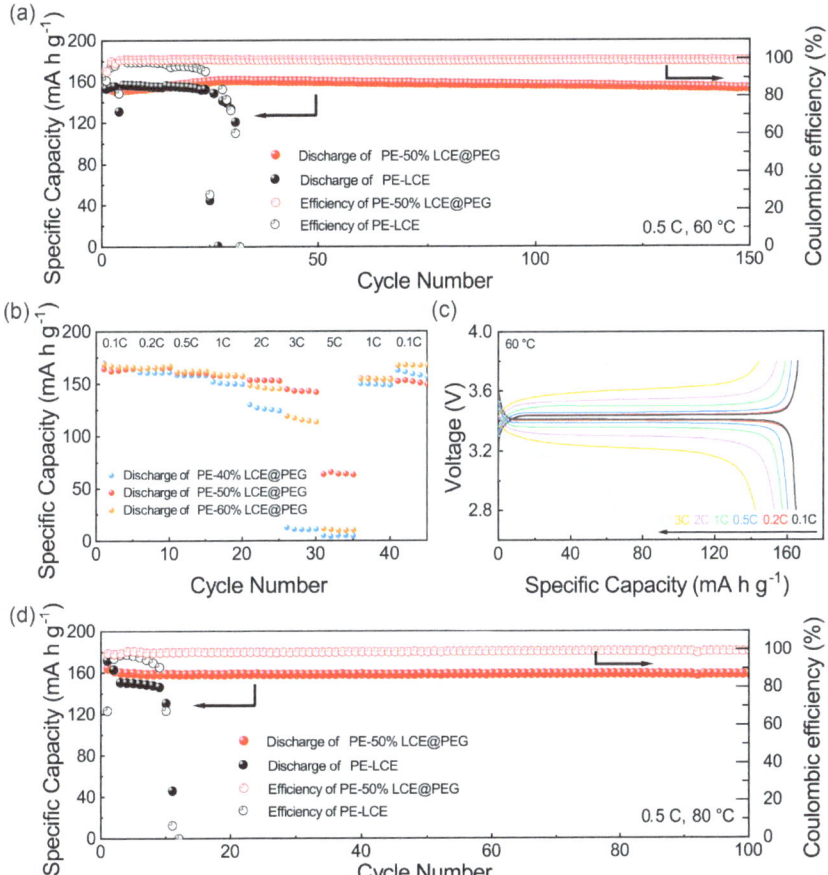

Figure 4. (**a**) Cyclic performances of the LiFePO$_4$//Li cells using PE−50%LCE@PEG and LCE under 0.5 C at 60 °C. (**b**) Rate capability of the LiFePO$_4$//Li batteries using PE−x%LCE@PEG (x = 40, 50, and 60) electrolytes at different rates at 60 °C. (**c**) Charge−discharge voltage profiles of LiFePO$_4$//Li battery using PE−50%LCE@PEG electrolyte at different rate under 60 °C. (**d**) Cyclic performances of the LiFePO$_4$//Li batteries using PE−50%LCE@PEG and LCE electrolyte under 0.5 C at 80 °C.

Furthermore, a series of flexibility and safety tests were performed on the LiFePO$_4$/PE−50%LCE@PEG/Li pouch cells. As shown in Figure 5, the pouch batteries with PE−50%LCE@PEG can continuously light up the yellow LED in the flat state, folded state, several fold and even cut, which displays the good reliability and safety of PE−50%LCE@PEG for flexible solid−state lithium batteries.

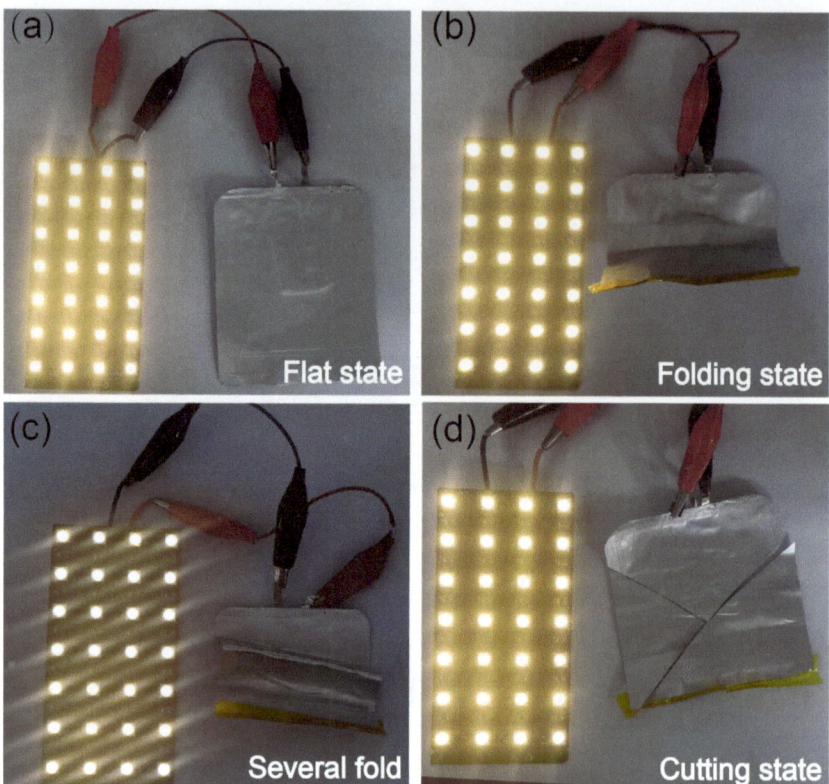

Figure 5. Optical images of the LED lamp powered by the pouch cell in various states: (**a**) flat state; (**b**) folding state; (**c**) several fold; (**d**) cutting state.

4. Conclusions

A series of PE−x%LCE@PEG (x = 40, 50, and 60) gel polymer electrolytes were successfully synthesized to realize high−temperature solid−state lithium batteries. The PE−50% LCE@PEG gel polymer electrolyte shows a high ionic conductivity of 1.73×10^{-4} S cm^{-1} at 60 °C with a high Li$^+$ transference number of 0.425. The critical current density of PE−50%LCE@PEG gel polymer electrolyte−based Li/Li symmetric cell is up to 1.1 mA cm^{-2}, and the cell exhibits cycling stability for 1200 h at 0.1 mA cm^{-2} and 0.1 mAh cm^{-2}. The assembled LiFePO$_4$/PE−50%LCE@PEG/Li solid−state batteries show outstanding cycling stability with no capacity decay over 150 cycles at 0.5 C under 60 °C. In addition, the cell exhibits a high initial reversible capacity of 160.3 mA h g^{-1} under a higher temperature of 80 °C at 0.5 C and exhibits excellent cycling stability for 100 cycles, demonstrating promising potential application for high−temperature solid−state lithium metal batteries.

Supplementary Materials: The following supporting information can be downloaded at: https://www.mdpi.com/article/10.3390/batteries9010028/s1, Figure S1: Optical images of PEGMEA, LiTFSI and initiator with varied amounts of LCE before (a) and after (b) the polymerization process. Figure S2. SEM images of (a) PE−40%LCE@PEG and (b) PE−60%LCE@PEG gel polymer electrolytes. Figure S3. EIS plots of PE−x%LCE@PEG (x = (a) 40, (b) 50, and (c) 60) gel polymer electrolytes at different temperatures (20~100 °C). Figure S4. Linear sweep voltammetry of LCE. Figure S5 Chronoamperometry of (a) Li/PE−40%LCE@PEG/Li, (b) Li/PE−50%LCE@PEG/Li, (c) Li/PE−60%LCE@PEG/Li symmetric cells at ambient temperature. The insets are the alternate current impedance spectra before and after polarization. Figure S6. Long−term cycling of symmetrical Li cells using the PE−40%LCE@PEG and PE−60%LCE@PEG electrolytes at 0.1 mA cm^{-2} and

0.1 mAh cm^{-2} under 60 °C. Figure S7. Cyclic performances of the LiFePO$_4$//Li batteries assembled with PE−40%LCE@PEG and PE−60%LCE@PEG under 0.5 C rate and 60 °C. Figure S8. Cyclic performances of the LiFePO$_4$//Li batteries assembled with PE−40%LCE@PEG and PE−60%LCE@PEG under 0.5 C rate and 80 °C. Figure S9. Cyclic performances of the LiFePO$_4$//Li battery with LCE and PE−x%LCE@PEG (x = 40, 50, and 60) electrolytes under 0.5 C rate and 30 °C. Figure S10. SEM images of the Li metal taken from (a) bare lithium, (b) LiFePO$_4$/PE−50%LCE@PEG/Li cell after 100 cycles under 0.5 C at 60 °C, (c) LiFePO$_4$/LCE/Li cell after 25 cycles under 0.5C at 60 °C.

Author Contributions: Conceptualization, X.X. and X.Y.; methodology, J.M. and Z.W.; validation, J.M., Z.W., J.W. and Z.G.; formal analysis, J.M., J.W. and Z.G.; investigation, J.M.; resources, X.Y.; data curation, J.M.; writing—original draft preparation, J.M.; writing—review and editing, X.Y.; visualization, J.M.; supervision, X.Y.; project administration, X.Y.; funding acquisition, X.Y. All authors have read and agreed to the published version of the manuscript.

Funding: The work was supported by the National Key R&D Program of China (Grant no. 2022YFB-3807700), the National Natural Science Foundation of China (Grant no. U1964205, U21A2075, 51872303, 51902321, 52172253, 52102326), Ningbo S&T Innovation 2025 Major Special Programme (Grant No. 2019B10044, 2021Z122), Zhejiang Provincial Key R&D Program of China (Grant No. 2022C01072), Jiangsu Provincial S&T Innovation Special Programme for carbon peak and carbon neutrality (Grant No. BE2022007) and Youth Innovation Promotion Association CAS (Y2021080).

Data Availability Statement: The data presented in this study are available upon request from the corresponding author.

Conflicts of Interest: The authors declare no conflict of interest.

References

1. Yamada, Y.; Furukawa, K.; Sodeyama, K.; Kikuchi, K.; Yaegashi, M.; Tateyama, Y.; Yamada, A. Unusual Stability of Acetonitrile-Based Superconcentrated Electrolytes for Fast-Charging Lithium-Ion Batteries. *J. Am. Chem. Soc.* **2014**, *136*, 5039–5046. [CrossRef] [PubMed]
2. Goodenough, J.B.; Park, K.S. The Li-Ion Rechargeable Battery: A Perspective. *J. Am. Chem. Soc.* **2013**, *135*, 1167–1176. [CrossRef] [PubMed]
3. Van Noorden, R. A BETTER BATTERY. *Nature* **2014**, *507*, 26–28. [CrossRef]
4. Voropaeva, D.Y.; Safronova, E.Y.; Novikova, S.A.; Yaroslavtsev, A.B. Recent progress in lithium-ion and lithium metal batteries. *Mendeleev Commun.* **2022**, *32*, 287–297. [CrossRef]
5. Guo, Y.; Wu, S.; He, Y.-B.; Kang, F.; Chen, L.; Li, H.; Yang, Q.-H. Solid-state lithium batteries: Safety and prospects. *eScience* **2022**, *2*, 138–163. [CrossRef]
6. Wang, Z.J.; Wang, Y.Y.; Zhang, Z.H.; Chen, X.W.; Lie, W.; He, Y.B.; Zhou, Z.; Xia, G.L.; Guo, Z.P. Building Artificial Solid-Electrolyte Interphase with Uniform Intermolecular Ionic Bonds toward Dendrite-Free Lithium Metal Anodes. *Adv. Funct. Mater.* **2020**, *30*, 2002414. [CrossRef]
7. Tarascon, J.M.; Armand, M. Issues and challenges facing rechargeable lithium batteries. *Nature* **2001**, *414*, 359–367. [CrossRef] [PubMed]
8. Hu, Z.L.; Xian, F.; Guo, Z.Y.; Lu, C.L.; Du, X.F.; Cheng, X.Y.; Zhang, S.; Dong, S.M.; Cui, G.L.; Chen, L.Q. Nonflammable Nitrile Deep Eutectic Electrolyte Enables High-Voltage Lithium Metal Batteries. *Chem. Mater.* **2020**, *32*, 3405–3413. [CrossRef]
9. Xu, K. Nonaqueous liquid electrolytes for lithium-based rechargeable batteries. *Chem. Rev.* **2004**, *104*, 4303–4417. [CrossRef] [PubMed]
10. Pieczonka, N.P.W.; Borgel, V.; Ziv, B.; Leifer, N.; Dargel, V.; Aurbach, D.; Kim, J.H.; Liu, Z.Y.; Huang, X.S.; Krachkovskiy, S.A.; et al. Lithium Polyacrylate (LiPAA) as an Advanced Binder and a Passivating Agent for High-Voltage Li-Ion Batteries. *Adv. Energy Mater.* **2015**, *5*, 1501008. [CrossRef]
11. Yang, C.P.; Yin, Y.X.; Zhang, S.F.; Li, N.W.; Guo, Y.G. Accommodating lithium into 3D current collectors with a submicron skeleton towards long-life lithium metal anodes. *Nat. Commun.* **2015**, *6*, 8058. [CrossRef]
12. Wood, S.M.; Pham, C.H.; Rodriguez, R.; Nathan, S.S.; Dolocan, A.D.; Celio, H.; de Souza, J.P.; Klavetter, K.C.; Heller, A.; Mullins, C.B. K+ Reduces Lithium Dendrite Growth by Forming a Thin, Less-Resistive Solid Electrolyte Interphase. *ACS Energy Lett.* **2016**, *1*, 414–419. [CrossRef]
13. Zhu, Z.Q.; Chen, X.D. Artificial interphase engineering of electrode materials to improve the overall performance of lithium-ion batteries. *Nano Res.* **2017**, *10*, 4115–4138. [CrossRef]
14. Chi, S.S.; Liu, Y.C.; Song, W.L.; Fan, L.Z.; Zhang, Q. Prestoring Lithium into Stable 3D Nickel Foam Host as Dendrite-Free Lithium Metal Anode. *Adv. Funct. Mater.* **2017**, *27*, 1700348. [CrossRef]

15. Deng, K.R.; Zeng, Q.G.; Wang, D.; Liu, Z.; Wang, G.X.; Qiu, Z.P.; Zhang, Y.F.; Xiao, M.; Meng, Y.Z. Nonflammable organic electrolytes for high-safety lithium-ion batteries. *Energy Storage Mater.* **2020**, *32*, 425–447. [CrossRef]
16. Shen, X.; Zhang, X.; Ding, F.; Huang, J.; Xu, R.; Chen, X.; Yan, C.; Su, F.; Chen, C.; Liu, X.; et al. Advanced Electrode Materials in Lithium Batteries: Retrospect and Prospect. *Energy Mater. Adv.* **2021**, *2021*, 1205324. [CrossRef]
17. Zhao, B.L.; Ma, L.X.; Wu, K.; Cao, M.X.; Xu, M.G.; Zhang, X.X.; Liu, W.; Chen, J.T. Asymmetric double-layer composite electrolyte with enhanced ionic conductivity and interface stability for all-solid-state lithium metal batteries. *Chin. Chem. Lett.* **2021**, *32*, 125–131. [CrossRef]
18. Zheng, J.; Tang, M.X.; Hu, Y.Y. Lithium Ion Pathway within $Li_7La_3Zr_2O_{12}$-Polyethylene Oxide Composite Electrolytes. *Angew. Chem.-Int. Ed.* **2016**, *55*, 12538–12542. [CrossRef]
19. Wang, Z.Y.; Shen, L.; Deng, S.G.; Cui, P.; Yao, X.Y. 10 μm-Thick High-Strength Solid Polymer Electrolytes with Excellent Interface Compatibility for Flexible All-Solid-State Lithium-Metal Batteries. *Adv. Mater.* **2021**, *33*, 2100353. [CrossRef] [PubMed]
20. Cheng, J.; Zhang, H.; Li, D.; Li, Y.; Zeng, Z.; Ji, F.; Wei, Y.; Xu, X.; Sun, Q.; Wang, S.; et al. Agglomeration-Free and Air-Inert Garnet for Upgrading PEO/Garnet Composite Solid State Electrolyte. *Batteries* **2022**, *8*, 141. [CrossRef]
21. Zhang, X.L.; Zhao, S.Y.; Fan, W.; Wang, J.N.; Li, C.J. Long cycling, thermal stable, dendrites free gel polymer electrolyte for flexible lithium metal batteries. *Electrochim. Acta* **2019**, *301*, 304–311. [CrossRef]
22. Zhou, H.Y.; Liu, H.D.; Li, Y.J.; Yue, X.J.; Wang, X.F.; Gonzalez, M.; Meng, Y.S.; Liu, P. In situ formed polymer gel electrolytes for lithium batteries with inherent thermal shutdown safety features. *J. Mater. Chem. A* **2019**, *7*, 16984–16991. [CrossRef]
23. Zhang, S.Z.; Xia, X.H.; Xie, D.; Xu, R.C.; Xu, Y.J.; Xia, Y.; Wu, J.B.; Yao, Z.J.; Wang, X.L.; Tu, J.P. Facile interfacial modification via in-situ ultraviolet solidified gel polymer electrolyte for high-performance solid-state lithium ion batteries. *J. Power Sources* **2019**, *409*, 31–37. [CrossRef]
24. Xu, D.; Su, J.M.; Jin, J.; Sun, C.; Ruan, Y.D.; Chen, C.H.; Wen, Z.Y. In Situ Generated Fireproof Gel Polymer Electrolyte with $Li_{6.4}Ga_{0.2}La_3Zr_2O_{12}$ As Initiator and Ion-Conductive Filler. *Adv. Energy Mater.* **2019**, *9*, 1900611. [CrossRef]
25. Cho, Y.G.; Hwang, C.; Cheong, D.S.; Kim, Y.S.; Song, H.K. Gel/Solid Polymer Electrolytes Characterized by In Situ Gelation or Polymerization for Electrochemical Energy Systems. *Adv. Mater.* **2019**, *31*, 1804909. [CrossRef] [PubMed]
26. Kim, D.; Liu, X.; Yu, B.Z.; Mateti, S.; O'Dell, L.A.; Rong, Q.Z.; Chen, Y. Amine-Functionalized Boron Nitride Nanosheets: A New Functional Additive for Robust, Flexible Ion Gel Electrolyte with High Lithium-Ion Transference Number. *Adv. Funct. Mater.* **2020**, *30*, 1910813. [CrossRef]
27. Liu, F.Q.; Wang, W.P.; Yin, Y.X.; Zhang, S.F.; Shi, J.L.; Wang, L.; Zhang, X.D.; Zheng, Y.; Zhou, J.J.; Li, L.; et al. Upgrading traditional liquid electrolyte via in situ gelation for future lithium metal batteries. *Sci. Adv.* **2018**, *4*, eaat5383. [CrossRef]
28. Liu, M.; Wang, Y.; Li, M.; Li, G.Q.; Li, B.; Zhang, S.T.; Ming, H.; Qiu, J.Y.; Chen, J.H.; Zhao, P.C. A new composite gel polymer electrolyte based on matrix of PEGDA with high ionic conductivity for lithium-ion batteries. *Electrochim. Acta* **2020**, *354*, 136622. [CrossRef]
29. Zhao, Q.; Liu, X.T.; Stalin, S.; Khan, K.; Archer, L.A. Solid-state polymer electrolytes with in-built fast interfacial transport for secondary lithium batteries. *Nat. Energy* **2019**, *4*, 365–373. [CrossRef]
30. Liu, X.C.; Ding, G.L.; Zhou, X.H.; Li, S.Z.; He, W.S.; Chai, J.C.; Pang, C.G.; Liu, Z.H.; Cui, G.L. An interpenetrating network poly(diethylene glycol carbonate)-based polymer electrolyte for solid state lithium batteries. *J. Mater. Chem. A* **2017**, *5*, 11124–11130. [CrossRef]
31. Zhou, Z.X.; Feng, Y.Y.; Wang, J.L.; Liang, B.; Li, Y.H.; Song, Z.X.; Itkis, D.M.; Song, J.X. A robust, highly stretchable ion-conducive skin for stable lithium metal batteries. *Chem. Eng. J.* **2020**, *396*, 125254. [CrossRef]
32. Fan, W.; Li, N.W.; Zhang, X.L.; Zhao, S.Y.; Cao, R.; Yin, Y.Y.; Xing, Y.; Wang, J.N.; Guo, Y.G.; Li, C.J. A Dual-Salt Gel Polymer Electrolyte with 3D Cross-Linked Polymer Network for Dendrite-Free Lithium Metal Batteries. *Adv. Sci.* **2018**, *5*, 1800559. [CrossRef] [PubMed]
33. Wang, Q.Y.; Xu, X.Q.; Hong, B.; Bai, M.H.; Li, J.; Zhang, Z.; Lai, Y.Q. Molecular engineering of a gel polymer electrolyte via in-situ polymerization for high performance lithium metal batteries. *Chem. Eng. J.* **2022**, *428*, 131331. [CrossRef]
34. Diederichsen, K.M.; McShane, E.J.; McCloskey, B.D. Promising Routes to a High Li+ Transference Number Electrolyte for Lithium Ion Batteries. *ACS Energy Lett.* **2017**, *2*, 2563–2575. [CrossRef]
35. Wang, Y.; Fu, L.; Shi, L.; Wang, Z.; Zhu, J.; Zhao, Y.; Yuan, S. Gel Polymer Electrolyte with High Li+ Transference Number Enhancing the Cycling Stability of Lithium Anodes. *ACS Appl. Mater. Interfaces* **2019**, *11*, 5168–5175. [CrossRef]
36. Xu, H.; Xie, J.; Liu, Z.; Wang, J.; Deng, Y. Carbonyl-coordinating polymers for high-voltage solid-state lithium batteries: Solid polymer electrolytes. *MRS Energy Sustain.* **2020**, *7*, 2. [CrossRef]
37. Lin, Z.; Guo, X.; Wang, Z.; Wang, B.; He, S.; O'Dell, L.A.; Huang, J.; Li, H.; Yu, H.; Chen, L. A wide-temperature superior ionic conductive polymer electrolyte for lithium metal battery. *Nano Energy* **2020**, *73*, 104786. [CrossRef]
38. Shen, L.; Deng, S.G.; Jiang, R.R.; Liu, G.Z.; Yang, J.; Yao, X.Y. Flexible composite solid electrolyte with 80 wt% $Na_{3.4}Zr_{1.9}Zn_{0.1}Si_2P_{0.8}O_{12}$ for solid-state sodium batteries. *Energy Storage Mater.* **2022**, *46*, 175–181. [CrossRef]

39. Zhou, W.D.; Wang, S.F.; Li, Y.T.; Xin, S.; Manthiram, A.; Goodenough, J.B. Plating a Dendrite-Free Lithium Anode with a Polymer/Ceramic/Polymer Sandwich Electrolyte. *J. Am. Chem. Soc.* **2016**, *138*, 9385–9388. [CrossRef]
40. Chang, Z.; Yang, H.J.; Zhu, X.Y.; He, P.; Zhou, H.S. A stable quasi-solid electrolyte improves the safe operation of highly efficient lithium-metal pouch cells in harsh environments. *Nat. Commun.* **2022**, *13*, 1510. [CrossRef]

Disclaimer/Publisher's Note: The statements, opinions and data contained in all publications are solely those of the individual author(s) and contributor(s) and not of MDPI and/or the editor(s). MDPI and/or the editor(s) disclaim responsibility for any injury to people or property resulting from any ideas, methods, instructions or products referred to in the content.

Review

Recent Progress and Perspectives of Solid State Na-CO$_2$ Batteries

Zelin Wang [1], Chunwen Sun [1,2,*], Liang Lu [1] and Lifang Jiao [3,*]

[1] School of Chemical & Environmental Engineering, China University of Mining and Technology-Beijing, Beijing 100083, China
[2] 21C Innovation Laboratory, Contemporary Amperex Technology Ltd. (21C LAB), Ningde 352100, China
[3] Key Laboratory of Advanced Energy Materials Chemistry (Ministry of Education), College of Chemistry, Nankai University, Tianjin 300071, China
* Correspondence: csun@cumtb.edu.cn (C.S.); jiaolf@nankai.edu.cn (L.J.)

Abstract: Solid state Na-CO$_2$ batteries are a kind of promising energy storage system, which can use excess CO$_2$ for electrochemical energy storage. They not only have high theoretical energy densities, but also feature a high safety level of solid-state batteries and low cost owing to abundant sodium metal resources. Although many efforts have been made, the practical application of Na-CO$_2$ battery technology is still hampered by some crucial challenges, including short cycle life, high charging potential, poor rate performance and lower specific full discharge capacity. This paper systematically reviews the recent research advances in Na-CO$_2$ batteries in terms of understanding the mechanism of CO$_2$ reduction, carbonate formation and decomposition reaction, design strategies of cathode electrocatalysts, solid electrolytes and their interface design. In addition, the application of advanced in situ characterization techniques and theoretical calculation of metal–CO$_2$ batteries are briefly introduced, and the combination of theory and experiment in the research of battery materials is discussed as well. Finally, the opportunities and key challenges of solid-state Na-CO$_2$ electrochemical systems in the carbon-neutral era are presented.

Keywords: Na-CO$_2$ battery; reaction mechanism; solid electrolyte; in situ characterization technology; theoretical calculation and simulation

Citation: Wang, Z.; Sun, C.; Lu, L.; Jiao, L. Recent Progress and Perspectives of Solid State Na-CO$_2$ Batteries. *Batteries* **2023**, *9*, 36. https://doi.org/10.3390/batteries9010036

Academic Editor: Claudio Gerbaldi

Received: 23 November 2022
Revised: 26 December 2022
Accepted: 30 December 2022
Published: 4 January 2023

Copyright: © 2023 by the authors. Licensee MDPI, Basel, Switzerland. This article is an open access article distributed under the terms and conditions of the Creative Commons Attribution (CC BY) license (https://creativecommons.org/licenses/by/4.0/).

1. Introduction

Metal–CO$_2$ batteries (such as Li/Na/Zn/K-CO$_2$ batteries) are a high energy density energy storage and power supply technology that enables CO$_2$ fixation and conversion [1,2]. Among various kinds of metal–CO$_2$ batteries, Na-CO$_2$ batteries have attracted more attention because of the abundant resources of sodium metal and similar physical and chemical properties to lithium. Na-CO$_2$ batteries exhibit better comprehensive performance, including high energy density (1.13 kWh kg^{-1}), and relatively high working voltage (2.35 V) [3]. In addition, the abundance of metal sodium is 1352 times that of metal lithium, so the cost is lower (the price of metal sodium is 20-times cheaper than that of metal lithium) [4,5]. In addition, a Na-CO$_2$ battery with the reaction of 4 Na + 3 CO$_2$ ↔ 2 Na$_2$CO$_3$ + C ($\Delta_r G^0_m$ = −905.6 kJ mol^{-1}) has a low reaction Gibbs free energy, which means that Na-CO$_2$ batteries may have a low charging voltage and inhibit electrolyte decomposition, which is conducive to improving round-trip efficiency and prolonging service life [6,7]. Compared with Li$^+$, Na$^+$ as a charge carrier has other advantages, for example, sodium has a larger ionic radius and atomic mass than lithium, but the Stokes radius of sodium is smaller than lithium, so it has higher mobility and ionic conductivity, resulting in a smaller polarization [8]. Therefore, Na-CO$_2$ batteries are widely considered to be a promising next generation energy storage power supply technology. However, the research of Na-CO$_2$ batteries is still in its infancy, and there are problems, for example, the reaction mechanism is still unclear, the types of cathodic materials and catalytic activity are relatively limited, the ionic conductivity and interfacial stability of the solid electrolyte still need to be improved, the impedance between the electrode and the interface is large and dendrites grow at the metal anode interface during the reaction. These problems lead to unsatisfactory electrochemical performance, such as cycling stability, overpotential, rate capability and specific full discharge capacity of Na-CO_2 batteries [9–11]. To solve these problems, it is necessary to comprehensively understand the reaction mechanism of Na-CO_2 battery and clarify the possible origins of the issues.

2. Mechanism of Na-CO_2 Electrochemistry

In 2011, Asaoka's group found that adding CO_2 to Li/Na-O_2 batteries could increase its discharge capacity and energy density, demonstrating the feasibility of metal–CO_2 batteries for the first time [12]. In 2013, Das et al. first designed a rechargeable O_2-assisted Na-CO_2 battery [13]. Under different partial pressures of CO_2, the discharge capacity of rechargeable Na-CO_2 (O_2) battery with tetraethlene glycol dimethyl ether (TEGDME) and ionic liquid (IL) electrolyte increased by 2.6 times and 2.1 times, respectively (Figure 1b,c), the optimal CO_2/O_2 ratio of the maximum discharge capacity ranges from 40% to 70% (Figure 1c) [13]. The ex situ Fourier transformed infrared (FTIR) spectroscopy test and X-ray powder diffraction (XRD) analysis showed that Na_2CO_3 and $Na_2C_2O_4$ exist simultaneously in TEGDME-based electrolyte; $Na_2C_2O_4$ is the main discharge product in IL electrolyte. These results proved that the electrolyte solvent may determine the final discharge product by affecting the intermediate stability.

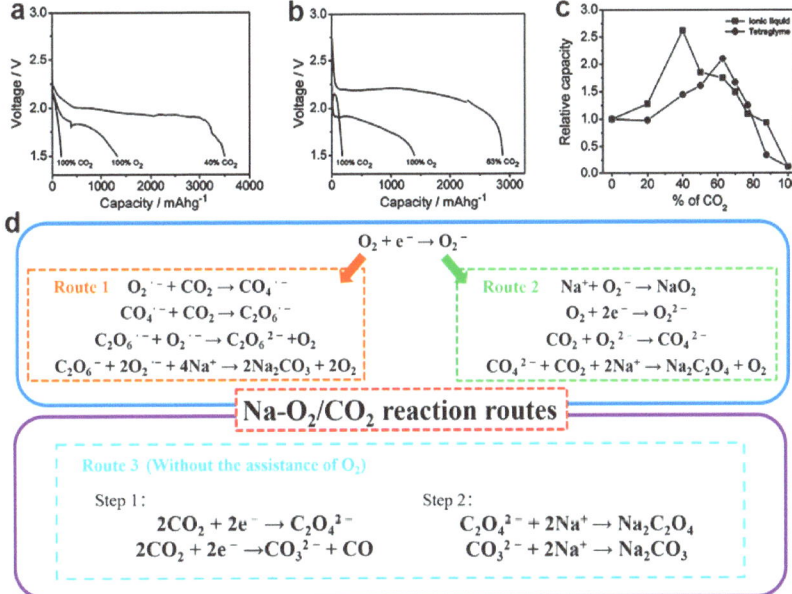

Figure 1. (**a**) Constant current discharge curves of Na-CO_2/O_2 battery based on IL electrolyte with mixed O_2/CO_2 supply. (**b**) Constant current discharge curves of Na-CO_2/O_2 battery with TEGDME-based electrolyte. (**c**) Relative capacity as a function of CO_2 concentration. Reproduced with permission from [13]. Copyright 2013, Elsevier Ltd. (**d**) Schematic of Na-O_2/CO_2 reaction routes.

In 2016, Hu et al. first proposed and demonstrated the electrochemical reaction mechanism of rechargeable Na-(pure) CO_2 batteries:

$$3CO_2 + 4Na \leftrightarrow 2Na_2CO_3 + C \tag{1}$$

which is inconsistent with the conjecture proposed by Das that $Na_2C_2O_4$ or Na_2CO_3 and CO is obtained in pure CO_2 atmosphere [6]. The Na-CO_2 battery was designed based on $NaClO_4$/TEGDME electrolyte and treated multi-walled carbon nanotube (MWCNT) cathode, which was successfully cycled 200 times in a pure CO_2 atmosphere, and the reaction mechanism was verified by a series of tests including XRD and X-ray photoelectron spectroscopy (XPS). Combined with the research results of Li-CO_2 system similar to Na-CO_2 system [14], it is reasonable to speculate that the reaction mechanism of rechargeable Na-pure CO_2 system is as follows: CO_2 molecules directly capture e^- to form $C_2O_4^{2-}$, the unstable $C_2O_4^{2-}$ undergoes a two-step disproportionation reaction to form CO_3^{2-} and C, and finally the discharge products Na_2CO_3 and C are produced as follows:

$$2CO_2 + 2e^- \rightarrow C_2O_4^{2-} \tag{2}$$

$$C_2O_4^{2-} \rightarrow CO_2^{2-} + CO_2 \tag{3}$$

$$C_2O_4^{2-} + CO_2^{2-} \rightarrow 2CO_3^{2-} + C \tag{4}$$

$$CO_3^{2-} + 2Na^+ \rightarrow Na_2CO_3 \tag{5}$$

The schematic diagram of Na-CO_2 battery structure is shown in Figure 2. Its electrochemical reaction route can be proposed as the following chemical equations:

$$\text{Cathode reactions: } 4Na^+ + 3CO_2 + 4e^- \rightarrow 2Na_2CO_3 + C \tag{6}$$

$$\text{Anode reactions: } Na \rightarrow Na^+ + e^- \tag{7}$$

$$\text{Overall reaction equation: } 4Na + 3CO_2 \rightarrow 2Na_2CO_3 + C \tag{8}$$

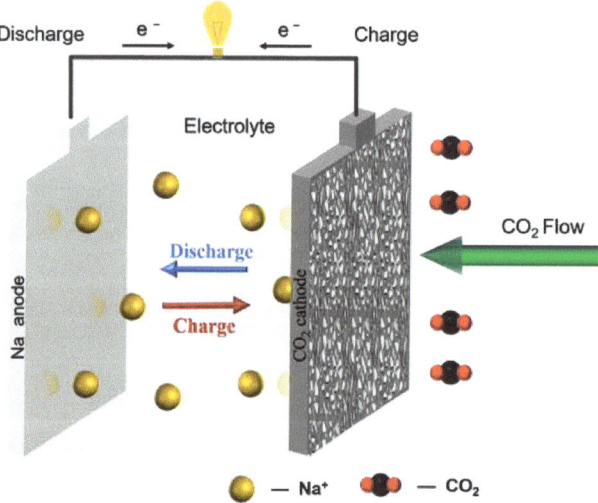

Figure 2. Schematic of a Na-CO_2 battery.

In addition, a rechargeable Na-O_2 (CO_2) battery with an IL-propylene carbonate-based electrolyte supplemented with 10% SiO_2 nanoparticles was reported by Archer et al. in 2014 [15]. The discharge product was detected as $NaHCO_3$, while CO_2 and O_2 were released during charging. Although it was speculated that the H in NaHCO3 might come from the introduction of trace amounts of H_2O during electrolyte preparation, there was no sufficient experimental evidence to prove it. Obviously, the exploration of the Na-CO_2 reaction mechanism is rather limited and further studies are needed.

3. Cathode Material/Catalysts of Na-CO$_2$ Batteries

The charging and discharging process of Na-CO$_2$ batteries is accompanied by the adsorption and desorption of CO$_2$, as well as the deposition and decomposition of the insulated discharge product Na$_2$CO$_3$ on the cathode surface. Inadequate reaction kinetics is the main obstacle, resulting in large overpotential, poor reversibility, poor rate performance and poor cycling stability, etc. [16]. Therefore, exploring an efficient, stable and low-cost electrocatalyst to facilitate CO$_2$ reduction and carbonate decomposition is one of the key issues in the development of this technology. The following issues should be considered during the research of cathode materials/catalysts [17–19]: to realize proper CO$_2$ adsorption properties, reasonably designed porous and macroporous structures are essential elements for designing cathode materials, so as to facilitate Na$^+$ and CO$_2$ diffusion, reduce the activation energy of the rate-controlling step and accommodate the insulation discharge product Na$_2$CO$_3$. In addition, the selection of efficient catalytic materials and the design of rich catalytic sites are the key factors to reduce the overpotential in the discharge process and improve the electrochemical performance. Finally, the factors such as abundant raw material resources, environmentally friendly preparation process and easy preparation are the prerequisites for the practical application of Na-CO$_2$ batteries. To realize the commercialization of Na-CO$_2$ batteries, the exploration of efficient catalytic materials is of broad significance. In the past decades, Na-CO$_2$ batteries have achieved rewarding results, especially in the design of efficient electrocatalysts. In this section, we review and discuss the research progress of cathode catalysts based on their chemical composition and microstructure from the three major categories: carbon and heteroatom-doped carbon materials, metal-loaded composite catalytic materials and single-atom catalysts, and then analyze their structure-activity relationship and the overall performance of Na-CO$_2$ batteries

3.1. Carbon Materials and Heteroatom-Doped Carbon Materials

Carbon materials have been widely used in various electrochemical energy storage devices, especially metal–O$_2$ batteries and metal–CO$_2$ batteries, due to their high electronic conductivity, large specific surface area, stable chemical and electrochemical properties, controllable pore structure and adjustable surface chemistry (defect engineering and heteroatom doping) [20–24]. Super P, Ketjen black (KB), activated carbon, carbon nanotubes, graphene, metal–organic framework materials (MOFs) and nitrogen-doped porous carbon materials have been widely used as cathodes for Li-CO$_2$ batteries, and they are also very suitable for Na-CO$_2$ batteries [13,15,25].

3.1.1. Commercial Carbon Materials

Commercial carbon materials can be used as cathode materials for metal–CO$_2$ batteries due to their good electronic conductivity, large surface area, relatively chemical stability, low cost, as well as their mature and scalable preparation process. Common commercial carbon materials, such as Ketjen black (KB) [12] and Super P [26,27], have been developed as porous cathode materials for Li-CO$_2$ batteries. However, the electrochemical performance of commercial carbon materials is not ideal owing to the inherent defects of low electrical conductivity, small pore volume, relatively small specific surface area and limited active sites [28].

Super P is first applied to the metal–CO$_2$ battery as the cathode material. Notably, the electrolyte plays a key role in determining the battery performance when the electrode materials are the same. As reported by Archer et al. Super P has almost no discharge capacity when it was used as a cathode material in ionic liquid electrolytes [26]; while Yang et al. applied a Super P cathode in combination with an ether-based electrolyte to a Li-CO$_2$ battery, showing a greatly enhanced discharge capacity at 100 mA g^{-1}, reaching 6062 mAh g^{-1} (further increase in discharge capacity with the addition of Ru metal, Ru@Super P, 8229 mAh g^{-1}) [27]. The reason for the significant difference in discharge capacity was ascribed to the electrolyte solvent could affect the stability of the intermediate discharge products and the formation mechanism of the final discharge products, thus the final discharge products and discharge capacities differed with different electrolytes as reported in the earliest studies on Na-$CO_2(O_2)$ batteries with Super P as the cathode material and ionic liquid electrolytes or ether-based electrolytes [13]. Therefore, the selection of a suitable electrolyte and cathode material plays a key role in determining the battery performance. Advanced characterization methods can be used to track these intermediate species and theoretical calculations can also further provide sufficient evidence.

3.1.2. Nanocarbon Materials

In addition to the above commercial activated carbon materials, nanostructured carbon materials (such as carbon nanotubes and graphene) are more widely explored in Na-CO_2 batteries [6,29,30]. They not only have novel structures, high electronic conductivity and high specific surface area, but also have excellent physical and chemical properties due to their unique quantum size effects and surface chemical states, so they have excellent electrochemical activity [31].

Chen's group prepared activated multi-walled carbon nanotube (a-MWCNT) cathode with three-dimensional tri-continuous porous structure (Figure 3a), high electronic conductivity and good wettability to electrolyte by boiling MWCNT with TEGDME at 100 °C and coating it on Ni network, which effectively improved its reactivity and reduced electrochemical polarization [6]. The rechargeable Na-CO_2 battery assembled with this material as cathode electrode catalyst and ether-based liquid electrolyte has a maximum reversible capacity of 60,000 mAh g^{-1} at 1 A g^{-1}, can be cycled 200 times and the initial discharge/charge voltage difference is only 0.6 V, and gradually increases to 1.3 V after 200 cycles (Figure 3b). Transmission electron microscopy (TEM) images (Figure 3b) shows grape-like discharge nanoparticles (50 nm) were randomly deposited on the t-MWCNT surface during the discharge process, and the grape-like discharge products disappeared after charging (Figure 4c), clearly indicating that the reaction is extremely reversible. A series of electrochemical test methods such as selected area electron diffraction (SAED), Raman spectroscopy (Figure 3d), XPS (Figure 3e), electron energy loss spectroscopy (EELS) (Figure 3f) similarly demonstrated the discharge products of Na-CO_2 batteries as Na_2CO_3 and C. Subsequently, this group reported solid-state Na-CO_2 batteries via acid-treated MWCNTs cathodes and NaF-modified anode [29]. The successful introduction of -COOH and -OH groups (Figure 3h) by acid treatment of MWCNTs before use can improve the adsorption capacity of CO_2 and thus the reaction kinetics of CO_2 electrochemical reduction was enhanced.

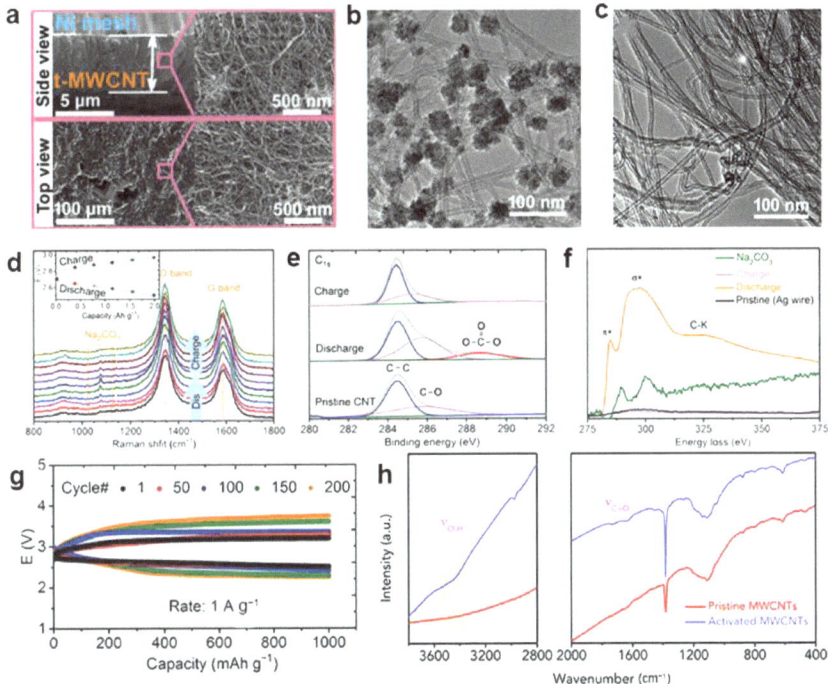

Figure 3. (**a**) SEM image of t-MWCNT cathode at room temperature, from top to side view. (**b**) TEM image. (**c**) HRTEM image. (**d**) In situ Raman spectra and corresponding discharge/charge product distributions at 11 selected points. (**e**) XPS of Ag wire cathode in different states. (**f**) EELS. (**g**) Complete discharge/charge curves at 1 A g^{-1}. Reproduced with permission [6]. Copyright 2016, John Wiley and Sons. (**h**) IR spectra of pristine and activated MWCNTS. Reproduced with permission [29]. Copyright 2019, Royal Society of Chemistry.

In general, carbon materials have remarkable characteristics such as large specific surface area, rich surface chemical properties, high intrinsic electronic conductivity, high chemical and electrochemical stability and low cost, and are the most commonly used cathode materials in Na-CO$_2$ batteries [17,29]. Although the incorporation of new nanostructured carbon materials such as graphene and carbon nanotubes has greatly improved the discharge capacity of Na-CO$_2$ batteries, their catalytic activity and resulting rate performance and cycling performance are still unsatisfactory, and require effective strategies, such as heteroatom doping to are needed to adjust the microstructure, porosity, defects and surface charge distribution.

3.1.3. Heteroatom-Doped Carbon Materials

It is widely accepted that the activation of CO$_2$ at the early stage of discharge depends mainly on the interaction of CO$_2$ with the chemically inert surface of the carbon material, i.e., the adsorption/desorption of CO$_2$. In the field of Li-CO$_2$ batteries, it has been demonstrated that heteroatom doping can modulate the charge distribution of nanostructured carbon material, strengthen nearby positively charged carbon atoms [32]. This has a significant impact on the adsorption mode and adsorption energy between the gas molecules (CO$_2$) and the carbon material surface, thus significantly enhancing the gas reduction kinetics and promoting the decomposition kinetics of the solid discharge products on the cathode surface of the metal–CO$_2$ batteries [20,33,34].

N element doping is an important strategy for catalyst design [35]. Due to the high electronegativity of nitrogen, N doping can break the charge neutrality of the carbon skeleton and cause charge redistribution, which gives the material excellent electronic and catalytic properties. For example, Hu et al. obtained nitrogen-doped carbon cathode nanomaterials with unique structures by calcining zeolite imidazolium salt framework (ZIF-8) at a certain temperature and then washing with dilute hydrochloric acid [25]. This material has higher CO_2 absorption and CO_2 adsorption properties, as well as better cycling stability. As shown in Figure 4b, the CO_2 absorption of the nitrogen-doped sample is much higher than that of carbon black, but its SSA is relatively small. According to the calculated results (Figure 4d), the surface of N-doped has stronger interaction with CO_2 bond, which can promote the reduction in CO_2 and the formation of discharge products. The electrochemical impedance spectrum shows that the resistance of the cell with the optimized nitrogen-doped nanocarbon (NC900) cathode increases only slightly after 80 cycles, which is much better than that of the cell with the carbon black cathode (about a six-fold increase after seven cycles), as shown in Figure 4c. Their assembled solid-state Na-CO_2 batteries Na | | liquid-free PEO-based polymer electrolyte | | optimized nitrogen-doped nanocarbon (NC900) cathode material exhibited better electrochemical performance: lower overpotential at 50 °C, higher discharge capacity of 10,500 mAh g^{-1}, the energy density of 180 Wh kg^{-1} and stable cycling for 320 h (at a capacity of 1000 mAh g^{-1}), as shown in Figure 4a. In addition, X-ray photoelectron spectroscopy (XPS) revealed that the signal of CO_3^{2-} was detected after the NC900 cathode was discharged and disappeared after charging, confirming the reversible formation and decomposition of Na_2CO_3 (Figure 4e). These excellent properties are attributed to the efficient catalytic effect of porous and highly conductive N-doped nanocarbons. Furthermore, the optimized nitrogen-doped nanocarbons facilitate the formation of sheet-like discharge products, which are easily decomposed into CO_2 after charging.

In 2019, Sun's group prepared nitrogen-doped single-walled carbon nanotubes (N-SWCNH) catalytic materials with a unique structure [36]. The excellent electrocatalytic performance of metal-free N-SWCNH for CO_2 reduction is mainly attributed to the unique structure of single-walled carbon nanotubes and nitrogen doping. The porous nature and unique conical structure of SWCNH provide sufficient storage space for discharge products; the highly dispersed nitrogen doping provides a large number of structural defect sites for CO_2 adsorption and electron transfer, which contributes to the electron affinity and CO_2 adsorption/desorption ability and improves catalyst activity and reversibility. They successfully prepared refillable hybrid Na-CO_2 batteries using N-doped single-walled carbon nanotubes (N-SWCNH) as the cathode catalyst and Na superionic conductor (NA-SICON) solid electrolyte as the separation medium for the hybrid electrolyte system. The use of aqueous electrolyte is also beneficial to the dissolution of discharge products, greatly improving the electrochemical reaction kinetics. As shown in Figure 4f,g, compared with the highly promising N-MWCNTs and Au NPs catalytic materials, N-SWCNH exhibited better performances and the prepared Na-CO_2 hybrid battery not only exhibited a low discharge/charge voltage difference of 0.49 V at a current density of 0.1 mA cm^{-1} (Figure 4f). It provides a high discharge capacity of 2293 mAh g^{-1} at a cut-off voltage and a current density of 0.2 mA cm^{-2}, as shown in Figure 4g. It can be cycled for more than 100 times at a current density of 0.1 mA cm^{-1} (Figure 4h).

In addition to N element doping, density function theory (DFT) calculations show that F doping and B doping also improve the performance [37]. In addition, multi-element co-doping, N/S co-doping [38] and N/B co-doping [39] are also used. Cheng et al. adopted the S/N co-doping strategy to improve the physical and chemical properties of the cathode material [40]. They designed electrophilic S vacancies and nucleophilic N-doped active centers on the surface of ReS_2, and used the synergistic coupling effect between heteroatoms to adjust the interaction of the catalyst with Li atoms and C/O atoms to show suitable adsorption during charging and discharging processes of Li-CO_2 batteries, respectively, thus reducing the activation energy of the rate-determining step and thus increasing the reaction rate.

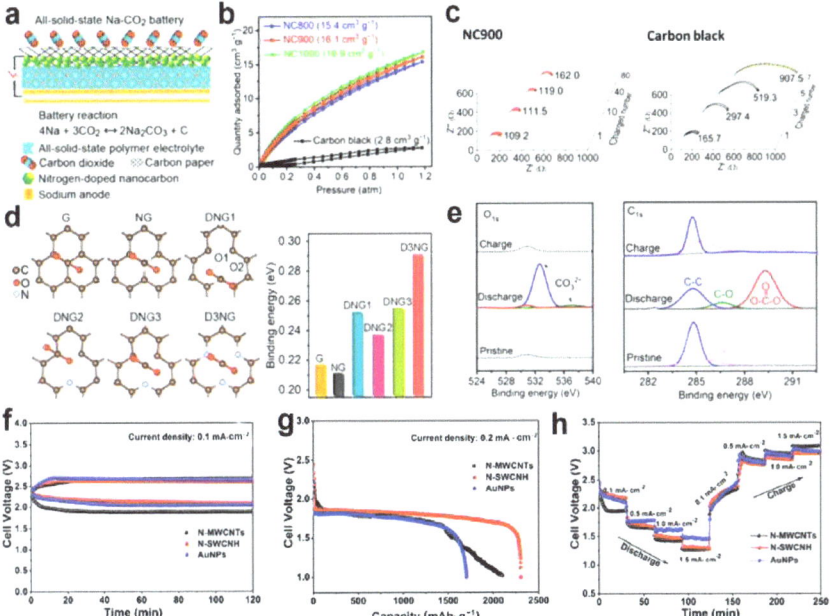

Figure 4. (**a**) Schematic of the structure of an all-solid-state Na-CO$_2$ cell with N-doped carbon cathode. (**b**) Low-pressure CO$_2$ adsorption isotherm of N-doped nanocarbon and carbon black at 273 K. The adsorption amount at 1 atm is shown in parentheses. (**c**) Nyquist plots of the NC900 cathode and carbon black cathode of the Na-CO$_2$ batteries in different charge states. (**d**) Density function theory (DFT) calculations of the interaction of undoped, graphitic N-doped and pyridine N-doped nanocarbon with CO$_2$ molecules and the adsorption of one CO$_2$ molecule on undoped and doped nanocarbon binding energy calculations. (**e**) XPS characterization of the discharged products. Reproduced with permission from [25]. Copyright 2020, American Chemical Society. (**f**–**h**) Electrochemical performance of Na-CO$_2$ batteries. (**f**) Discharge–charge voltage curves of the cells with N-MWCNTs, AuNPs and N-SWCNH as catalysts at 0.1 mA cm^{-2} current density. (**g**) Discharge capacity curves at a current density of 0.2 mA cm^{-2}. (**h**) Rate performances at different current densities. Reproduced with permission from [36]. Copyright 2020, Elsevier Ltd.

In summary, the design of three-dimensional (3D) structures and heteroatom doping is an effective strategy to facilitate the diffusion of reactants and improve the catalytic activity of carbon nanomaterials for CO$_2$ reduction and Na$_2$CO$_3$ decomposition. However, the optimal design for carbon materials cannot yield the desired high catalytic activity. Therefore, an in-depth analysis of the active center of heteroatom doping and the related CO$_2$ reduction mechanism with the help of advanced characterization is beneficial for exploring the methods to achieve uniform doping and controlled preparation.

3.2. Metal-Loaded Composites

Despite the demonstrated applicability of nitrogen-doped carbon materials, the catalytic activity for the reversible reaction of the cathode in Na-CO$_2$ batteries is quite limited [41]. Therefore, various carbon loaded metal composites were designed as cathodes for activity modulation and thus to improve the performance of Na-CO$_2$ batteries. The results show that metal-loaded composite catalysts appear to be a good choice for reducing the battery charging potential and improving the electrochemical performance of the battery compared to pure carbon materials [38,42–47].

3.2.1. Precious Metals and Their Composites

Due to the inherent electronic configuration of half-filled anti-bonds and high electrical conductivity of noble metals, noble metals and their composites usually have the advantages of good adsorption, low resistance and low overpotential as electrocatalysts [2,48].

Among all precious metal elements, ruthenium (Ru) is one of the most widely studied precious metal catalysts [27,49]. For example, Guo et al. obtained Ru@KB composites by in situ reductions of $RuCl_3$ on porous Ketjen Black [16]. As shown in Figure 5a–c, compared with pure KB as the cathode of Na-CO_2 battery, Ru@KB can significantly improve the discharge capacity up to 11,537 mAh g^{-1}, cycle stability for more than 130 cycles and coulomb efficiency of 94.1% of Na-CO_2 battery. Their electrochemical performance and ex situ characterization confirmed that ruthenium nanoparticles can significantly reduce the charging overpotential, promote the reversible reaction between Na_2CO_3 and carbon, and further improve the cycling stability of Na-CO_2 batteries. Besides the excellent catalytic activity of metallic Ru, its corresponding oxide RuO_2 has also been shown to have significant catalytic ability for reversible metal–CO_2 batteries. In addition, Pt [50], Ag, Ir [51,52], Au [53], Pd [54] and other noble metals and their oxides have been shown to be very effective in bifunctional catalysis for metal–CO_2 batteries, and need further be explored.

Compared with pure carbon-based materials, the composite by loaded with noble metal-based catalysts help to reduce the active energy required for CO_2 reduction and decomposition of discharging products, exhibit excellent catalytic activity in facilitating the electrochemical reaction of Na-CO_2 batteries, allowing for smaller overpotentials and higher energy efficiency, and significantly increase the discharge/charge capacity of the batteries. Unfortunately, the expensive cost and limited resources seriously hinder the commercial application of noble metal-based catalysts. Therefore, reasonable strategies such as developing noble metal-based single-atom catalysts or other inexpensive and abundant transition metal-based catalysts should be explored to meet their industrial development.

3.2.2. Transition Metals and Their Composites

Transition metals Ni [21], Co [43,55], Mn [56], Cu [57] and Fe [58] are supported on carbon-based materials with high specific surface area and high electronic conductivity due to their unique adjustable structure and multivalent characteristics, providing rich active sites for electrochemical reactions. In recent years, single metal composites, alloy type composites and transition metal oxide composites of transition metals have been widely reported. Moreover, due to the advantages of rich reserves and low cost of transition metals, transition metal-based composites are a feasible solution for future controlled scale up production [17].

In 2020, Xu et al. obtained an efficient active material (Co/Co_9S_8 @SNHC) for hybrid system Na-CO_2 batteries by anchoring Co/Co_9S_8 active nanoparticles on biomass-derived S and N-doped graded porous carbon via a microporous/mesoporous domain-limited synthesis strategy, as shown in Figure 5d [45]. The Na-CO_2 battery with Co/Co_9S_8 @SNHC not only exhibits a low overcharge potential of ~0.32 V and a charge/discharge voltage difference of only 0.65 V (Figure 5e). As shown in Figure 5f, it also shows better rate performance, cycle stability (cycled for more than 200 cycles at a current density of 0.1 mA cm^{-1}) and higher specific area discharge capacity (~18.9 mA cm^{-2}). These excellent electrochemical properties are attributed to the meso- and mesoporous-limited domains of the biomass carbon skeleton, which not only effectively inhibit the agglomeration of Co/Co_9S_8 nanoparticles, but also provide diffusion channels for CO_2 and Na^+ as well as sufficient space for storing the discharge products. In addition, the effective synergistic interactions between the effective catalytically active sites (Co/Co_9S_8, C-N, C-S bonds) and the defect-rich carbon interfaces (S, N doping) similarly prevent the agglomeration and separation of Co/Co_9S_8 nanoparticles, enhance the catalytic activity and improve the stability.

In addition to monometallic composites, bimetallic composites are also potential candidates for CO$_2$ reduction reaction (CO$_2$RR) and CO$_2$ electroreduction reaction (CO$_2$ER). In 2021, Xu et al. obtained bimetallic nitrogen-doped carbon materials of Fe-Cu-N-C with dense bimetallic active sites as catalysts for Na-CO$_2$ batteries with mixed air by introducing Fe^{3+} and Cu^{2+} regulated in situ pyrolytic growth of carbon nanotubes via solid-phase reactions [58]. They suggested that the excellent electrocatalytic activity of Fe-Cu-N-C is attributed to the synergy between the N-doped carbon framework with more defects and a large number of active sites in the Fe-N$_x$, Cu-N$_x$ and Fe/Fe$_3$C nanocrystals. Moreover, Fe^{3+} is the key to catalyze the conversion of g-C$_3$N$_4$ to CNT conformation, while Cu^{2+} gives the carbon nanotubes a good structure and uniform diameter. As shown in Figure 5g,h, the Fe-Cu-N-C materials synthesized at a pyrolysis temperature of 700 °C exhibit an ultra-low voltage gap of 0.44 V and a cycle efficiency of 83.2%, as well as a large discharge capacity of 8411 mAh g^{-1} and a long-term cycling performance of 1550 cycles (over 600 h).

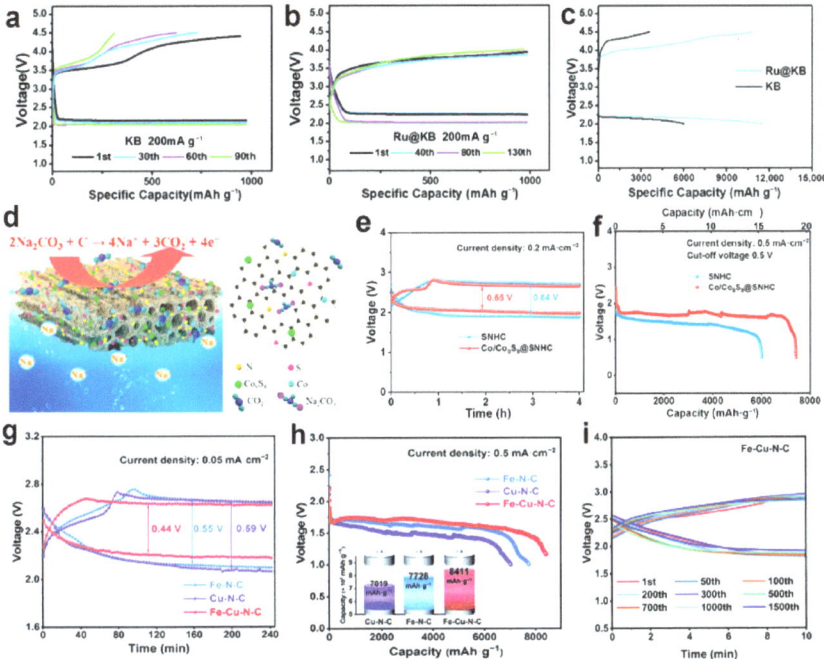

Figure 5. Na-CO$_2$ battery cycling behavior of KB and Ru@KB cathode at 200 mA g^{-1}: (**a**) KB cathode. (**b**) Ru@KB composite cathode. (**c**) Discharge–charge profiles of Na-CO$_2$ batteries with KB and Ru@KB composite cathodes at a current density of 100 mA g^{-1} in the first cycle. Reproduced with permission from [16]. Copyright 2019, Royal Society of Chemistry. (**d**) Schematic illustration of catalytic cathode of hybrid Na-CO$_2$ battery. (**e**) Discharge–charge voltage curves at a current density of 0.2 mA cm^{-2}. (**f**) Discharge capacity curves of hybrid Na-CO$_2$ batteries with Co/Co$_9$S$_8$@SNHC or SNHC at a current density of 0.5 mA cm^{-2}. Reproduced with permission from [45]. Copyright 2021, Elsevier Ltd. (**g–i**) Electrochemical performances of the as-obtained Fe-Cu-N-C, catalysts: (**g**) Discharge–charge voltage curves at 0.05 mA cm^{-2}; (**h**) Discharge–charge cycling curves; (**i**) Discharge–charge curves of Fe-Cu-N-C. Reproduced with permission from [58]. Copyright 2021, Royal Society of Chemistry.

Due to the excellent catalytic activity, low cost and simple preparation methods, transition metal oxides such as NiO [59,60], MnO [37], MnO$_2$ [61], have been extensively studied in the field of Na-CO$_2$ batteries. Fang et al. obtained in situ grown (CMO@CF) composites of Co$_2$MnO$_4$ on carbon fibers by a simple hydrothermal method and high temperature annealing [55] which achieved a reversible charge/discharge process and remained stable after 75 cycles at 200 mA g^{-1}. In addition, XRD, XPS, Raman spectroscopy and SEM characterization demonstrated that the high catalytic activity of CMO@CF electrodes is mainly due to the homogeneous morphological, chemical and structural stability and the hybridized Co^{2+}/Co^{3+} and Mn^{2+}/Mn^{3+} redox pairs. ZnCo$_2$O@CNT materials were synthesized by the same hydrothermal method as that of preparing ZnCo$_2$O$_4$ porous carbon nanorod cathodes [43]. The discharge capacity of 500 mAh g^{-1} at 100 mA g^{-1} can be charged and discharged stably for at least 150 cycles. Theoretical calculations show that there is a strong adsorption energy for CO$_2$, Na and Na$_2$CO$_3$ on three surfaces of ZnCo$_2$O$_4$, namely the [001] surface of ZnCo$_2$O$_4$, the [111] surface with only Co atoms exposed and the [111] surface with Co and Zn atoms exposed. In addition, the exposed Co atoms on these three surfaces of the density of states (DOS) calculated surface are the real catalytically active sites for the CO$_2$ electrochemical reaction.

Apparently, the transition metal-based composites combine the high SSA and electronic conductivity of porous carbon materials and the unique multivalent characteristics of transition metals, which enable an effective catalytic effect in the application of Na-CO$_2$ batteries, which are considered as the most attractive catalytic cathodes. Therefore, in future research, the catalytic mechanism of transition metal materials should be understood in depth, and more attention should be paid to and design of composite carbon-based materials with synergistic effects of transition metals, transition metal oxides and multi-transition metals, and their controlled and large-scale production should be realized.

3.2.3. Other Types of Composite Materials

In addition to metal nanoparticles loaded on porous carbon materials, several composites have been applied to metal–CO$_2$ batteries, such as molybdenum-based electrode materials [62], metal–organic complexes, polymers and self-supporting freestanding cathodes.

Molybdenum carbide (Mo$_x$C) has a d-band electronic structure similar to that of precious metals, especially metallic palladium, and is considered to be a "similar catalyst to precious metals". In early studies, Mo$_2$C was first applied to lithium–oxygen batteries and was shown to improve the Coulombic efficiency and cycle life of the batteries [63]. After that, Chen's group prepared Mo$_2$C/CNT composites by a simple carbon thermal reduction method and applied them to Li-CO$_2$ batteries [64]. In this work, the Li$_2$C$_2$O$_4$ intermediate was stabilized by forming a Li$_2$C$_2$O$_4$-Mo$_2$C species on the surface of the catalyst. Finally, amorphous Li$_2$C$_2$O$_4$ final discharge products with thin film morphology were obtained instead of Li$_2$CO$_3$ species. In this case, Mo$_2$C makes the amorphous Li$_2$C$_2$O$_4$ components readily decomposable at very low potentials below 3.5 V. Moreover, it shows excellent cycling properties. However, the specific mechanism of Mo$_2$C catalyzing the formation of amorphous Li$_2$C$_2$O$_4$ remains unclear yet, and some advanced characterization tools as well as theoretical simulation calculations are needed in this regard to strongly elucidate the effects of Mo$_2$C on the performances of CO$_2$ batteries. In addition, MoS$_2$ is also an effective electrocatalyst for the decomposition of Li$_2$CO$_3$ [65]. In fact, we believe that it is a worthwhile effort to study the crystal structure of Mo$_x$C and to develop new molybdenum-based catalysts/materials for Na-CO$_2$ batteries.

Metal–organic backbone (MOF) is a crystalline porous organic-inorganic hybrid material with periodic network structure self-assembled by inorganic metal centers (usually metal ions or metal clusters) and bridging organic ligands [24]. Furthermore, MOFs not only combine the stiffness of inorganic materials with the flexibility of organic materials, but also have a porous structure and a metal–nonmetal structure, offering attractive prospects in the field of metal–air batteries research [66–68]. Successful applications of MOFs have been reported for Li-CO$_2$ batteries [24] and Na-CO$_2$ batteries [25]. The results show that MOFs have significant potential to increase the discharge capacity and reduce the polarization potential of Li-CO$_2$ batteries. However, the electronic conductivity of MOF is still insufficient, and the catalytic efficiency of the cathode needs to be further improved. We believe that the use of conductive MOFs with high electronic conductivity, along with metal ion doping to synergistically promote catalytic activity, can be attempted. In addition, MOF-derived single-atom catalysts are expected to be a promising new class of MOF-based materials in the laboratory due to their excellent activity in ORR [46,69].

Due to the insulator nature of discharge product, Na_2CO_3, of the $Na-CO_2$ batteries higher external voltages are typically required to decompose Na_2CO_3 during the charging process. However, high charging voltages typically lead to degradation of the battery components (e.g., electrolytes, carbon additives and binders), resulting in a decrease in overall battery performance [14]. Therefore, the development of carbon-free and binder-free cathodes is also an important task.

3.3. Single-Atom Catalysts

Large-scale applications of precious metals are hindered by their high price and scarce resources. Non-precious metal materials are promising candidates for noble metals, but with lower catalytic performance. In order to reduce the use of noble metals and further enhance the effective activity of metal atoms, single-atom catalysts (SACs) have emerged and attracted extensive attention [70]. Compared with nanomaterials, SACs can maximize the use of metal atoms; in addition, due to quantum confinement effects, these sub-nanometer catalysts have unexpected catalytic properties and can be used for various energy conversion reactions, and their applications in metal–CO_2 batteries are also highly anticipated [71–73]. Despite these advantages, SACs still suffer from some drawbacks, such as synthesis and characterization difficulties.

When metal particles are dispersed on the support at the atomic/cluster level, the properties of the catalyst, such as surface free energy, unsaturated coordination environment, quantum size effect and metal–support interaction change drastically, and SACs are easily agglomerated during preparation and application, leading to catalyst deactivation (Figure 6) [69]. Theoretically, increasing the loading of single-atom catalysts and avoiding atomic agglomeration are mainly achieved by increasing the surface area of the carrier and enhancing the interaction of the metal carriers [73]. Researchers have obtained "bottom-up" and "top-down" synthesis strategies for SACs [74]. Typical "bottom-up" strategies include atomic layer deposition (ALD), wet chemical methods, electrochemical deposition and chemical vapor deposition [75–78]. Atomic layer deposition (ALD) allows for precise control of the film thickness by evaporating the active precursor into a vapor and depositing it layer by layer on a carrier in a self-limiting manner using sophisticated equipment [79,80]. Wet chemical methods consisting of wet impregnation and co-precipitation are considered a promising technique for the synthesis of SACs because of their simplicity and the possibility of mass production of SACs [81]. However, it should be noted that the SACs obtained by this strategy usually exhibit the disadvantages of low metal loading and easy aggregation. In addition, electrochemical deposition has proven to be a universal method for the preparation of single-atom catalysts. Single-atom catalysts on various substrates such as transition group metals and oxides, sulfides, selenides and carbon materials were obtained using electrochemical deposition by Zeng et al. [82]. In contrast to the bottom-up strategy, top-down strategies are generally ball milling, high-temperature atom capture and pyrolysis [83–85]. Ball milling is a simple and versatile method that enables large-scale preparation by breaking/reconstructing the bonding, thus efficiently producing SA [83]. The pyrolysis strategy often uses various metal–organic complex precursors with molecular confinement and organic ligand coordination characteristics to obtain uniformly dispersed M-N-C site catalysts by pyrolysis, which are considered to be the most promising non-precious metal-based ORR catalysts [86]. Despite the consumption of organic reagents, it shows clear advantages: simpler preparation process, controllable tuning of the performance of SACs by controlling the pyrolysis environment and inducing defects [87,88].

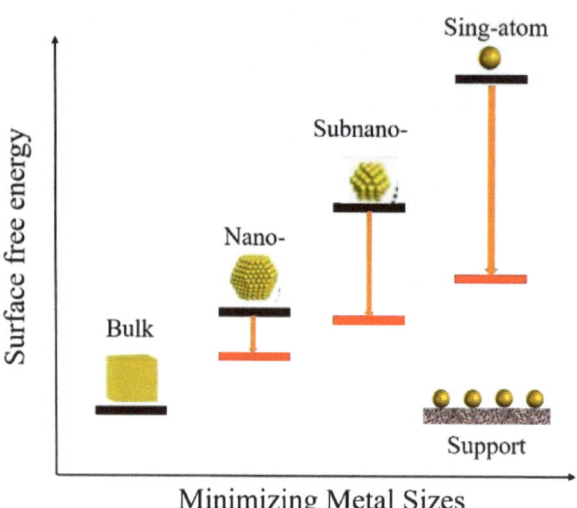

Figure 6. Schematic representation of surface free energy and specific activity of supported catalysts as a function of particle size.

With the development of nanotechnology and characterization science, advanced characterization techniques provide rich information and reliable evidence for studying the composition, structure and performance relationships of catalysts, which plays a crucial role in the rapid development of single-atom catalysts [89,90]. In addition, theoretical calculations can provide a reasonable reaction model for uniformly dispersed active sites and calculate the activation energy of the reaction. Combining advanced characterization techniques and data analysis of theoretical calculations, detailed information on the geometry and electronic structure of the active sites of single-atom catalysts can be obtained, thus elucidating the structure and performance relationships of single-atom materials in the catalytic process and guiding the design of single-atom catalysts [91,92]. The contents of characterization technology and theoretical calculation will be elaborated in the following section.

SACs are a promising redox electrocatalyst. However, the application of SACs in Na-CO_2 batteries is relatively new and much research is needed to explore the potential of SACs for practical applications in the next generation of rechargeable batteries. Theoretically, SACs ensure 100% atom utilization efficiency, which is very valuable for saving metal resources, especially precious metals. Recently, Zhu et al. successfully prepared single-atom Pt deposited on nitrogen-doped carbon nanotubes (Pt@NCNT) as a cathode material for Na-CO_2 batteries [44]. The single-atom dispersed Pt catalytic site, with a unique electronic structure and low coordination environment (Figure 7a,b), can achieve high activity and high selectivity due to the well-dispersed and fully exposed active site. Compared with the air batteries with pure NCNT as the positive electrode, the Pt single-atom catalyst can effectively improve the discharge reaction rate. The schematic diagram of Figure 7c illustrates the reaction mechanism based on Pt-SA in the discharge/charge electrochemical process of Na-CO_2 battery.

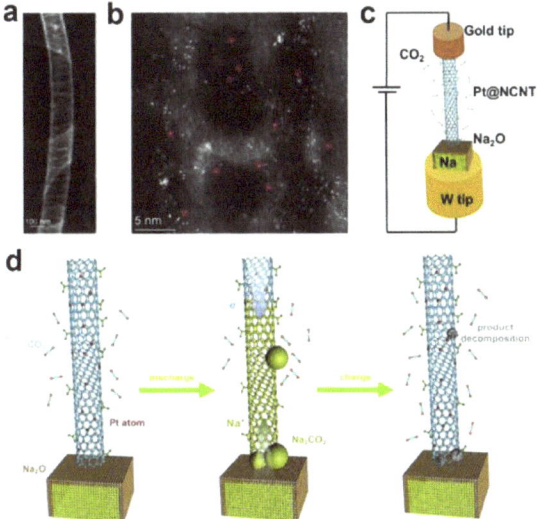

Figure 7. STEM-HAADF images of Pt@NCNT at low magnification (**a**) and high magnification (**b**). Red, circle in 7b shows the presence of single Pt atoms. (**c**) Schematic diagram of Na-CO$_2$ nanobatteries constructed in ETEM. (**d**) Schematic diagram of the discharge/charge electrochemical processes in the Na-CO$_2$ nanobattery. Reproduced with permission from [44]. Copyright 2020, Elsevier Ltd.

Given the two important factors of catalyst performance and cost, non-precious metal SACs with similar activity to noble metal SACs are more attractive [93,94]. Future work should target the combination of nitride or carbide supports with non-precious metal single atoms, which may provide unique electronic interactions with the metal and thus improve anode electrochemical catalytic performance [95]. In addition, SACs applied to Na-CO$_2$ batteries require harsh operating conditions, and the development of industrial fabrication methods for low-cost, stable, highly metal-loaded SACs is critical.

4. Progress of Solid Electrolyte and Interface Research

Many Na-CO$_2$ batteries reported still use the conventional organic liquid electrolyte [6,43], because Na-CO$_2$ batteries operate in a semi-open system, there are great safety hazards in actual application, such as the inherent volatility and flammability of organic solvents, and potential problems such as reduced battery life due to interfacial side reactions under high pressure [7]. Replacing the liquid electrolyte with a solid electrolyte can not only effectively solve the above safety hazards, but also inhibit the growth of dendrites and prevent CO$_2$ from corroding the metal Na anode during the long-term cycling process [11]. In addition, through scientific and reasonable structural design, solid electrolyte can also provide wider electrochemical window, higher energy density and longer cycle life [96–98].

For the solid electrolyte of the Na-CO$_2$ batteries, in addition to meeting the necessary energy requirements for a solid sodium ion conductor, e.g., high ionic conductivity, non-conductivity, interface compatibility, simplicity of preparation, low cost and environmental friendliness, it needs to be highly resistant to superoxide and peroxide to facilitate the smooth Na-CO$_2$ reaction [11]. It is difficult for the reported sodium ion solid electrolyte for Na-CO$_2$ batteries to meet all the above requirements at the same time, further research is needed [99,100]. In this section, we provide a brief overview of several important applications of solid electrolyte in the Na-CO$_2$ batteries, and analyze the research on solid electrolyte interface. Sodium-based electrolytes can be classified into three categories according to the chemical composition of the electrolyte and the transport mechanism of sodium ion, namely inorganic solid electrolytes (ISEs), polymer electrolytes (PEs) and composite polymer electrolytes (CPEs).

4.1. Inorganic Solid Electrolytes

Inorganic solid electrolytes (ISEs) are usually characterized by high ionic conductivity ($10^{-5} \sim 10^{-2}$ S cm^{-1}), high ionic mobility number, mechanical rigidity, non-flammability and non-flow characteristics due to their structural properties [101]. In order to understand the reasons for the high ionic conductivity of ISEs and to further improve its ionic conductivity, it is necessary to understand its ionic transport mechanisms and properties. The three principal migration mechanisms shown in Figure 8a are vacancy leap, gap-site leap and linkage leap [102]. According to the mechanism of vacancy leap and gapsite leap ion transport, the ionic transport is mainly related to the activation energy of the material and the number of defects in the vacancy, so ion doping can effectively improve the ionic conductivity of the material [103–105]. The ionic transport in linkage leap is not directly jumping between vacancies, but knocking the ions of adjacent sites to make them migrate to vacancies, so the ionic transport potential of linkage leap mechanism is lower than that of vacancy or gap site leap, so increasing the concentration of sodium ions in ISEs can improve the ionic conductivity [106–109]. ISEs can be classified according to their components and crystal structure types: NASICON, Na-beta-Al$_2$O$_3$, sulfide electrolytes and complex hydride electrolytes.

4.1.1. NASICON Structured Electrolytes

Na$_{1+x}$Zr$_2$Si$_x$P$_{3-x}$O$_{12}$ ($0 \leq x \leq 3$) was proposed by Hong and Goodenough in 1976 and consists of NaZr$_2$(PO$_4$)$_3$ and Na$_4$Zr$_2$(SiO$_4$)$_3$ solid solution; the NASICON-structured fast ion conductor material with a 3D transport channel for Na$^+$ was obtained by partially replacing the pentavalent P in NaZr$_2$(PO$_4$)$_3$ by the tetravalent Si and introducing Na$^+$ to balance the charge [110,111]. Among them, Na$_{1+x}$Zr$_2$Si$_x$P$_{3-x}$O$_{12}$ ($0 \leq x \leq 3$) fast ionic conductors are CO$_2$-stable ISEs. Due to its high ionic conductivity, good thermal and chemical stability, Na Superionic conductor (NASICON) structure electrolytes is the most widely studied inorganic material in the solid electrolyte of Na-CO$_2$ batteries [36,42,112,113]. Kim et al. [114] demonstrated that after NASICON materials contact with water, H$_3$O$^+$ occupies part of the Na$^+$ sites, causing a shift in the NASICON peak position, as revealed by XRD analysis. Although NASICON reacts with water, it does not cause the fracture and decomposition of the NASICON ceramic sheet, and there is no drastic decay in ionic conductivity.

According to the crystal structure analysis of this series of materials by Hong (Figure 8b) [111], the crystal structure of the material is monoclini with space group C2/c when $1.8 \leq x \leq 2.2$ at room temperature; when x is outside this range, the crystal structure becomes tripartite (rhombic structure) with space group R-3c; with the change in temperature, the two crystal structures can be converted, and the phase transition temperature depends on the specific components of the material, usually at 150~200 °C. In the triangular structure (R-3c), SiO$_4$ or PO$_4$ tetrahedra are connected to ZrO$_6$ octahedra at the top corners, forming the three-dimensional skeleton for Na$^+$ transport. Therefore, the most commonly used method to increase the ionic conductivity is doping. Heterogeneous element doping is used to enhance ionic conductivity by changing the local structure of the crystal, making the Na$^+$ transport channel larger and lowering the ion diffusion potential barrier [107]. Among Na$_{1-x}$Zr$_2$Si$_x$P$_{3-x}$O$_{12}$ ($0 \leq x \leq 3$), Na$_3$Zr$_2$Si$_2$PO$_{12}$ (x = 2) with the monoclinic phase structure is more stable. It exhibits the highest ionic conductivity of ~10^{-4} and ~10^{-1} S cm^{-1} at room temperature and 300 °C, respectively [110]. As shown in Figure 8c, Song et al. investigated the crystal structure and ionic conductivity of NASICON doped with different contents of alkaline earth metal ions (Mg, Ca, Sr, Ba) [105]. It was found that Mg ion doping improved its conductivity most significantly, and the ionic conductivity of Na$_{3.1}$Zr$_{1.95}$Mg$_{0.05}$Si$_2$PO$_{12}$ could reach 3.5×10^{-3} S cm^{-1} at room temperature. However, Ba^{2+} ion substituted compounds by the elements with larger ionic radii exhibit narrower bottlenecks than the original compounds, suggesting that the ionic radius of the substituent plays an important role [105]. Lu et al. prepared $Na_{3.2}Zr_{1.9}Mg_{0.1}Si_2PO_{12}$ matrix materials by doping Mg^{2+} with Zr sites, showing an ionic conductivity of 1.16 mS cm^{-1} at room temperature in order to obtain high-performance Na-CO_2 battery [113] (Figure 8d). In addition, heterogeneous elements with similar ionic radii to Zr (including calcium ions [115] and aluminum ions [116]) doped NASICON-based composite electrolytes can significantly improve the ionic conductivity of NZSP by changing the Na^+ concentration and crystal localization structure, thus improving the Na^+ transport channels [117].

Another way to improve the ionic conductivity of NASICON materials is to reduce the grain boundary resistance by adjusting the grain size and grain boundary chemistry [100]. Ihlefeld et al. investigated the size effect on the grain boundary resistance of $Na_{1+x}Zr_2Si_xP_{3-x}O_{12}$ (0.25 < x < 1.0) [118]. The results showed that changing the Si/P ratio, increasing the process temperature and decreasing the annealing temperature all reduced its grain size and thus the ionic conductivity is increased. Hu et al. synthesized La^{3+} doped NASICON by adding $La(CH_3COO)_3$ to the precursor through a self-growth strategy [103]. XRD results showed that several new phases of $Na_3La(PO_4)_2$, La_2O_3 and $LaPO_4$ appeared at the grain boundaries of the final electrolyte material (Figure 8e). The newly appeared phases adjusted the chemical composition of the grain boundaries and increased the ionic conductivity at the grain boundaries.

4.1.2. Na-Beta-Al_2O_3

Na-beta-Al_2O_3 has become one of the most studied sodium ion solid electrolytes due to its high ionic conductivity and suitable mechanical properties, and was first applied in solid-state Na-S batteries [119–121]. Na-beta-Al_2O_3 has two crystal structure types as shown in Figure 8f, both are layered structures made of alternating stacks of spinel and sodium-conducting layers, with Na^+ conducting in two dimensions between two adjacent spinel stacks, called sodium-conducting layers. one crystal structure of Na-beta-Al_2O_3 is a hexagonal crystal system structure made of two spinel structures stacked with space group $P6m_3/mmc$, labeled beta-Al_2O_3, composed of Na_2O-(8~11) Al_2O_3; the other is a tripartite crystal structure, consisting of three spinel structures stacked in the space group R3m, labeled β''-Al_2O_3, composed of Na_2O-(5~7) Al_2O_3.

A higher proportion of β''-Al_2O_3 phase is usually desired in the material because the ionic conductivity of the β''-Al_2O_3 phase is higher than that of the β-Al_2O_3 phase. This is not difficult to analyze: the two adjacent layers of spinel structure are connected by O^{2-} in the Na^+ conducting sodium layer, forming an Al-O-Al bond and Na^+ two-dimensional ion transport along the ab plane. For the β-Al_2O_3 phase, O^{2-} in the Na^+ conduction layer has a higher electrostatic gravitational force on the surrounding Na^+ and can accommodate a smaller amount of Na^+, while for the β''-Al_2O_3 phase, O^{2-} in the Na^+ conduction layer has a lower electrostatic gravitational force on the surrounding Na^+ and can accommodate more Na^+, thus the ionic conductivity of the β''-Al_2O_3 phase is higher than that of the β-Al_2O_3 phase. Single-crystal β''-Al_2O_3 can exhibit ionic conductivity up to 1 S cm^{-1} at 300 °C, while polycrystalline structures exhibit 0.002 S cm^{-1} at room temperature and 0.2–0.4 S cm^{-1} at 300 °C. The lower ionic conductivity exhibited by polycrystalline structures is due to the high grain boundary resistance in polycrystalline β''-Al_2O_3. However, pure β''-Al_2O_3 is a thermodynamically sub-stable phase, which decomposes into Al_2O_3 and β-Al_2O_3 at 1500 °C and has relatively poor mechanical properties (200 MPa). Na-beta-Al_2O_3 is generally prepared by solid-phase method; however, the solid-phase synthesized Na-β''-Al_2O_3 powder has residual $NaAlO_2$ at the grain boundaries of the β''-Al_2O_3 phase and β-Al_2O_3 phase, which is unstable in air and easily reacts with CO_2 and H_2O [122,123]. Virkar et al. used steam-assisted method to obtain dense ceramic flakes by sintering Y-ZrO_2 and α-Al_2O_3 at 1450 °C to enhance their chemical stability [124].

The research of this material focuses on achieving stabilization in CO_2 atmosphere and reducing the impedance of grains and grain boundaries by suitable preparation methods, enhancing the ratio of β''-Al_2O_3 and reducing unwanted by-products. Doping is usually used to stabilize the β″-Al$_2$O$_3$ phases, such as Li$^+$, Mg^{2+}, Ni^{2+} and Ti^{4+}; in addition, the overall mechanical strength can be enhanced by synthesizing mixed crystals of β″-Al$_2$O$_3$ and β-Al$_2$O$_3$ or by adding zirconium oxide [125].

4.1.3. Sulfide Electrolytes

Compared with oxide solid electrolytes, sulfides as electrolyte materials have higher ionic conductivities and lower grain boundary resistances, which is due to the intrinsic characteristics of sulfur [126]. S is less electronegative than O, which is less binding to Na$^+$ and facilitates the free movement of Na$^+$, thus it has lower grain boundary resistance; S has a larger ionic radius than O, and S replaces O to expand the lattice structure and form channels that facilitate the diffusion of Na$^+$ and thus has a high ionic conductivity [106,127]. In addition, sulfide electrolytes also have the advantages of mild synthesis conditions, good mechanical strength, good ductility, etc. However, sulfide is unstable in humid air, easy to absorb water and easy to decompose with water in air, releasing toxic H$_2$S gas [128–130]. Therefore, it is necessary to improve the ionic conductivity and air stability by designing new sulfide-based electrolytes for the application in metal–CO$_2$ batteries.

Na$_3$PS$_4$ has two crystal structures, the tetragonal phase (P-42$_1$c, $a = b = 6.9520$ Å, $c = 7.0757$ Å) and the cubic phase (I-43m, $a = b = c = 7.0699$ Å) [131,132]. The Figure 8g shows that in the cubic phase, Na$^+$ is distributed in two distorted tetrahedral interstitial sites with space group I-43m, while in the tetragonal phase, Na$^+$ is distributed in a tetrahedral site and an octahedral site with space group P-42$_1$c. Typically, Na$_3$PS$_4$ exists as a tetragonal phase, which can be transformed into a cubic phase at about 530 K. The cubic phase monomer has the lowest activation energy and has been studied the most. In 2012, the Hayashi and colleagues reported glass-ceramic sulfide electrolytes with an ionic conductivity of 2 × 10^{-4} S cm^{-1} at room temperature, and the high ionic conductivity can be attributed to the stability of cubic Na$_3$PS$_4$ in microcrystalline glass electrolytes at room temperature [126]. It was found that the ionic conductivity of several crystalline phases of Na$_3$PS$_4$ sulfide electrolytes, has the following pattern: ionic conductivity of the cubic phase > ionic conductivity of the tetragonal phase > ionic conductivity of the glass-ceramic phase > ionic conductivity of the glass phase. Therefore, how to obtain a stable cubic phase Na$_3$PS$_4$ crystal structure at room temperature is one of the methods to improve the ionic conductivity of sulfide solid electrolytes.

Tuning the size of unit cell/channel by introducing Na vacancies, gaps or modulating the interaction between Na$^+$ and anion skeleton in the lattice by means of elemental doping is an important method to improve the ionic conductivity of sulfide electrolytes. The doping of P sites with homologous As^{5+} of larger ionic radius expands the lattice and introduces Na vacancies while increasing the distance between Na-S [128,133]. In the tetragonal phase of Na$_3$PS$_4$, replacing S^{2-} with negative monovalent F$^-$, Cl$^-$, Br$^-$ and I$^-$ can also introduce Na vacancies according to the charge balance theory, thus increasing the migration probability of Na$^+$ from one site to neighboring sites and thus increasing the ionic conductivity [134,135]. De Klerk and Wagemaker investigated the effect of Na$^+$ vacancies on the ionic conductivity of Na$_3$PS$_4$ by molecular dynamics (MD) simulations: a significant increase at 300 K compared to pure cubic Na$_3$PS$_4$ (0.17 S cm^{-1}) [136]. Experimental results show that the ionic conductivity of chlorine-doped tetragonal sulfide (Na$_{2.9375}$PS$_{3.9375}$Cl$_{0.0625}$) at 303 K is 1.14 × 10^{-1} S cm^{-1}, which is much higher than that of the pristine tetragonal Na$_3$PS$_4$ (5 × 10^{-5} S cm^{-1}) [135]. The first-principles study shows that the Na$^+$ gap defect structure can also increase the Na$_3$PS$_4$ carrier density, i.e., in the cubic phase of Na$_3$PS$_4$, doping with tetravalent ions M^{4+} (M = Si, Sn, Ti, Ge) replaces the P^{5+} sites, while more Na$^+$ will be introduced to maintain the electroneutrality and broaden the Na$^+$ channel size, thus reducing the gap migration barrier and increasing its ionic conductivity [127]. The silica sulfide doped microcrystalline glass electrolytes (94Na$_3$PS$_4$·6Na$_4$SiS$_4$) also show higher ionic conductivity (7.4 × 10^{-4} S cm^{-1}) compared to the undoped state [120]. In addition, the interaction between Na$^+$ and the anionic backbone has a significant effect on the ionic conductivity of the electrolyte. Theoretical calculations show that Se doping can increase the polarizability of the anionic framework, smooth the lattice and lowering the activation potential barrier [137].

However, to apply sulfide electrolytes in metal–air batteries, it is important to make them stable in air. In 2016, Liang et al. proposed an air-stable sulfide electrolyte [138]. When the P-site of Na_3PS_4 is completely replaced by Sb to obtain Na_3SbS_4, the unit cell and channels are expanded to further improve the ionic conductivity of Na_3PS_4, exhibiting a high ionic conductivity of 1 mS cm^{-1} at 25 °C and a good compatibility with sodium metal anodes. More importantly, Na_3SbS_4 is stable when exposed to O_2 and H_2O, which can be well explained by the theory of soft and hard acids and bases [138,139]. In Na_3PS_4, the interaction between O^{2-} and P^{5+} (hard acid) is stronger, so S^{2-} (soft base) is easily replaced by O^{2-} (hard acid), while in Na_3SbS_4, Sb^{5+} (soft acid) and S^{2-}. The interaction between them is strong and not easy to be destroyed by O_2. When exposed to air, pure Na_3SbS_4 tends to form Na_3SbS_4-xH_2O and H_2O may reversibly lose after heating at 150 °C for 1 h (Figure 8h). Despite the attractive stability and ionic conductivity of Na_3SbS_4 (~10^{-3} S cm^{-1}), the toxicity of Sb should be considered. At present, the research on sodium sulfide ionic solid electrolytes is still at the early stage and further studies are needed to improve their ionic conductivity, chemical stability and electrochemical stability.

4.1.4. Complex Hydride Electrolytes

In addition to the inorganic solid electrolytes mentioned above, complex hydride electrolytes are also good conductors of Na ions. In 2012, Orimo and colleagues first reported complex hydrides as Na$^+$ ion solid electrolytes [140]. The reported complex hydride electrolytes have high ionic conductivities and ion mobility numbers, but complex hydrides are usually prone to water absorption, difficult to maintain stably in air. The electrochemical window is also not meet the requirement. If its chemical stability can be improved by modification, it may provide a new solution for the application of Na-CO_2 batteries.

Complex hydrogen compounds consist of a metal cation Na$^+$ and a complex anion consisting of a central atom and a ligand hydrogen atom, which have been reported as $Na_2(BH_4)(NH_2)$ [141], $Na_2B_{12}H_{12}$ [142,143], $NaCB_{11}H_{12}$ [144], $NaCB_9H_{10}$ [145], $Na_2(B_{12}H_{12})_{0.5}(B_{10}H_{10})_{0.5}$ [146,147], $Na_2(CB_9H_{10})(CB_{11}H_{12})$ [144], $Na_2B_{10}H_{10}$ [148] and Na_3OBH_4 [149], etc. For example, $Na_2B_{12}H_{12}$ exhibits high ionic conductivity (>0.1 S cm^{-1} at 573 K) due to its high-temperature disordered bulk-centered cubic phase (cation vacancy rich structure). However, due to the high phase transition temperature, the complex hydride electrolytes of large anions cannot meet the requirements of practical applications in different fields. It is necessary to reduce or eliminate the phase transition temperature of complex hydrides. It has been shown that the phase transition temperature can be significantly lowered after the introduction of C by chemical modification of the anion [145,150], for example, $NaCB_{11}H_{12}$ vs. $Na_2B_{12}H_{12}$: 380 and 529 K; $NaCB_9H_{10}$ vs. $Na_2B_{10}H_{10}$: 290 and 380 K, with ionic conductivity of 7×10^{-2} S cm^{-1} at room temperature (Figure 8i). In addition, mixing different anions helps to reduce or eliminate the transition temperature due to the introduction of geometric hindrance [140]. Tang and colleagues found that the transition temperature can also be effectively reduced by ball milling by reducing the grain size and disordering, and that a more homogeneous mixing of atoms is less prone to phase transitions [151]. Further studies should focus on investigating the stability and ionic conductivity of composite hydrides and applying them to solid-state Na-CO_2 batteries.

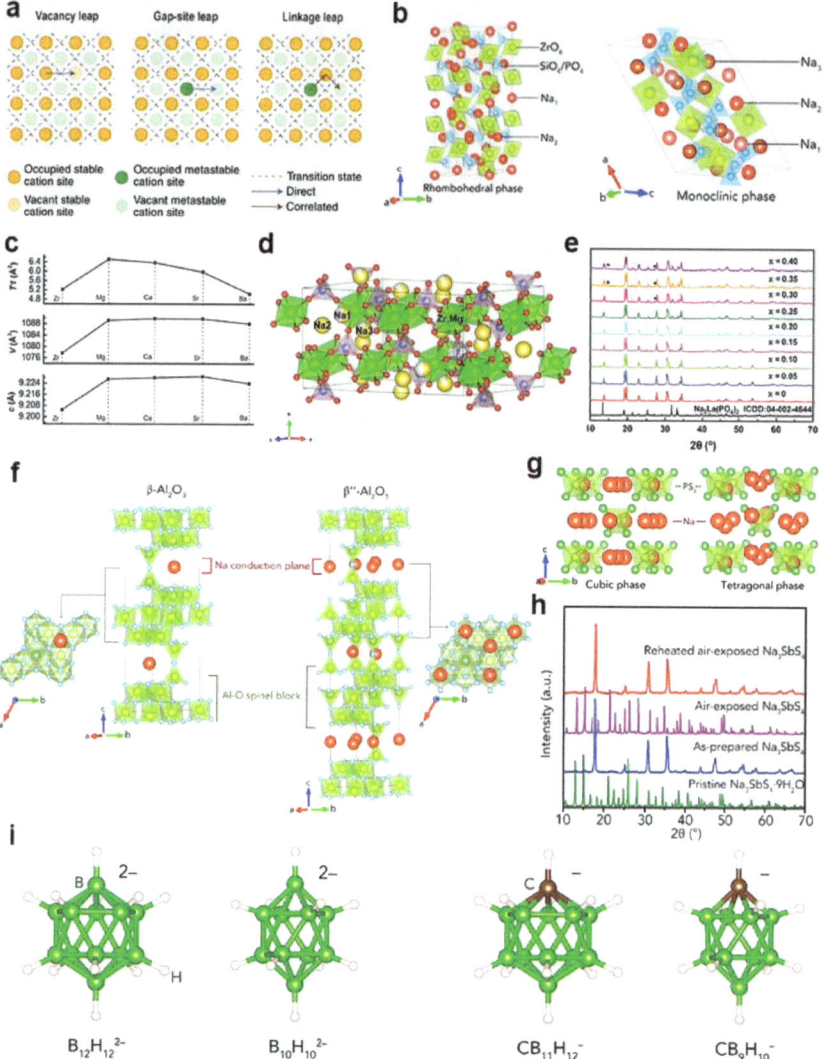

Figure 8. (a) Arrows indicate the three typical migration mechanisms: vacancy site leap, gap site leap and linkage leap. The circles indicate cations in stable (green) and substable (orange) positions in the model lattice. The dashed lines indicate the transition states of cation jumps imposed by the anion skeleton (not explicitly shown). Reproduced with permission from [102]. Copyright 2019, Springer Nature. (b) Crystal structures of representative NASICON ($Na_3Zr_2Si_2PO_{12}$) with rhombic and monoclinic crystalline phase. Reproduced with permission from [111]. Copyright 1976, Elsevier Ltd. (c) Lattice parameter, volume of unit cell and area of the bottleneck (marked as T1 in the inset) of $Na_3Zr_2Si_2PO_{12}$ and $Na_{3.1}Zr_{1.95}M_{0.05}Si_2PO_{12}$ (M = Mg, Ca, Sr, Ba). Reproduced with permission from [105]. Copyright 2016, Springer Nature. (d) Simplified view of the structure of $Na_{3.2}Zr_{1.9}Mg_{0.1}Si_2PO_{12}$ approximately along the [101] direction. Reproduced with permission from [113]. Copyright 2022, John Wiley and Sons. (e) X-ray diffraction patterns. The X-ray diffraction patterns of the nominal composition $Na_{3+x}La_xZr_{2-x}Si_2PO_{12}$ (x = 0, 0.05, 0.10, 0.15, 0.20, 0.25, 0.30, 0.35, 0.40) solid electrolytes. The peaks in the circled by the dotted box represent the $Na_3La(PO_4)_2$ minor phase. The peaks marked by the black solid spheres and black asterisks denote the La_2O_3 and

LaPO$_4$, respectively. Reproduced with permission from [103]. Copyright 2017, John Wiley and Sons. (**f**) Crystal structures of β-Al$_2$O$_3$ and β″-Al$_2$O$_3$. Reproduced with permission from [100]. Copyright 2018, Elsevier Ltd. (**g**) Crystal structures of Na$_3$PS$_4$ with cubic and tetragonal phase. Reproduced with permission from [131]. Copyright 2017, Reproduced with permission from. (**h**) XRD patterns of pristine Na$_3$SbS$_4$·9H$_2$O, as-prepared Na$_3$SbS$_4$, air-exposed Na$_3$SbS$_4$ and reheated air-exposed Na$_3$SbS$_4$ (150 °C for 1 h under vacuum). Reproduced with permission from [138]. Copyright 2016, John Wiley and Sons. (**i**) Relative geometries of the B$_{12}$H$_{12}{}^{2-}$, B$_{10}$H$_{10}{}^{2-}$, CB$_{11}$H$_{12}{}^{-}$ and CB$_9$H$_{10}{}^{-}$ anions. Reproduced with permission from [148]. Copyright 2014, John Wiley and Sons.

4.2. Polymer Electrolytes

The concept of polymer electrolytes (PEs) was first proposed in 1973 when Fenton et al. discovered that alkali metal salts dissolved in polyethylene oxide were found to form conductive complexes [152]. Subsequently, Feuillade et al. introduced organic plasticizers into the polymer–salt binary system to obtain a quasi-solid electrolyte [153]. Compared with inorganic solid electrolytes, electrolytes with polymer matrix usually have good flexibility, easy processing and high tolerance to vibration, shock and mechanical deformation and better interfacial contact and compatibility between electrodes and electrolytes.

For the ion mobility number and ionic conductivity of PEs, the methods of increasing amorphous regions and fixing anions are usually used. Soluble metal salts with anion stabilizing effect are generally preferred [11], such as NaCF$_3$SO$_3$, NaPF$_6$ and NaClO$_4$; in addition, a certain percentage of inorganic fillers are also added to enhance the amorphous region and ionic transport of organic electrolytes, thus improving ionic conductivity, mechanical strength and electrochemical properties. Polymer electrolytes are classified into solvent-free polymer electrolytes (SPEs) and gel polymer electrolytes (GPEs) based on their composition and physical form, and then explains the ion transport mechanisms, basic properties and their applications in Na-CO$_2$ batteries [99].

4.2.1. Solvent-Free Polymer Electrolytes

Solvent-free polymer electrolytes (SPEs) are usually consisted of only polymer matrix and lithium salt as solute without adding liquid solvent as plasticizer, and can be easily fabricated by solvent casting, thermoforming or extrusion techniques. SPEs that have been reported for Na-CO$_2$ batteries are mainly of the polyethylene oxide (PEO) type [25,30]. The carbonate electrolytes are susceptible to nucleophilic attack by superoxide radicals and are not suitable for metal–air batteries, while other types of SPEs are still to be developed.

PEO is the earliest and most studied, and its chemical structure is H-(O-CH$_2$-CH$_2$)$_n$-OH, a polyether compound with the advantages of good chemical stability, good compatibility with alkali metal negative electrode, good flexibility, good water solubility, low density, good viscoelasticity and easy film formation, and tolerance to superoxide radicals, which is a more suitable organic material for Na-CO$_2$ battery electrolyte. However, the current PEO-based solid electrolytes during application are often troubled by problems such as: the relatively high degree of room temperature crystallization of PEO (PEO chains are mainly crystallized at 65 °C), resulting in low room temperature ionic conductivity (~10^{-8} S/cm), thus requiring operation at higher temperatures; low upper limit of electrochemical stability potential ($\leq$4.2 V), thus preventing the use of high voltage cathode materials; poor dimensional thermal stability (softening point of 55–64 °C); low mechanical strength ($\leq$10 MPa).

According to the ionic transport mechanism of PEO-based PEs (Figure 9a), the metal salts dissociate and delocalize metal cation (Li$^+$, Na$^+$), metal ions with polar groups on the polymer chain, such as O and N. Under the action of electric field and above the glass transition temperature, the polymer molecular chain segments in the amorphous region (amorphous region) are able to vibrate, and with the movement of the polymer chain, the metal cations are continuously complexed and dissociated with the groups on the polymer chain segments. Metal cation jumps from one coordination site (usually composed of more than three electron-giving groups) to the next, thus enabling the transfer of metal ions [154]. However, since the movement of chain segments transfers both metal cation and anions, the metal cation migration number of PEs is usually less than 0.5 and their ionic conductivity is generally inferior to that of ISEs [155–157].

The following aspects can be used to improve the comprehensive performance of PEO-based polymer electrolytes: (1) to facilitate the migration of metal ions, the polymer-cation interaction must be a compromise between sufficient strength (ensuring salt solubility through cation solventization) and sufficient instability (facilitating ion hopping from one coordination site to another); (2) selecting a polymer matrix with a dielectric constant a polymer matrix with a large dielectric constant, a high dielectric constant facilitates effective charge separation of the metal salt and thus a high Na^+ concentration; (3) the higher backbone flexibility and motility of the PEO chain facilitates the segmental movement of the polymer chain and (4) the high molecular weight of the polymer matrix is also desirable to obtain a polymer electrolyte with better mechanical strength.

In 2018, Chen's team obtained a high-performance $PEO/NaClO_4/SiO_2$ all-solid polymer electrolyte by adding inorganic filler nano-SiO_2 and assembled an all-solid flexible $Na-CO_2$ batteries with sodium as the anode, $PEO/NaClO_4/SiO_2$ as the electrolyte and multi-walled carbon nanotubes as the cathode [30]. In the all-solid polymer electrolyte, PEO acts as a Na ion conductor to transport Na between the Na anode and the gas cathode, and the addition of nano-SiO_2 can reduce the crystallinity and promote the dissociation of $NaClO_4$, thus improving the ionic conductivity, in addition to improving its mechanical strength and thermal stability. The highest sodium ionic conductivity of 6.4×10^{-4} S cm^{-1} and the highest ion transfer number of 0.56 were obtained at an operating temperature of 70 °C when the SiO_2 content in the polymer electrolyte was 3 wt% (200 μm thickness). Assembled flexible batteries with good bendability (21,000 times), foldability and shape customizability, as well as in the bending state (0~360°), stable operation time (80 h), high cycling stability (240 cycles with −0.4 V increase in overpotential), high capacity (450 mAh g^{-1}) and high energy density (173 Wh kg^{-1}) were obtained in electrochemical tests. Furthermore, Sun et al. developed a non-aqueous solution of expanded perfluorosulfonic ethylene carbonate and propylene carbonate (EC-PC) with a hexagonal structure of $Na_{0.67}Ni_{0.23}Mg_{0.1}Mn_{0.67}O_2$ as the cathode, and metal sodium as the anode. It proved to be a safe and durable all-solid-state Na-ion battery [158].

The research on SPEs for $Na-CO_2$ batteries is obviously insufficient, and subsequent studies need to further explore and investigate the Na^+ transport mechanism of SPEs to elucidate the intrinsic connection between Na^+ and polymer matrix, and to further develop polymer electrolyte matrix that can be applied to $Na-CO_2$ battery systems by simple blending, copolymerization, hyperbranching or cross-linking methods to find electrolyte materials with better overall performance.

4.2.2. Gel Polymer Electrolytes

Gel polymer electrolytes (GPEs) is a semi-solid electrolyte in a gel state, consisting of a polymer matrix, electrolyte salt and plasticizer. For GPEs, ion transport is the result of synergistic transport between the solid polymer matrix and the liquid electrolyte, and their room temperature ionic conductivity is higher than that of SPEs due to the presence of a certain amount of solvent; however, the safety in batteries is poorer than that of SPEs due to the presence of organic solvents. In addition, GPEs have both polymeric properties and are more flexible and easier to process than glass-ceramic type solid electrolytes. Therefore, GPEs exhibit good mechanical properties and good compatibility with electrodes. In quasi-solid polymer electrolytes, sodium ion transport occurs mainly in the liquid plasticizer containing dissolved lithium salts, while the polymer matrix provides mechanical strength to the GPEs and keeps it quasi-solid, thus minimizing the safety risk caused by leakage of liquid components. During battery charging and discharging, the plasticizer in the GPEs reacts on the electrode surface to form a solid electrolyte interface (SEI) film, similar to a liquid electrolyte. In contrast, electrochemically inert polymeric matrices are typically not involved in SEI formation.

PVDF-based gel electrolytes are the most widely studied matrix materials for GPEs due to their good film formation, large dielectric constant, high glass transition temperature and strong electron-absorbing groups. In 2017, Hu et al. [159] reported a quasi-solid Na-CO_2 battery with an applied PVDF-HFP-4% SiO_2/$NaClO_4$-TEGDME CPE with advantages such as high ionic conductivity (1.0 mS cm^{-1}), strong toughness, non-flammability (automatically extinguished within 1 s even if ignited for 5 s) and operated at a capacity of 1000 mAh g^{-1}, 400 cycles at 200 °C, as shown in Figure 9b,c. The product has shown higher safety and good performance.

In addition to the above PVDF-based gel polymers, PEO-based and cyano polymer-based GPEs are widely used in lithium batteries, and related sodium ion gel electrolytes and their applications in sodium batteries are yet to be developed.

PEO-based GPEs have a high degree of crystallization at room temperature and low ionic conductivity at room temperature, making them difficult to apply in practical batteries. Suitable plasticizers can be added, such as polyethylene glycol [153] or crown ether [160], to reduce the crystallinity and enhance the room temperature ionic conductivity. Co-blending [161], cross-linking [162] or the addition of inorganic fillers to improve the mechanical strength and thermal stability.

Cyano (e.g., polyacrylonitrile) [153] cyano (such as polyacrylonitrile and cyanoethyl polyvinyl ether) is a polar group with a high dipole moment [163], and is a strong electron-absorbing group with a dielectric constant of about 30. Cyano has high oxidation resistance and its introduction into the polymer matrix can increase the oxidative decomposition voltage of the electrolyte. Feuillade et al. [153] prepared polyacrylonitrile electrolytes with ionic conductivity close to 1×10^{-3} S cm^{-1} at room temperature. In polyacrylonitrile gel electrolytes, the ionic conductivity increases with increasing plasticizer and salt content. However, due to the strong polarity of the cyano group, there is a large passivation effect on the metal negative electrode leading to an increase in the interfacial resistance, thus requiring an effective negative interface protection.

4.3. Composite Polymer Electrolytes

Composite polymer electrolytes (CPEs) combining the soft mechanical properties of organic solid electrolytes and the high ionic conductivity of inorganic solid electrolytes offer a promising direction for highly stable Na-CO_2 battery electrolytes [97,115,164]. PEs have a generally low ionic conductivity. Inorganic powders incorporated into solid polymer electrolytes to obtain composite solid electrolytes can effectively reduce the orderly arrangement of polymer chains and prevent polymer crystallization; and more ion transport channels are formed in the surface region of nanoparticles, which helps to improve ionic conductivity. In 1988, Skaarup et al. added fast ionic conductor particles Li_3N as fillers to polyethylene oxide (PEO)-$LiCF_3SO_3$ matrix to obtain a composite polymer electrolyte, which not only improved its mechanical strength, but also had a higher ionic conductivity than the pure polymer electrolyte [165]. More interestingly, Wieczorek et al. found that the addition of nonionic fillers such as Al_2O_3 also improved the ionic conductivity of PEO-based polymer electrolytes, mainly due to the increase in the amorphous phase [166]. In addition, the addition of inorganic filler can improve the mechanical properties of electrolyte and maintain good flexibility, which improves the interfacial contact between electrolyte and electrode.

Inorganic fillers are mainly divided into two major categories: one is inert fillers, which do not have ion transport capability by themselves, such as Al_2O_3 [167], SiO_2 and TiO_2 [168], etc.; the other is active fillers, which have ion transport capability by themselves, such as Na_2SiO_3 [169], NASICON [170], β''-Al_2O_3 [171] and other inorganic solid electrolytes. Sun's group prepared a high performance PVDF-HFP-$Na_{3.2}Zr_{1.9}Mg_{0.1}Si_2PO_{12}$ organic-inorganic composite solid electrolyte [113]. The $Na_{3.2}Zr_{1.9}Mg_{0.1}Si_2PO_{12}$ inorganic material with an ionic conductivity of 1.16 mS cm^{-1} at room temperature was obtained by replacing the Zr ion in $Na_3Zr_2Si_3PO_{12}$ with Mg^{2+}, which was then combined with structurally stable PVDF-HFP with excellent mechanical properties as a solid electrolyte for Na-CO_2 batteries.

After 120 cycles at cut-off capacities of 200 mA g^{-1} and 500 mAh g^{-1}, the intermediate gap voltage was below 2 V. The Coulomb efficiency, multiplicative performance and cycling performance of the batteries were significantly improved.

In addition to the above method of compounding inorganic particles directly with polymers, composite solid electrolytes can also be prepared by compounding functional polymers, functional inorganic particles, or polymer/inorganic particles together with a polymer matrix. For example, blending cellulose in sodium ionomer solid polymers [172] or glass fiber [173] can effectively enhance the mechanical properties of the electrolyte film; or adding functional inorganic particles to immobilize the anions on the polymer matrix chains and inhibit the formation of polarization centers, thus promoting the movement of cations [174].

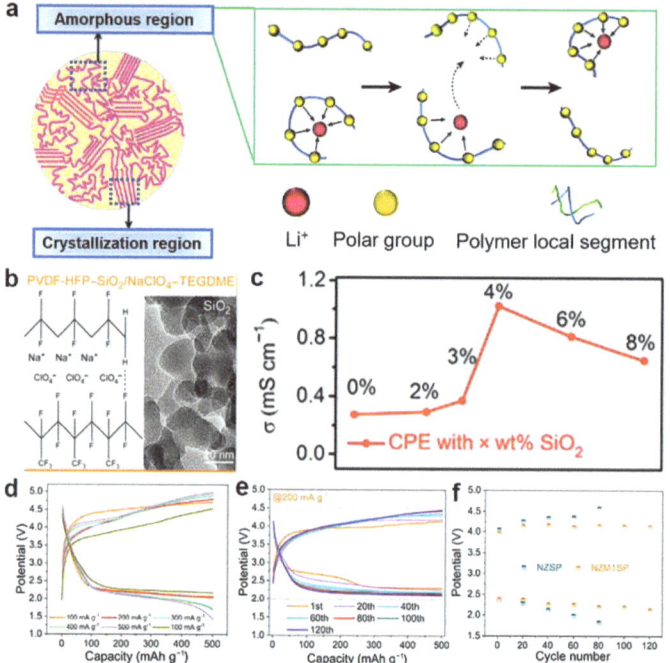

Figure 9. (**a**) Schematic illustration of Li ion transport mechanism in PEO-based SPEs. Reproduced with permission from [154]. Copyright 2016, Royal Society of Chemistry. (**b**) Composition of the CPE. Inset: Transmission electron microscopy (TEM) image of fumed SiO$_2$. (**c**) Ionic conductivity of CPE with various contents of SiO$_2$. Reproduced with permission from [159]. Copyright 2017, the American Association for the Advancement of Science. (**d**) Charge/discharge curves tested at different current densities. (**e**) Cut-off capacity of NZM1SP-PVDF-HFP electrolytes at 200 mA g^{-1} for 500 mAh g^{-1}. (**f**) Corresponding changes in intermediate charging and discharging voltages of cells with NZSP-PVDF-HFP and NZM1SP-PVDFHFP electrolytes. Reproduced with permission from [113]. Copyright 2022, John Wiley and Sons.

4.4. Interfaces of Solid Electrolytes

During the electrochemical cycling of a solid-state Na-CO$_2$ batteries, the interface problems between the solid electrolyte and the electrode include two main ones: interfacial close contact and interfacial compatibility (no chemical reaction).

First, the close contact between the interfaces will facilitate the rapid transport of electrons or ions. In the conventional liquid electrolyte battery, the electrolyte has extremely high fluidity and good wettability to the electrode material, which ensures a low contact impedance inside the battery. In contrast, in solid-state batteries, the effective contact area is greatly reduced due to the point to point solid–solid contact between the electrolyte and the electrode, thus the interfacial impedance between the solid electrolyte and the electrode is very high (Figure 10a). Therefore, the current research work on solid-state batteries is mainly focused on improving the contact between solid electrolyte and electrode material and reducing the interfacial impedance. Sun et al. developed a monolithic symmetric cell for solid-state sodium batteries (SSSBs). A three-dimensional (3D) electronic and ionic conductive network is formed by integrating sodium anodes into NZSP-type monolithic structure. The interfacial resistance of the monolithic symmetric cell was significantly reduced, and exhibited a stable sodium plating/strip cycle with a low polarization of over 600 h [175].

Tong et al. artificially improved the interfacial stability and interfacial contact on the positive side of NASICON by obtaining a butanedionitrile-based plastic crystal electrolyte in situ on the cathode interface side (Figure 10b). This interfacial layer can be reversibly deformed with the change in particle volume, which reduces the interfacial charge transfer resistance and NASICON electrolyte is protected from successive interfacial side reactions, and thus the prepared solid-state Na-CO_2 battery can stably cycle for more than 50 cycles at a limited capacity of 500 mAh g^{-1} [29]. In addition, solution casting method and in situ polymerization are also effective measures to enhance their interfacial contact.

In addition, chemical and electrochemical compatibility between different components will contribute to the stable operation of solid-state batteries (Figure 10c). Usually, the electrolyte and electrode will react at the interface, and dendrites may also form at the interface during the electroplating and stripping process, leading to short-circuiting of the battery [176,177]. Effective measures to solve the above problems include: (1) Optimizing the composition of the solid electrolyte or electrode, where the electrolyte reacts with the electrode to form an intermediate phase, but the interfacial products are stable and can reduce the interfacial impedance. For example, Tong et al. found that on the anode interface side, $Na_3Zr_2Si_2PO_{12}$ reacts with Na metal to form a thin passivation layer (indicated by NaO$^-$), which not only prevents further reaction between electrolyte and Na metal, but also improves the interfacial contact between electrolyte and anode. (2) Optimizing the composition of electrode, by reducing the local current density and homogenizing the electric field distribution, so as to guide the sodium ion fluxes and improve the problem of battery short circuit caused by interface dendrites. Based on this idea, Sun et al. modified the surface oxidation functional groups of three-dimensional carbon cloth to make the carbon cloth sodium-friendly, and then injected molten sodium to obtain a stable Na @ CC anode which inhibited dendrite growth. The composite anode greatly improves the reversibility and safety of the electrode [178]. (3) To introduce artificial interfacial layers to reduce the interfacial impedance caused by the chemical or electrochemical reaction of the intermediate phase (Figure 10d). Goodenough et al. [179] heated sodium metal and NASICON to 380 °C, chemically reacted to get stable interface layer and then reduced to room temperature, and the negative metal Na showed good wettability to NASICON, and the symmetric batteries of sodium achieved good cycling at 65 °C and at current densities of 0.15 mA cm^{-2} and 0.25 mA cm^{-2}. Sun et al. added 1,1,1,3,3,3-hexafluoroisopropylmethyl ether (HFPM) to 0.1 M NaPF$_6$ in 1,2-dimethoxyethane (DME) and fluoroethylene carbonate (FEC) electrolyte to form a new fluorine-containing organic layer, effectively stabilized the surface of Na anode, and showed excellent cycling performance [180].

Recently, a new approach can solve the above interfacial close contact and interfacial compatibility simultaneously. For example, Sui et al. and others constructed a thin self-reinforcing GPE in situ by polymerizing 1,3-dioxolane (DOL) in the nanofiber skeleton [181]. The framework is composed of polydopamine modified PVDF-HFP (PDA/PVDF-HFP) nanofiber membrane (Figure 10e). It has good affinity with PDOL and DOL, and can improve ionic conductivity. The DOL precursor solution can be firmly absorbed in the porous membrane, making the PVDF-HFP chain well swelled and forming good contact with the electrode interface, and be more stable to the negative metal Na [182]. After polymerization, there is no residual organic liquid GPE to ensure the full safety of the battery. Similarly, Sun et al. injected methylmethacrylate (MMA) into porous $Na_3Zr_2Si_2PO_{12}$-polymer vinylidene fluoride-hexafluoropropylene (NZSP-PVDF-HFP) composite membrane and obtained in situ polymerization polymethylmethacrylate (PMMA)-filled composite electrolyte membrane with excellent electrochemical performance [183].

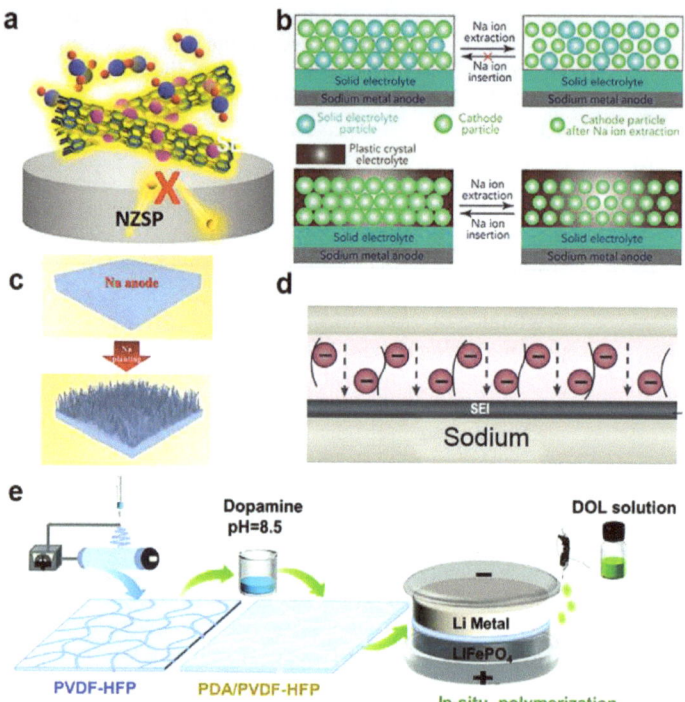

Figure 10. (**a**) Schematic diagram of the "point-to-point" contact between the porous cathode and the NZSP. Reproduced with permission from [42]. Copyright 2021, Elsevier Ltd. (**b**) SN wetting on the porous cathode surface. Reproduced with permission from [29]. Copyright 2019, Royal Society of Chemistry. (**c**) Na dendrite growth process. (**d**) Contact model of ceramic solid electrolyte with metallic sodium in the process of sodium plating. Reproduced with permission from [184]. Copyright 2020, Elsevier Ltd. (**e**) Preparation of 3D PDOL@PDA/PVDF-HFP gel polymer electrolyte via in situ polymerization. Reproduced with permission from [182]. Copyright 2022, John Wiley and Sons.

5. In Situ Characterization Techniques and Theoretical Calculations

The electrochemical performance of metal–air batteries is closely related to the composition and structure of each component and the process of battery fabrication. A detailed understanding of the mechanism of action of each component is required to assemble a battery with excellent performance [185]. The combination of theoretical calculations and characterization techniques can help to better understand the electrochemical reaction mechanisms and the constitutive relationships of batteries materials in the cell system, which in turn can guide the design of safer batteries with higher energy density and longer life [186,187].

5.1. In Situ Characterization Techniques

During charge and discharge, the transport of Na^+ ions between the cathode and anode usually induces various changes in the electrode materials and the electrode/electrolyte interface, such as material volume expansion and contraction, structural phase changes, morphological evolution and surface reconstruction phenomena. With the help of in situ characterization technology, the structural development process of Na-CO_2 batteries can be monitored in real time on multiple scales, providing a strong theoretical basis for studying the structural evolution of Na-CO_2 batteries [188–191]. This section summarizes typical in situ characterization techniques, including in situ diffraction (XRD, NPD and PDF), in situ microscopy (SEM, TEM and AFM), and in situ spectroscopy (FTIR, Raman and XAS), and gives some representative application examples. The key importance of each technique is highlighted in the study of metal–air batteries [185,189,191–193].

5.1.1. In Situ Diffraction Technique

The diffraction technique is a common method for characterizing the composition of crystalline materials and phases. The basic idea is as follows: any phase has its own characteristic diffraction spectrum; the diffraction patterns of any two phases cannot be identical; the diffraction peaks of multi-phase samples are the simple superposition of the phases [194,195]. Therefore, the position and intensity of the peaks are closely related to the composition, arrangement and type of atoms in the crystal. Typical in situ diffraction techniques include in situ XRD, in situ NPD and in situ PDF.

XRD originates from the interference of X-rays scattering crystals to produce diffraction patterns. In situ XRD can be used to monitor information on phase changes, lattice parameter changes and intermediate phase formation of the batteries' material during charge and discharge [194,196]. The development of in situ XRD technology offers the possibility to explore the reaction mechanism of metal–CO_2 batteries [197–199]. The battery structure for in situ XRD is shown in Figure 11a [196]. Chen and colleagues used in situ XRD to observe the growth of Na_2O, the discharge product of Na-O_2 batteries in different electrolytes [198]. It proved that the electrolyte has an important influence on the reaction mechanism of Na-O_2 batteries, i.e., NaO_2 is the main discharge product in high and medium donor number electrolytes, but is rapidly converted to Na_2O_2 in low donor number electrolytes.

In addition to the conventional laboratory X-ray light source, the synchrotron X-ray light source can provide higher intensity and larger photon energy, which can collect higher signal-to-noise diffraction data in a short period of time and thus better study the lattice changes and structural evolution of battery materials during charging and discharging, reflecting more real-time change information [200].

Similar to in situ XRD techniques, in situ neutron diffraction allows for unique information for detecting the structural evolution of materials because neutrons are more sensitive to lighter elements (e.g., H [201,202], Li [97,203], O [204–206], Na [207], Al [208]) and can easily distinguish elements with similar atomic numbers (e.g., Fe and Mn) [209]. Specific applications include component determination, crystallinity, lattice constants, ion diffusion, expansion tensor, bulk modulus, phase transition and structure refinement and determination. However, the disadvantage of neutron diffraction is that it requires a strong neutron source, often a large sample size and a long data collection time [210]. Using NPD technology, Sun et al. found that there is an opportunity to create more sodium ion diffusion paths and improve the sodium ion diffusion capacity by applying perturbations, such as doping elements [113]. However, in situ XRD or NPD methods to study the long-range crystal structure of materials are unable to determine phase transitions in amorphous electrode materials or local structural changes during battery cycling.

Unlike the above diffraction techniques, the pair distribution function (PDF) can provide information about the local structure in the material, and can study both crystalline and amorphous materials [211]. The pair distribution function is also based on X-ray or neutron light source tests, but it provides information on the structure of the material at the atomic scale, especially on the atom–atom interactions in the structure. Fourier transform allows to obtain a real-space pair distribution function map, based on peak positions that can directly correspond to adjacent atomic spacings in real space; the peak intensity corresponds to the relative abundance of adjacent atomic spacings. For amorphous materials, local structural information, such as bond lengths and coordination numbers, can be extracted from the pair distribution familiarity map; for crystalline materials, the corresponding structural information can be fitted by taking the corresponding crystal structure model [212,213]. For crystalline materials, the corresponding crystal structure model can be adopted to fit the corresponding structural information. Unfortunately, there are few reports on the use of in situ PDF for characterization, and we believe that these studies will play a more important role in the design and development of solid state Na-CO_2 batteries in the future.

In situ diffraction techniques are important to obtain information on the structural changes in battery materials. The application of in situ diffraction techniques such as in situ XRD, in situ NPD and in situ PDF in the study of metal–air batteries are emphasized. However, some microscopy and spectroscopy techniques are still needed to obtain more comprehensive information about the structure and composition of cell materials. For example, while synchrotron radiation-based X-ray imaging techniques can provide morphological and chemical information over a wide length scale from tens to hundreds of nanometers. Although synchrotron X-ray imaging techniques can provide morphological and chemical information on scales from tens of nanometers to a few millimeters, imaging at the atomic length scale (submicron) still requires transmission electron microscopy (TEM) techniques. In addition, spectroscopic techniques can also provide important complementary information for structural analysis of cellular materials.

5.1.2. In Situ Microscopy Techniques

In situ microscopy techniques, such as Scanning Electron Microscopy (SEM), Transmission Electron Microscope (TEM) and Atomic Force Microscope (AFM), allow visual observation of particle size and morphological changes on the material surface and is often used to observe the fine structure of the electrochemical reaction process in metal–air batteries and thus to determine the (catalytic) reaction mechanism. In situ microscopy combined with other spectroscopic techniques, such as energy dispersive spectroscopy (EDS) and electron energy loss spectroscopy (EELS), gives information on the chemical composition, electronic structure and chemical bonding of the sample. TEM-EDS is essential for understanding the chemical composition and microstructure of materials. Scanning transmission electron microscopy-electron energy loss spectroscopy (STEM-EELS) also allows for selective imaging of various parts of the electron energy loss spectrum to obtain information on the structure, interfacial features, diffusion pathways and electronic structure at the atomic scale in the sample.

SEM uses a focused electron beam to scan and image the surface of a specimen point by point, and obtains information about the surface morphology of the test specimen by accepting, amplifying and displaying imaging of these signals. Usually, SEM collects secondary electrons from the sample surface, and its lining degree reflects the surface morphology and roughness of the sample. The spatial resolution of SEM can reach 10 nm, and the actual resolution is limited by the conductivity of the sample and the environment of the electron microscope cavity, with some time-resolved capability for in situ characterization of the battery. Neelam et al. studied the degradation mechanism of two sulfide-based solid electrolytes, b-Li_3PS_4 and Li_6PS_5Cl, during the operation of batteries by in situ SEM [214]. As shown in Figure 11b, compared with Li_6PS_5Cl, b-Li_3PS_4 shows huge plating, faster dendrite formation, cracks and ultimately cell failure.

TEM has extremely high spatial resolution, can realize the observation of diffraction patterns in tiny regions (several nanometers), and is suitable for the study of crystal structure of microcrystals, surfaces and thin films. Because electrons are easily scattered or absorbed by objects with low penetration, TEM must prepare ultra-thin samples, usually in thickness of 50–100 nm; on the other hand, the intense irradiation of the electron beam tends to damage the sample and bring artifacts. The in situ TEM technique has been applied to the study of the reaction mechanism and catalyst of metal–CO_2 batteries [44,215–217]. Zhu et al. directly studied the morphological changes in the cathode surface of single Pt atom-loaded nitrogen-doped carbon nanotubes (Pt@NCNT) during the charging and discharging of Na-CO$_2$ batteries using in situ ambient transmission electron microscopy (TEM) under CO$_2$ atmosphere (Figure 11c) [44]. Han et al. investigated the mechanism of the reversible redox reaction in Na-O$_2$ batteries with CuS cathodes using in situ TEM [215,216]. The results show that Na$_2$O$_2$ is the main final ORR product, which uniformly covers the whole linear cathode (Figure 11d). In addition, Huang et al. studied the reaction process of Na-O$_2$/CO$_2$ battery in situ using spherical aberration-corrected environmental transmission electron microscopy, and characterized the structure and composition of the discharging and partially charging products by using annular dark field images (ADF) and EELS [187]. Figure 12a,b showed the ADF diagrams characterizing the cathode surface: the discharge and charge products had similar spherical profiles in the low-loss and core-loss spectra. As shown in Figure 12c,d, EELS showed that: only Na and Na$_2$O$_2$ are present in the discharge and charge spheres, and the shell layer shows the formulation of Na$_2$CO$_3$.

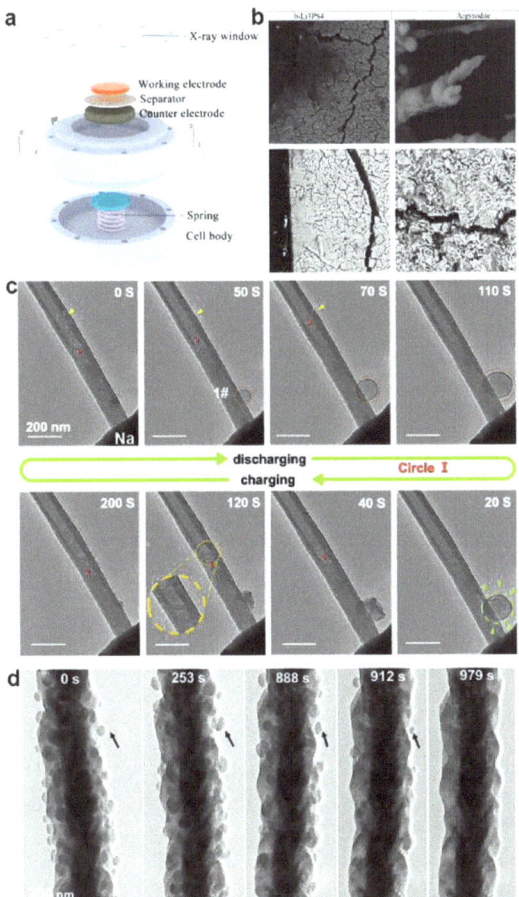

Figure 11. (**a**) Schematic diagram of a typical battery device for in-situ XRD testing;. Reproduced with permission from [196]. Copyright 2019, John Wiley and Sons. (**b**) Plating, cracking and dendrite formation in b-Li$_3$PS$_4$ and Li$_6$PS$_5$C. Reproduced with permission from [214]. Copyright 2021, Cambridge University Press. (**c**) Morphology evolution of individual platinum-doped NCNTs (Pt@NCNTs) during electrochemical discharge and charge of Na-CO$_2$ nanobatteries. Reproduced with permission from [44]. Copyright 2020, Elsevier Ltd. (**d**) In situ TEM micrographs of the OER process in Na-O$_2$ batteries. Reproduced with permission from [215]. Copyright 2020, American Chemical Society.

Compared with SEM and TEM, AFM can more easily simulate cell environment conditioning, and be used for in situ operational characterization to monitor the microscopic morphology of the electrode surface in real time, provide physicochemical information of the electrode surface at the nanoscale, and provide an experimental basis for the optimization of electrode materials and electrolyte modification. In addition, unlike scanning electron microscopy which can only provide two-dimensional images, AFM can provide true three-dimensional images. At the same time, scanning electron microscopy needs to operate under high vacuum conditions, while AFM can work under atmospheric pressure or even in liquid environments. Cort'es et al. used improved AFM without putting AFM in glove box, visualized the formation and decomposition process on thin Li_2O_2 carbon cathode as shown in Figure 12e. This equipment provides a technical support for future studies of new cathode materials [218].

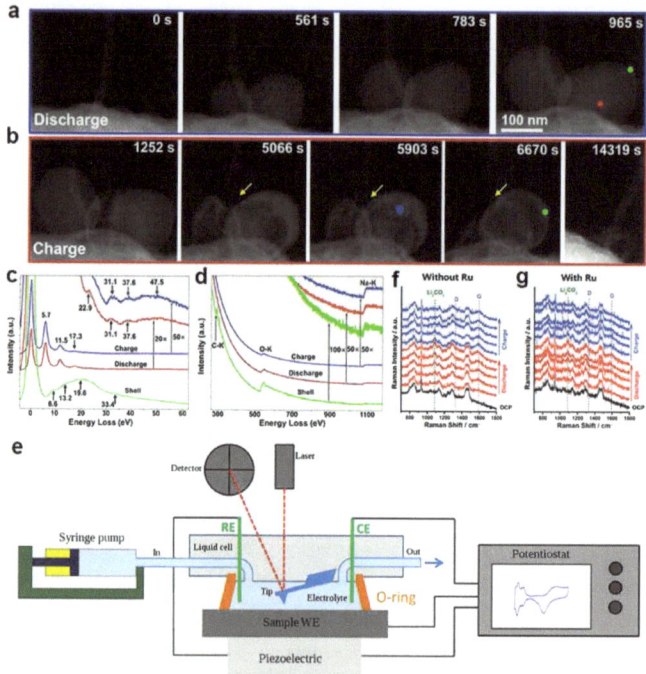

Figure 12. Annular dark field images (ADF) and EELS characterization of the structure and composition of the discharge and partial charging products: (**a,b**) ADF images of the time-dependent structural evolution of the air cathode of the Na-O_2/CO_2 batteries. During the discharge reaction, two balls emerged at 561 s at the CNT-Na substrate-O_2/CO_2 triple point, which then grew under a constant −3 V bias, showing the core-shell structure of the ball. (**c,d**) Low-loss and core-loss EELS spectra. Reproduced with permission from [187]. The low-loss and core-loss spectra of both the discharged (red profile) and charged (blue profile) balls show similar profile: in the Low-Loss region, there are three multiple plasma peaks at 5.7, 11.5 and 17.3 eV respectively, and another three plasma peaks at 22.9, 31.1 and 37.6 eV, and combined with side absorption peaks of O-K and Na-K in the Core-Loss region, indicating the presence of only Na and Na_2O_2 exist in the discharge sphere and charge sphere. The Core-Loss of the shell layer shows the presence of Na, C and O, indicating the formation of Na_2CO_3.Copyright 2020, American Chemical Society. (**e**) Schematic diagram of the flow electrochemical atomic force microscope (FE-AFM). Reproduced with permission from [218]. Copyright 2021, Elsevier Ltd. (**f,g**) In situ SERS characterization using $LiCF_3SO_3$-TEGDME with and without Ru catalyst electrodes during discharge and recharge in CO_2. Reproduced with permission from [27]. Copyright 2017, Royal Society of Chemistry.

5.1.3. In Situ Spectroscopy Technique

Spectroscopic techniques such as XPS, FTIR, Raman and XAS are sensitive to changes in the local chemical environment, have elemental resolution and are suitable for compositional analysis of crystalline and amorphous phases.

X-ray photoelectron spectroscopy (XPS) is based on the photoelectric effect, using the interaction of X-rays with the surface of the sample to generate photoelectrons, and using an energy analyzer to analyze the kinetic energy of the photoelectrons, the X-ray photoelectron spectrum is thus obtained. Based on the measured kinetic energy of photoelectrons, it is possible to determine which elements are present on the surface and to know the content of an element on the surface based on the intensity of photoelectrons with a certain energy, with an error of about 20%. It is possible to determine both the relative concentration of elements and the relative concentration of different oxidation states of the same element. XPS not only determines the composition of all elements on the surface except H and He, but also gives information on the chemical state of each element, with high energy resolution and a certain spatial resolution (currently on the micron scale) and temporal resolution (on the minute scale). XPS also gives information on the surface, tiny areas and depth distribution information. In situ atmospheric pressure XPS (APXPS) is an effective tool to study the chemical state changes on the surface of solid electrolytes and catalysts. Wang et al. used synchrotron in situ APXPS to study the redox reaction of CO_2 on the surface of porous carbon electrode in ionic liquid electrolytes [219]. The results showed that the reduction in pure CO_2 at porous carbon electrodes has no electrochemical activity at room temperature. In contrast, when O_2 (CO_2:O_2 = 2:1) is added to the ionic liquid electrolyte of Li-CO_2 batteries, CO_2 is reduced to low-valent amorphous carbon and Li_2O_2/Li_2O. Compared with the reaction of Li_2CO_3, the charging reaction of amorphous carbon, Li_2O_2 and Li_2O is faster. The main reason why the ultimately discharge product is not Li_2CO_3 is that the strong solvation between ionic liquid electrolyte and Li^+ stabilizes the intermediate anion of metastable CRR reaction.

In situ Fourier infrared reflectance absorption spectroscopy (FTIR) is particularly advantageous in identifying different functional groups and components during electrochemical reactions, and thus in situ FTIR techniques can be used to analyze reaction intermediates and determine reaction mechanisms. Although the application of in situ FTIR in metal–air batteries has not been widely reported, it can be expected to play an important role in future research in this field.

In situ Raman spectroscopy is based on the detection of laser-induced vibrational, rotational and other low-frequency leap patterns in electrode materials, and is a fingerprint spectrum of the structure of matter [192]. Raman spectroscopy is suitable for vibration mode measurements of symmetric vibrations, non-polar groups and homoatomic bonds, such as S=S, S-S, N≡N, C=C and O=O. In addition to identifying substance types and chemical components, Raman spectroscopy mainly measures molecular vibration frequencies to quantitatively understand intermolecular forces and intra-molecular forces, and to infer information such as symmetry, geometry and arrangement of atoms in molecules. However, conventional Raman spectroscopy signals are relatively weak, and surface-enhanced Raman spectroscopy (SERS) is able to detect lower concentrations of species in the surface layers of bulk materials.

Zhou et al. used in situ SERS to study the catalytic activity of Ru in promoting CO_2 reduction to Li_2CO_3 and C on ruthenium-free sputtered gold electrodes and gold-ruthenium electrodes [27]. By observing figures f and g, it can be seen that the peak Raman intensity corresponding to Li_2CO_3 at 1080 cm^{-1} decreases during the charging process and completely disappears at the end of the charging process. Comparing Figure 12f,g, the difference is mainly the Raman peak corresponding to the G-band of carbon at 1580 cm^{-1}. As shown in Figure 12f, the peak value of the carbon G band corresponding to 1580 cm^{-1} in the gold electrode hardly changes during the whole charging process. In the Au-Ru electrode (Figure 12g), the peak intensity of the G-band in carbon (1580 cm^{-1}) decreases during the charging process and disappears completely at the end of the charging process, which is similar to the Raman peak of Li_2CO_3. This result shows that the ruthenium catalyst can significantly promote the reaction between Li_2CO_3 and carbon during charging (charging reaction), while the absence of ruthenium leads to the self-decomposition of Li_2CO_3 and leads to an irreversible discharge–recharging cycle of the battery.

However, SERS can only characterize information from a few atomic layers on the surface. In contrast, X-ray absorption spectroscopy (XAS) can effectively capture deeper signal variations in bulk materials. By analyzing and fitting the data to the X-ray absorption spectra, the most accurate characterization of the sample as a whole can be obtained, including interatomic distances, number and type of neighboring atoms and information on the average oxidation state of the elements in the coordination environment [189,220]. In addition, unlike in situ diffraction techniques, in situ XAS can provide real-time element-specific information on crystalline and amorphous evolution and phase transitions. XAS can be divided into X-ray absorption near edge structure (XANES) spectroscopy and extended X-ray absorption fine structure (EXAFS) [221]. EXAFS is commonly used in the range of 150–2000 eV to obtain quantitative local structural information such as bond length, coordination number and disorder of the central and coordination atoms. On the other hand, XANES can quickly identify the chemical state of elements in the low kinetic energy range of 5–150 eV. In addition, XANES can be used for time-resolved experiments and high-temperature in situ chemistry experiments because of its violent vibrations, short spectrum acquisition time and temperature dependence. The self-reinforced catalysis of $LiCoO_2$ as an electrocatalyst for $Li-O_2$ batteries was reported by Zhou et al. [222]. The evolution of Co electrons and local structure was investigated using XAFS, illustrating that the intercalation/extraction of Li^+ in $LiCoO_2$ not only induces changes in Co valence and regulates the electron/crystal structure, but also the surface disorder, lattice strain and local symmetry, thus promoting bidirectional catalysis. The in situ technique XAFS has not been reported.

Photoelectronic information (XPS), which is sensitive to the material surface, can detect the changes in the element composition and chemical state information on the battery surface in real time. Raman spectroscopy is derived from the inelastic scattering of incident laser by the molecules on the surface of the object. In situ Raman spectroscopy can detect the changes in material composition and structure on the electrode or solid electrolyte surface of the all-solid battery. In situ XAS plays an irreplaceable role in the rapid and high precision analysis of the elements and their valence states and their respective distributions in solid state batteries. With the construction of more and more advanced synchronous light sources, in situ spectroscopy will play an increasingly important role in the research of solid state metal–air batteries and other energy materials.

Since each characterization method has its own advantages and disadvantages, combining the advantages of different characterization techniques to study the physicochemical changes and failure mechanisms of solid $Na-CO_2$ electrode materials and interfaces during operation can provide an important information for further understanding and optimization of material performance, and provide strong support for subsequent improvement of battery performance including energy density, cycle life and safety. With the development of modern technology, these in situ characterization techniques and data acquisition and analysis systems will be further improved and intelligent, and more in situ characterization techniques will be available for more systematic real-time inspection of all-solid-state metal–air batteries to guide the design of solid-state metal–air batteries.

5.2. Theoretical Calculations and Simulations

Theoretical calculations and simulations also play important roles in the development of battery technology. With the rapid development and wide application of computer simulation techniques in quantum and atomic scale-based materials science research, it provides the possibility to understand the reaction mechanism of microscopic metal–air batteries and to develop and design new battery materials efficiently. For catalytic materials, theoretical calculations and simulations can calculate adsorption energy, migration energy, charge distribution, electronic structure and even defect states to understand the reaction mechanism; for electrode materials, theoretical calculations and simulations can calculate the energy band gap, sodium storage voltage, bulk deformation and charge compensation mechanism of positive and negative electrodes and electrolyte materials. In addition, theoretical calculations and simulations can study the electrochemical window, ionic conductivity, ion mobility and ion diffusion barrier and transport mechanism. With the dramatic increase in computer computing power, methods such as high-throughput computing and machine learning have also started to develop rapidly, making the research and development of battery materials tend to be more efficient and intelligent. The combination of theoretical calculations and experiments helps to use the deep understanding of electrochemical theory to design and control electrochemical reactions in turn, thus advancing the battery research. This section outlines some common computational and simulation methods based on quantum mechanical theory in battery research, and briefly introduces the application of materials genetic engineering and machine learning in battery materials development.

5.2.1. Theoretical Calculation and Simulation Methods Based on Quantum Mechanics

Density Function Theory (DFT) was developed by Kohn, the 1998 Nobel Prize Chemistry awardee, which enables efficient solution of the many-body Schrödinger equation. In recent years, theoretical calculations based on density DFT and molecular dynamics (MD) simulations have greatly contributed to the understanding of the reaction mechanism of metal–air batteries at the atomic scale and to the efficient design of battery materials. The calculation of various properties of materials by inputting structural information of crystals has provided a theoretical basis for understanding the reaction mechanism, catalytic reaction kinetics and ionic transport in electrolytes in metal–air batteries.

Due to the multi-electron composite process of oxygen reduction reaction (ORR) and oxygen evolution reaction (OER), an in-depth understanding of the catalytic reaction process of cathode catalysts is essential to improve the reaction kinetics by changing the growth path and morphological evolution of the cathode surface [223]. In recent years, first-principles calculations have become a routine approach to understand complex catalytic phenomena and experiments at the electronic level, and have been further extended to find and design novel catalysts [25,61,224]. Jiang et al. used DFT calculations to investigate the contribution of point defects on carbon surfaces to the reaction of non-aqueous $Li-O_2$ batteries [225]. In this work, five representative defect structures were considered, including SV (point defects), DV5-8-5 (two pentagons and one octagon), DV555-777 (three pentagons and three heptagons), DV5555-6-7777 (four pentagons, one hexagon and four heptagons) and SW (Stone–Wales) defects. The calculated free energy results (Figure 13a–f) showed that the DV555-777, DV5555-67777 and SW type defects can increase the discharge voltage and decrease the charge voltage, while promoting the ORR and OER processes. In addition, the discharge voltage of DV555-777 is the highest, and the charging voltage of DV5555-6-7777 and SW is the lowest. Therefore, non-aqueous $Li-O_2$ batteries require carbon materials containing DV5555-6-7777 and SW type defects. Deng et al. obtained porous self-supporting cathodes for $Li-CO_2$ batteries by directly anchoring two-dimensional (2D) cobalt-doped CeO_2 nanosheets on graphene aerogels. Combined with the experimental results and DFT calculations, it was proved that co-doped CeO_2 nanosheets could effectively improve the conductivity and adsorption capacity of CO_2. In addition, they obtained the possible reaction path of non-aqueous $Li-O_2$ batteries from the perspective of thermodynamics, and the analysis results are shown in Figure 13g [224]. Recently, Sun et al. constructed a new autocatalytic system for lithium–air batteries using an in situ electrochemical doping strategy combined with theoretical computational simulations and systematically investigated its reaction mechanism and battery performance. The work was based on density function theory (DFT) for the theoretical study, and it was found that the batteries' performance could be significantly enhanced by doping the discharge product Li_2O_2 with metal ions, and then the theoretical calculations were used to guide the experimental screening of the most effective doping structure [226].

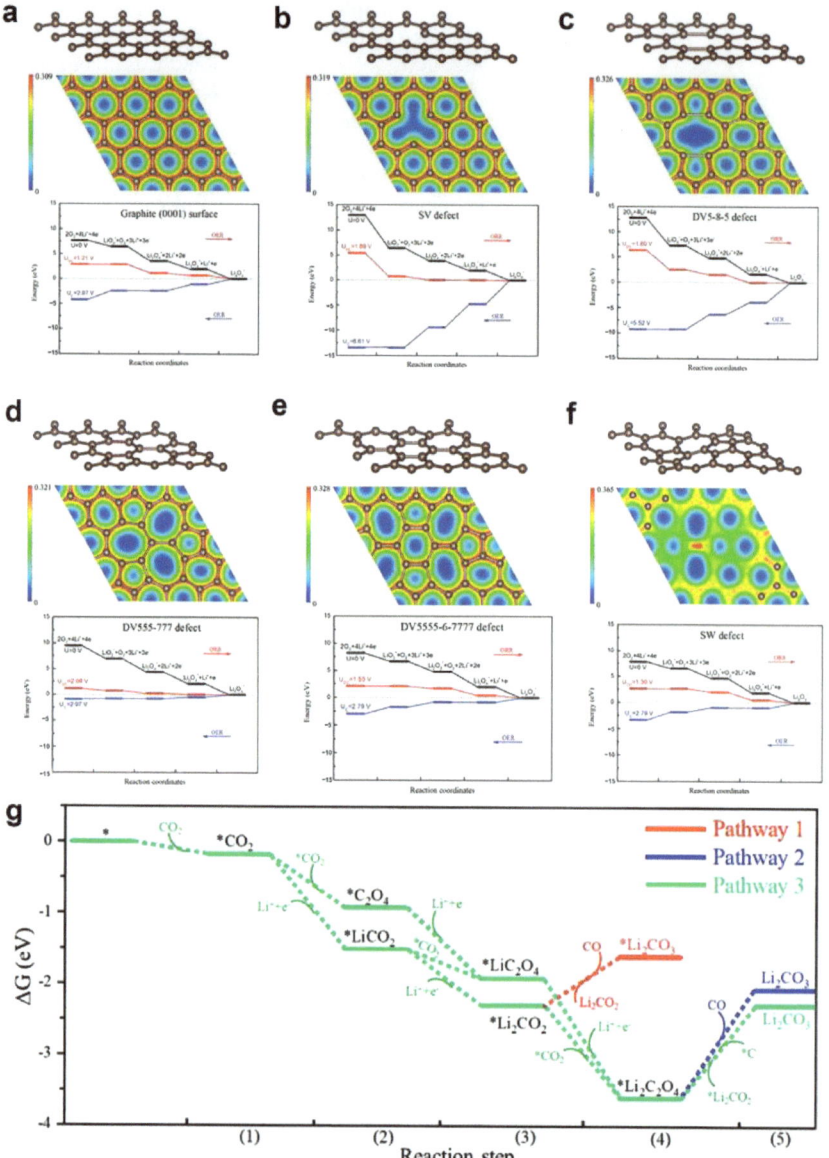

Figure 13. Optimized structures (top), electron density profiles (middle) and free energy diagrams (bottom) of (**a**) graphite (0001) surface. (**b**) SV defects. (**c**) DV5-8-5 defects. (**d**) DV555-777 defects. (**e**) DV5555-6-7777 defects and (**f**) SW defects. The brown balls represent carbon atoms and the electron density is in $|e|$ Bohr^{-3}. Reproduced with permission from [225]. Copyright 2016, American Chemistry Society. (**g**) Free energy diagrams of co-doped CeO_2 (110) surface CO_2 to Li_2CO_3 were obtained based on DFT calculations. Asterisk * represents activated reactive intermediates. Reproduced with permission from [224]. Copyright 2021, Elsevier Ltd.

For solid electrolytes, polymeric solid electrolytes are difficult to establish standard models for calculations due to their complex structures. In contrast, inorganic solid electrolytes, such as $Na_3Zr_2Si_2PO_{12}$, are easier to model due to their simple structure, and their corresponding theoretical simulations and calculations are more studied [107,227]. Park et al. investigated the effect of adding excess Na to standard NASICON and the mechanism of Na^+ ion transport [107]. Both experimental and theoretical calculations demonstrated that the main mechanisms of Na^+ ion transport in the NASICON structure are, grain boundary diffusion at low temperatures and grain diffusion at high temperatures, respectively (Figure 14a). In addition, it was found that after adding 10% excess Na, the expansion of the bottleneck of polycrystalline particle sodium diffusion channel was conducive to improving the bulk conductivity of NASICON. DFT calculation results show that by adding 10% excess Na, the minimum bottleneck area was expanded from 4.9922 Å to 5.4086 Å; the activation energy was significantly reduced. The minimum bottleneck area is the key factor to determine the conductivity of Na ion in NASICON electrolytes, which is basically consistent with the experimental results. Figure 14b shows that the ionic conductivity of NAICON has been significantly improved after adding excessive sodium, resulting from the enlargement of the bottleneck areas in the Na diffusion channels of polycrystalline grains.

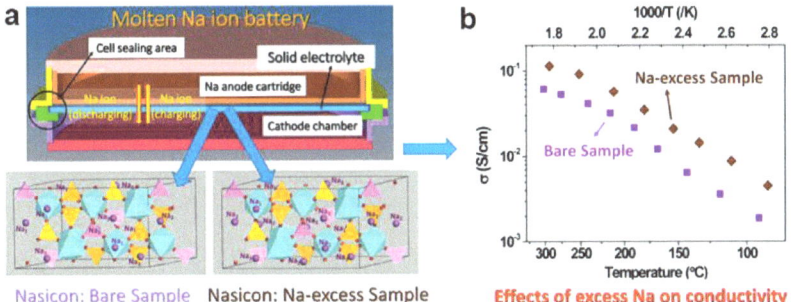

Figure 14. (**a**) Schematics of the cross−section of a molten Na battery and NASICON structures (bare sample and Na-excess sample); (**b**) measured total ionic conductivity for bare and Na excess sample. Reproduced with permission from [107]. Copyright 2016, American Chemical Society.

The combination of DFT calculations and MD simulations has a strong potential in facilitating the development of metal–air batteries: the mass transfer dynamics of gas molecules and metal ions throughout battery system can be studied at the atomic scale. Notably, most of the current theoretical calculations are based on predictions under 0 K and vacuum conditions. However, powerful predictive tools should model the actual battery environment and conduct extensive experiments to verify theoretical simulations and improve accuracy.

5.2.2. Introduction to Material Genetic Engineering and Machine Learning

Materials genetic engineering is a new concept and method for materials research that has emerged in recent years and is at the forefront of materials science and engineering in the world today. Materials genetic engineering is a technological integration and synergy of high-throughput materials computing, high-throughput materials synthesis and characterization experiments and databases, which can at least double the speed of materials discovery, manufacturing and application, and reduce the cost by at least half. Materials genetic engineering is a leap forward in materials science and technology, replacing the traditional trial-and-error method of multiple sequential iterations with high-throughput parallel iterations, gradually changing from "experience-guided experiments" to a new model of "theoretical prediction and experimental verification" in materials research, and finally realizing materials design on demand.

There are some important steps in the process of high-throughput computational screening of materials [228]. First, a database of experimental crystal structures containing a large number of materials is needed as the seven points for high-throughput screening; second, structural information is input in bulk and data are screened; then the screened crystal structure information is generated into a data format that can be calculated using density functional theory and put into the server for calculation; then the calculation files are output and kept together to form a database after the calculation is completed; finally, the data from the database are used to analyze the various properties of the material, which can be used to guide the material design or to filter the data for the next cycle.

In recent years, the combination of machine learning and materials genetic engineering has led to the advancement of materials informatics and promoted the development of materials science. Currently, the use of data-driven machine learning algorithms to build material performance prediction models and then apply them to material screening and new material development has attracted much attention from research community. Machine learning can be seen as an umbrella term for a class of algorithms that can mine potential laws from large amounts of experimental data, acquire new knowledge or skills, build corresponding data analysis models and allow machines to repeatedly analyze the corresponding content and reorganize existing knowledge structures through the input of new data to continuously improve their performance. In the field of materials development, machine learning can show great potential and advantages, especially in the design of new catalysts, organic and inorganic battery materials, etc. [229]. A machine learning toolkit for genetic engineering improves the performance of solid-state $Na-CO_2$ batteries by searching for high-performance solid electrolytes.

Materials genetic engineering and machine learning aim to establish a new model of data-driven, computational simulation and theoretical prediction first, experimental validation second, thus replacing the existing empirical and experimental-based materials R&D paradigm, which is a powerful tool for the future development of metal–air batteries.

6. Outlook

Metal–CO_2 batteries are considered as a promising clean technology because of their potential for high energy density energy storage and power supply, as well as their ability to convert and immobilize CO_2 as well as mitigate global warming trends. After nearly a decade of development, although the research of $Na-CO_2$ battery has made great progress, compared with other energy storage systems, the research of $Na-CO_2$ battery is still in the infancy stage, facing the challenges of unclear electrochemical reaction mechanism, slow catalytic reaction of CO_2 gas at the cathode and the discharge product carbonate has difficulties in reversible transformation due to problems such as good thermodynamic stability, limited research on electrolyte materials, lack of suitable electrolytes and interface incompatibility. To solve these problems and realize the practical application of solid-state $Na-CO_2$ batteries operating at room temperature, we propose the following aspects for further research. First, the analysis of the electrochemical mechanism of $Na-CO_2$ batteries, especially the formation and decomposition process of the discharge products in the rate-determined step, is of great significance for the study of $Na-CO_2$ batteries, and should be considered in conjunction with in situ experimental studies, such as the electrochemical scanning microscope system combined with Fourier transform infrared spectroscopy, to further rationalize the design of the $Li/Na-CO_2$ cell system by analyzing the carbonate formation and decomposition pathways. Secondly, the sluggish reaction kinetics is a non-negligible challenge for $Na-CO_2$ batteries, which will lead to large overpotential and poor reversibility. Therefore, efficient catalysts with low cost, reasonable structure and high catalytic ability need to be developed to fill this gap. Considering the cost and performance, heteroatom doping, noble metal single-atom catalysts and transition metal complex catalysts are very promising.

Third, the applications of solid electrolytes in $Na-CO_2$ batteries are rather limited, and there is an urgent need for researchers to design and develop more solid electrolyte materials with high performance (e.g., high ionic conductivity and good chemical/electrochemical stability) based on fundamental principles and theoretical calculations, and to create stable interfaces with good contacts. In fact, ion transport in electrolytes and at interfaces is a multi-scale process, including atomic scale, micro and mesoscopic scale and device scale [102]. Monitoring the ion conduction and interfacial reactions at different scales, together with atomic-scale characterization techniques and theoretical modeling, are important for the design and fabrication of high-performance batteries. To improve the interfacial contact and form a three-dimensional (3D) electronic and ionic conducting network monolithic cell architecture may be a good choice. Finally, how to scale up the production of high-performance solid electrolytes by a simple, scalable and cost-effective method is also an important consideration for practical applications.

Finally, with the rapid development of in situ characterization technology and high performance computing, more comprehensive real-time inspection, data acquisition and analysis systems will be available and with the powerful predictive capabilities of computer simulation technology (especially theoretical calculations and simulations, genetic engineering of materials and machine learning), the development of battery materials will tend to be more efficient and intelligent, which is a powerful tool for the future development of metal–air batteries.

In summary, Na-CO_2 batteries demonstrate great application potential in CO_2 fixation as well as energy storage. They involve a cross multidiscipline. We believe that the development of practical metal–CO_2 batteries will bring new hope for achieving the strategic goal of carbon neutralization and carbon peaking.

Author Contributions: Proposing and supervising this project, C.S.; Writing—review and editing, Z.W., L.L., L.J. and C.S. All authors have read and agreed to the published version of the manuscript.

Funding: This work was supported by Foundation of Key Laboratory of Advanced Energy Materials Chemistry (Ministry of Education), Nankai University, 21C Innovation Laboratory, Contemporary Amperex Technology Ltd. by project No. 21C-OP-202212, the Foundation of State Key Laboratory of High-efficiency Utilization of Coal and Green Chemical Engineering (Grant No. 2022-K15), China University of Mining & Technology (Beijing), Beijing National Laboratory for Condensed Matter Physics, and the National Natural Science Foundation of China (No. 51672029 and 51372271).

Institutional Review Board Statement: Written informed consent has been obtained from the patient(s) to publish this paper.

Informed Consent Statement: Informed consent was obtained from all subjects involved in the study.

Conflicts of Interest: The authors declare no conflict of interest.

References

1. Xie, J.F.; Zhou, Z.; Wang, Y.B. Metal-CO_2 Batteries at the Crossroad to Practical Energy Storage and CO_2 Recycle. *Adv. Funct. Mater.* **2020**, *30*, 1908285. [CrossRef]
2. Wang, F.; Li, Y.; Xia, X.H.; Cai, W.; Chen, Q.G.; Chen, M.H. Metal-CO_2 Electrochemistry: From CO_2 Recycling to Energy Storage. *Adv. Energy Mater.* **2021**, *11*, 2100667. [CrossRef]
3. Xia, C.; Black, R.; Fernandes, R.; Adams, B.; Nazar, L.F. The critical role of phase-transfer catalysis in aprotic sodium oxygen batteries. *Nat. Chem.* **2015**, *7*, 496–501. [CrossRef] [PubMed]
4. Palomares, V.; Serras, P.; Villaluenga, I.; Hueso, K.B.; Carretero-González, J.; Rojo, T. Na-ion batteries, recent advances and present challenges to become low cost energy storage systems. *Energy Environ. Sci.* **2012**, *5*, 5884–5901. [CrossRef]
5. Li, W.J.; Han, C.; Wang, W.L.; Gebert, F.; Chou, S.L.; Liu, H.K.; Zhang, X.H.; Dou, S.X. Commercial Prospects of Existing Cathode Materials for Sodium Ion Storage. *Adv. Energy Mater.* **2017**, *7*, 1700274. [CrossRef]
6. Hu, X.F.; Sun, J.C.; Li, Z.F.; Zhao, Q.; Chen, C.C.; Chen, J. Rechargeable Room-Temperature Na-CO_2 Batteries. *Angew. Chem. Int. Ed.* **2016**, *55*, 6482–6486. [CrossRef]
7. Jena, A.; Tong, Z.Z.; Chang, H.; Hu, S.F.; Liu, R.S. Capturing carbon dioxide in Na-CO_2 batteries: A route for green energy. *J. Chin. Chem. Soc.* **2020**, *68*, 421–428. [CrossRef]
8. Pham, T.A.; Kweon, K.E.; Samanta, A.; Lordi, V.; Pask, J.E. Solvation and Dynamics of Sodium and Potassium in Ethylene Carbonate from ab Initio Molecular Dynamics Simulations. *J. Phys. Chem. C* **2017**, *121*, 21913–21920. [CrossRef]
9. Mu, X.W.; Pan, H.; He, P.; Zhou, H.S. Li-CO_2 and Na-CO_2 Batteries: Toward Greener and Sustainable Electrical Energy Storage. *Adv. Mater.* **2020**, *37*, e1903790.

10. Xie, Z.J.; Zhang, X.; Zhang, Z.; Zhou, Z. Metal-CO$_2$ Batteries on the Road: CO$_2$ from Contamination Gas to Energy Source. *Adv. Mater.* **2017**, *29*, 1605891. [CrossRef]
11. Zheng, Z.; Wu, C.; Gu, Q.; Konstantinov, K.; Wang, J. Research Progress and Future Perspectives on Rechargeable Na-O$_2$ and Na-CO$_2$ Batteries. *Energy Environ. Mater.* **2021**, *4*, 158–177. [CrossRef]
12. Takechi, K.; Shiga, T.; Asaoka, T. A Li-O$_2$/CO$_2$ battery. *Chem. Commun.* **2011**, *47*, 3463–3465. [CrossRef]
13. Das, S.K.; Xu, S.M.; Archer, L.A. Carbon dioxide assist for non-aqueous sodium-oxygen batteries. *Electrochem. Commun.* **2013**, *27*, 59–62. [CrossRef]
14. Liu, B.; Sun, Y.L.; Liu, L.Y.; Chen, J.T.; Yang, B.J.; Xu, S.; Yan, X.B. Recent advances in understanding Li-CO$_2$ electrochemistry. *Energy Environ. Sci.* **2019**, *12*, 887–922. [CrossRef]
15. Xu, S.M.; Lu, Y.Y.; Wang, H.S.; Abruña, H.D.; Archer, L.A. A rechargeable Na-CO$_2$/O$_2$ battery enabled by stable nanoparticle hybrid electrolytes. *J. Mater. Chem. A* **2014**, *2*, 17723–17729. [CrossRef]
16. Guo, L.N.; Li, B.; Thirumal, V.; Song, J.X. Advanced rechargeable Na-CO$_2$ batteries enabled by a ruthenium@porous carbon composite cathode with enhanced Na$_2$CO$_3$ reversibility. *Chem. Commun.* **2019**, *55*, 7946–7949. [CrossRef]
17. Sui, D.; Chang, M.J.; Wang, H.Y.; Qian, H.; Yang, Y.L.; Li, S.; Zhang, Y.S.; Song, Y.Z. A Brief Review of Catalytic Cathode Materials for Na-CO$_2$ Batteries. *Catalysts* **2021**, *11*, 603. [CrossRef]
18. Zhou, J.W.; Li, X.L.; Yang, C.; Li, Y.C.; Guo, K.K.; Cheng, J.L.; Yuan, D.W.; Song, C.H.; Lu, J.; Wang, B. A Quasi-Solid-State Flexible Fiber-Shaped Li-CO$_2$ Battery with Low Overpotential and High Energy Efficiency. *Adv. Mater.* **2019**, *31*, e1804439. [CrossRef]
19. Yang, C.; Guo, K.K.; Yuan, D.W.; Cheng, J.L.; Wang, B. Unraveling Reaction Mechanisms of Mo$_2$C as Cathode Catalyst in a Li-CO$_2$ Battery. *J. Am. Chem. Soc.* **2020**, *142*, 6983–6990. [CrossRef]
20. Zhang, Z.; Zhang, Q.; Chen, Y.N.; Bao, J.; Zhou, X.L.; Xie, Z.J.; Wei, J.P.; Zhou, Z. The First Introduction of Graphene to Rechargeable Li-CO$_2$ Batteries. *Angew. Chem.* **2015**, *127*, 6550–6653. [CrossRef]
21. Zhang, Z.; Wang, X.G.; Zhang, X.; Xie, Z.J.; Chen, Y.N.; Ma, L.P.; Peng, Z.Q.; Zhou, Z. Verifying the Rechargeability of Li-CO$_2$ Batteries on Working Cathodes of Ni Nanoparticles Highly Dispersed on N-Doped Graphene. *Adv. Sci.* **2017**, *5*, 1700567. [CrossRef] [PubMed]
22. Zhang, X.; Zhang, Q.; Zhang, Z.; Chen, Y.N.; Xie, Z.J.; Wei, J.P.; Zhou, Z. Rechargeable Li-CO$_2$ batteries with carbon nanotubes as air cathodes. *Chem. Commun.* **2015**, *51*, 14636–14639. [CrossRef] [PubMed]
23. Sun, J.Y.; Zhao, N.; Li, Y.Q.; Guo, X.X.; Feng, X.F.; Liu, X.S.; Liu, Z.; Cui, G.L.; Zheng, H.; Gu, L.; et al. A Rechargeable Li-Air Fuel Cell Battery Based on Garnet Solid Electrolytes. *Sci. Rep.* **2017**, *7*, 41217. [CrossRef]
24. Li, S.W.; Dong, Y.; Zhou, J.W.; Liu, Y.; Wang, J.M.; Gao, X.; Han, Y.Z.; Qi, P.F.; Wang, B. Carbon dioxide in the cage: Manganese metal–organic frameworks for high performance CO$_2$ electrodes in Li-CO$_2$ batteries. *Energy Environ. Sci.* **2018**, *11*, 1318–1325. [CrossRef]
25. Hu, X.F.; Joo, P.H.; Matios, E.; Wang, C.L.; Luo, J.M.; Yang, K.S.; Li, W.Y. Designing an All-Solid-State Sodium-Carbon Dioxide Battery Enabled by Nitrogen-Doped Nanocarbon. *Nano Lett.* **2020**, *20*, 3620–3626. [CrossRef] [PubMed]
26. Xu, S.M.; Das, S.K.; Archer, L.A. The Li-CO$_2$ battery: A novel method for CO$_2$ capture and utilization. *RSC Adv.* **2013**, *3*, 6656–6660. [CrossRef]
27. Yang, S.X.; Qiao, Y.; He, P.; Liu, Y.J.; Cheng, Z.; Zhu, J.J.; Zhou, H.S. A reversible lithium-CO$_2$ battery with Ru nanoparticles as a cathode catalyst. *Energy Environ. Sci.* **2017**, *10*, 972–978. [CrossRef]
28. Ha, T.A.; Pozo-Gonzalo, C.; Nairn, K.; MacFarlane, D.R.; Forsyth, M.; Howlett, P.C. An investigation of commercial carbon air cathode structure in ionic liquid based sodium oxygen batteries. *Sci. Rep.* **2020**, *10*, 7123. [CrossRef]
29. Lu, Y.; Cai, Y.C.; Zhang, Q.; Liu, L.J.; Niu, Z.Q.; Chen, J. A compatible anode/succinonitrile-based electrolyte interface in all-solid-state Na-CO$_2$ batteries. *Chem. Sci.* **2019**, *10*, 4306–4312. [CrossRef]
30. Wang, X.C.; Zhang, X.J.; Lu, Y.; Yan, Z.H.; Tao, Z.L.; Jia, D.Z.; Chen, J. Flexible and Tailorable Na-CO$_2$ Batteries Based on an All-Solid-State Polymer Electrolyte. *ChemElectroChem* **2018**, *5*, 3628–3632. [CrossRef]
31. Li, Y.L.; Wang, J.J.; Li, X.F.; Geng, D.S.; Banis, M.N.; Li, R.Y.; Sun, X.L. Nitrogen-doped graphene nanosheets as cathode materials with excellent electrocatalytic activity for high capacity lithium-oxygen batteries. *Electrochem. Commun.* **2012**, *18*, 12–15. [CrossRef]
32. Ci, L.J.; Song, L.; Jin, C.H.; Jariwala, D.; Wu, D.X.; Li, Y.J.; Srivastava, A.; Wang, Z.F.; Storr, K.; Balicas, L.; et al. Atomic layers of hybridized boron nitride and graphene domains. *Nat. Mater.* **2010**, *9*, 430–435. [CrossRef] [PubMed]
33. Sui, D.; Xu, L.Q.; Zhang, H.T.; Sun, Z.H.; Kan, B.; Ma, Y.F.; Chen, Y.S. A 3D cross-linked graphene-based honeycomb carbon composite with excellent confinement effect of organic cathode material for lithium-ion batteries. *Carbon* **2020**, *157*, 656–662. [CrossRef]
34. Li, Y.; Chen, M.H.; Liu, B.; Zhang, Y.; Liang, X.Q.; Xia, X.H. Heteroatom Doping: An Effective Way to Boost Sodium Ion Storage. *Adv. Energy Mater.* **2020**, *10*, 2000927. [CrossRef]
35. Wang, H.; Jia, J.; Song, P.; Wang, Q.; Li, D.; Min, Y.; Qian, C.; Wang, L.; Li, Y.; Ma, C.; et al. Efficient Electrocatalytic Reduction of CO$_2$ by Nitrogen-Doped Nanoporous Carbon/Carbon Nanotube Membranes: A Step Towards the Electrochemical CO$_2$ Refinery. *Angew. Chem. Int. Ed.* **2012**, *18*, 12–15.
36. Xu, C.F.; Zhang, K.W.; Zhang, D.; Chang, S.L.; Liang, F.; Yan, P.F.; Yao, Y.C.; Qu, T.; Zhan, J.; Ma, W.H.; et al. Reversible hybrid sodium-CO$_2$ batteries with low charging voltage and long-life. *Nano Energy* **2020**, *68*, 104318. [CrossRef]

37. Li, S.W.; Liu, Y.; Zhou, J.W.; Hong, S.S.; Dong, Y.; Wang, J.M.; Gao, X.; Qi, P.F.; Han, Y.; Wang, B. Monodispersed MnO nanoparticles in graphene-an interconnected N-doped 3D carbon framework as a highly efficient gas cathode in Li-CO_2 batteries. *Energy Environ. Sci.* **2019**, *12*, 1046–1054. [CrossRef]
38. Dong, H.Y.; Jin, C.; Gao, Y.C.; Xiao, X.L.; Li, K.; Tang, P.P.; Li, X.N.; Yang, S.T. Nitrogen and sulfur co-doped three-dimensional graphene@NiO composite as cathode catalyst for the Li-O_2 and Li-CO_2 batteries. *Mater. Res. Express* **2019**, *6*, 115616. [CrossRef]
39. Sun, Z.M.; Wang, D.; Lin, L.; Liu, Y.H.; Yuan, M.W.; Nan, C.Y.; Li, H.F.; Sun, G.B.; Yang, X.J. Ultrathin hexagonal boron nitride as a van der Waals' force initiator activated graphene for engineering efficient non-metal electrocatalysts of Li-CO_2 battery. *Nano Res.* **2022**, *15*, 1171–1177. [CrossRef]
40. Chen, B.; Wang, D.S.; Tan, J.Y.; Liu, Y.Q.; Jiao, M.L.; Liu, B.L.; Zhao, N.Q.; Zou, X.L.; Zhou, G.M.; Cheng, H.M. Designing Electrophilic and Nucleophilic Dual Centers in the ReS_2 Plane toward Efficient Bifunctional Catalysts for Li-CO_2 Batteries. *J. Am. Chem. Soc.* **2022**, *144*, 3106–3116. [CrossRef]
41. Shen, J.R.; Wu, H.T.; Sun, W.; Wu, Q.B.; Zhen, S.Y.; Wang, Z.H.; Sun, K.N. Biomass-derived hierarchically porous carbon skeletons with in situ decorated IrCo nanoparticles as high-performance cathode catalysts for Li-O_2 batteries. *J. Mater. Chem. A* **2019**, *7*, 10662–10671. [CrossRef]
42. Tong, Z.Z.; Wang, S.B.; Fang, M.H.; Lin, Y.T.; Tsai, K.T.; Tsai, S.Y.; Yin, L.C.; Hu, S.F.; Liu, R.S. Na-CO_2 battery with NASICON-structured solid-state electrolyte. *Nano Energy* **2021**, *85*, 105972. [CrossRef]
43. Thoka, S.; Tong, Z.Z.; Jena, A.; Hung, T.F.; Wu, C.C.; Chang, W.S.; Wang, F.M.; Wang, X.C.; Yin, L.C.; Chang, H.; et al. High-performance Na-CO_2 batteries with $ZnCo_2O_4$@CNT as the cathode catalyst. *J. Mater. Chem. A* **2020**, *8*, 23974–23982. [CrossRef]
44. Zhu, Y.; Feng, S.; Zhang, P.; Guo, M.; Wang, Q.; Wu, D.; Zhang, L.; Li, H.; Wang, H.; Chen, L.; et al. Probing the electrochemical evolutions of Na-CO_2 nanobatteries on Pt@NCNT cathodes using in-situ environmental TEM. *Energy Storage Mater.* **2020**, *33*, 88–94. [CrossRef]
45. Xu, C.; Zhan, J.; Wang, Z.; Fang, X.; Chen, J.; Liang, F.; Zhao, H.; Lei, Y. Biomass-derived highly dispersed Co/Co_9S_8 nanoparticles encapsulated in S, N-co-doped hierarchically porous carbon as an efficient catalyst for hybrid Na-CO_2 batteries. *Mater. Today Energy* **2021**, *19*, 100594. [CrossRef]
46. Jin, T.; Han, Q.Q.; Jiao, L.F. Binder-Free Electrodes for Advanced Sodium-Ion Batteries. *Adv. Mater.* **2020**, *32*, 1806304. [CrossRef]
47. Zhai, Y.J.; Han, P.; Yun, Q.B.; Ge, Y.Y.; Zhang, X.; Chen, Y.; Zhang, H. Phase engineering of metal nanocatalysts for electrochemical CO_2 reduction. *eScience* **2022**, *2*, 467–485. [CrossRef]
48. Lu, J.; Lee, Y.J.; Luo, X.; Lau, K.C.; Asadi, M.; Wang, H.H.; Brombosz, S.; Wen, J.; Zhai, D.; Chen, Z.; et al. A lithium-oxygen battery based on lithium superoxide. *Nature* **2016**, *529*, 377–382. [CrossRef]
49. Qiao, Y.; Xu, S.M.; Liu, Y.; Dai, J.Q.; Xie, H.; Yao, Y.G.; Mu, X.W.; Chen, C.J.; Kline, D.J.; Hitz, E.M.; et al. Transient, in situ synthesis of ultrafine ruthenium nanoparticles for a high-rate Li-CO_2 battery. *Energy Environ. Sci.* **2019**, *12*, 1100–1107. [CrossRef]
50. Zhang, S.; Wen, Z.; Rui, K.; Shen, C.; Lu, Y.; Yang, J. Graphene nanosheets loaded with Pt nanoparticles with enhanced electrochemical performance for sodium-oxygen batteries. *J. Mater. Chem. A* **2015**, *3*, 2568–2571. [CrossRef]
51. Xing, Y.; Yang, Y.; Li, D.; Luo, M.; Chen, N.; Ye, Y.; Qian, J.; Li, L.; Yang, D.; Wu, F.; et al. Crumpled Ir Nanosheets Fully Covered on Porous Carbon Nanofibers for Long-Life Rechargeable Lithium-CO_2 Batteries. *Adv. Mater.* **2018**, *30*, e1803124. [CrossRef] [PubMed]
52. Wu, G.; Li, X.; Zhang, Z.; Dong, P.; Xu, M.; Peng, H.; Zeng, X.; Zhang, Y.; Liao, S. Design of ultralong-life Li-CO_2 batteries with IrO_2 nanoparticles highly dispersed on nitrogen-doped carbon nanotubes. *J. Mater. Chem. A* **2020**, *8*, 3763–3770. [CrossRef]
53. Wang, X.; Xie, J.; Ghausi, M.A.; Lv, J.; Huang, Y.; Wu, M.; Wang, Y.; Yao, J. Rechargeable Zn-CO_2 Electrochemical Cells Mimicking Two-Step Photosynthesis. *Adv. Mater.* **2019**, *31*, e1807807. [CrossRef]
54. Ma, W.; Liu, X.; Li, C.; Yin, H.; Xi, W.; Liu, R.; He, G.; Zhao, X.; Luo, J.; Ding, Y. Rechargeable Al-CO_2 Batteries for Reversible Utilization of CO_2. *Adv. Mater.* **2018**, *30*, e1801152. [CrossRef] [PubMed]
55. Fang, C.; Luo, J.; Jin, C.; Yuan, H.; Sheng, O.; Huang, H.; Gan, Y.; Xia, Y.; Liang, C.; Zhang, J.; et al. Enhancing Catalyzed Decomposition of Na_2CO_3 with Co_2MnO_x Nanowire-Decorated Carbon Fibers for Advanced Na-CO_2 Batteries. *ACS Appl. Mater. Interfaces* **2018**, *10*, 17240–17248. [CrossRef]
56. Dong, L.Z.; Zhang, Y.; Lu, Y.F.; Zhang, L.; Huang, X.; Wang, J.H.; Liu, J.; Li, S.L.; Lan, Y.Q. A well-defined dual Mn-site based Metal-organic framework to promote CO_2 reduction/evolution in Li-CO_2 batteries. *Chem. Commun.* **2021**, *57*, 8937–8940. [CrossRef]
57. Zhang, Z.; Zhang, Z.; Liu, P.; Xie, Y.; Cao, K.; Zhou, Z. Identification of cathode stability in Li-CO_2 batteries with Cu nanoparticles highly dispersed on N-doped graphene. *J. Mater. Chem. A* **2018**, *6*, 3218–3223. [CrossRef]
58. Xu, C.; Zhan, J.; Wang, H.; Kang, Y.; Liang, F. Dense binary Fe-Cu sites promoting CO_2 utilization enable highly reversible hybrid Na-CO_2 batteries. *J. Mater. Chem. A* **2021**, *9*, 22214–22128. [CrossRef]
59. Zhang, P.; Zhang, J.; Sheng, T.; Lu, Y.; Yin, Z.; Li, Y.; Peng, X.; Zhou, Z.; Li, J.; Wu, Y.; et al. Synergetic Effect of Ru and NiO in the Electrocatalytic Decomposition of Li_2CO_3 to Enhance the Performance of a Li-CO_2/O_2 Battery. *ACS Catalysis* **2019**, *10*, 1640–1651. [CrossRef]
60. Xiao, X.; Zhang, Z.; Yu, W.; Shang, W.; Ma, Y.; Zhu, X.; Tan, P. Ultrafine Co-Doped NiO Nanoparticles Decorated on Carbon Nanotubes Improving the Electrochemical Performance and Cycling Stability of Li-CO_2 Batteries. *ACS Appl. Energy Mater.* **2021**, *4*, 11858–11866. [CrossRef]

61. Ge, B.; Sun, Y.; Guo, J.; Yan, X.; Fernandez, C.; Peng, Q. A Co-Doped MnO$_2$ Catalyst for Li-O$_2$ Batteries with Low Overpotential and Ultrahigh Cyclability. *Small* **2019**, *15*, e1902220. [CrossRef] [PubMed]
62. Wang, W.X.; Xiong, F.Y.; Zhu, S.H.; Chen, J.H.; Xie, J.; An, Q.Y. Defect engineering in molybdenum-based electrode materials for energy storage. *eScience* **2022**, *2*, 278–294. [CrossRef]
63. Kwak, W.; Lau, K.; Shin, C.; Amine, K.; Curtiss, L.; Sun, Y. A Mo$_2$C/Carbon Nanotube Composite Cathode for Lithium-Oxygen Batteries with High Energy Efficiency and Long Cycle Life. *ACS Nano* **2015**, *9*, 4129–4137. [CrossRef] [PubMed]
64. Hou, Y.; Wang, J.; Liu, L.; Liu, Y.; Chou, S.; Shi, D.; Liu, H.; Wu, Y.; Zhang, W.; Chen, J. Mo$_2$C/CNT: An Efficient Catalyst for Rechargeable Li-CO$_2$ Batteries. *Adv. Funct. Mater.* **2017**, *27*, 1700564. [CrossRef]
65. Asadi, M.; Sayahpour, B.; Abbasi, P.; Ngo, A.T.; Karis, K.; Jokisaari, J.R.; Liu, C.; Narayanan, B.; Gerard, M.; Yasaei, P.; et al. A lithium-oxygen battery with a long cycle life in an air-like atmosphere. *Nature* **2018**, *555*, 502–506. [CrossRef]
66. Senthil Kumar, R.; Senthil Kumar, S.; Anbu Kulandainathan, M. Highly selective electrochemical reduction of carbon dioxide using Cu based metal organic framework as an electrocatalyst. *Electrochem. Commun.* **2012**, *25*, 70–73. [CrossRef]
67. Millward, A.R.; Yaghi, O.M. Metal-Organic Frameworks with Exceptionally High Capacity for Storage of Carbon Dioxide at Room Temperature. *J. Am. Chem. Soc.* **2005**, *127*, 17998–17999. [CrossRef]
68. Wu, D.; Guo, Z.; Yin, X.; Pang, Q.; Tu, B.; Zhang, L.; Wang, Y.G.; Li, Q. Metal-organic frameworks as cathode materials for Li-O$_2$ batteries. *Adv. Mater.* **2014**, *26*, 3258–3262. [CrossRef]
69. Chen, Y.; Ji, S.; Chen, C.; Peng, Q.; Wang, D.; Li, Y. Single-Atom Catalysts: Synthetic Strategies and Electrochemical Applications. *Joule* **2018**, *2*, 1242–1264. [CrossRef]
70. Qiao, B.; Wang, A.; Yang, X.; Allard, L.F.; Jiang, Z.; Cui, Y.; Liu, J.; Li, J.; Zhang, T. Single-atom catalysis of CO oxidation using Pt$_1$/FeO$_x$. *Nat. Chem.* **2011**, *3*, 634–641. [CrossRef]
71. Wang, Y.; Liu, Y.; Liu, W.; Wu, J.; Li, Q.; Feng, Q.; Chen, Z.; Xiong, X.; Wang, D.; Lei, Y. Regulating the coordination structure of metal single atoms for efficient electrocatalytic CO$_2$ reduction. *Energy Environ. Sci.* **2020**, *13*, 4609–4624. [CrossRef]
72. Cheng, Y.; Yang, S.; Jiang, S.P.; Wang, S. Supported Single Atoms as New Class of Catalysts for Electrochemical Reduction of Carbon Dioxide. *Small Methods* **2019**, *3*, 1800440. [CrossRef]
73. Cheng, Y.; Zhao, S.; Johannessen, B.; Veder, J.P.; Saunders, M.; Rowles, M.R.; Cheng, M.; Liu, C.; Chisholm, M.F.; De Marco, R.; et al. Atomically Dispersed Transition Metals on Carbon Nanotubes with Ultrahigh Loading for Selective Electrochemical Carbon Dioxide Reduction. *Adv. Mater.* **2018**, *30*, e1706287. [CrossRef]
74. Shah, S.S.A.; Najam, T.; Bashir, M.S.; Peng, L.; Nazir, M.A.; Javed, M.S. Single-atom catalysts for next-generation rechargeable batteries and fuel cells. *Energy Storage Mater.* **2022**, *45*, 301–322. [CrossRef]
75. Xi, J.; Sun, H.; Wang, D.; Zhang, Z.; Duan, X.; Xiao, J.; Xiao, F.; Liu, L.; Wang, S. Confined-interface-directed synthesis of Palladium single-atom catalysts on graphene/amorphous carbon. *Appl. Catal. B Environ.* **2018**, *225*, 291–297. [CrossRef]
76. Liu, D.; Wu, C.; Chen, S.; Ding, S.; Xie, Y.; Wang, C.; Wang, T.; Haleem, Y.A.; ur Rehman, Z.; Sang, Y.; et al. In situ trapped high-density single metal atoms within graphene: Iron-containing hybrids as representatives for efficient oxygen reduction. *Nano Res.* **2018**, *11*, 2217–2228. [CrossRef]
77. Wang, T.; Zhao, Q.; Fu, Y.; Lei, C.; Yang, B.; Li, Z.; Lei, L.; Wu, G.; Hou, Y. Carbon-Rich Nonprecious Metal Single Atom Electrocatalysts for CO$_2$ Reduction and Hydrogen Evolution. *Small Methods* **2019**, *3*, 1900210. [CrossRef]
78. Jiang, K.; Liu, B.; Luo, M.; Ning, S.; Peng, M.; Zhao, Y.; Lu, Y.R.; Chan, T.S.; de Groot, F.M.F.; Tan, Y. Single platinum atoms embedded in nanoporous cobalt selenide as electrocatalyst for accelerating hydrogen evolution reaction. *Nat. Commun.* **2019**, *10*, 1743. [CrossRef]
79. Sun, S.; Zhang, G.; Gauquelin, N.; Chen, N.; Zhou, J.; Yang, S.; Chen, W.; Meng, X.; Geng, D.; Banis, M.N.; et al. Single-atom Catalysis Using Pt/Graphene Achieved through Atomic Layer Deposition. *Sci. Rep.* **2013**, *3*, 1775. [CrossRef]
80. Liu, X.; Jia, S.; Yang, M.; Tang, Y.; Wen, Y.; Chu, S.; Wang, J.; Shan, B.; Chen, R. Activation of subnanometric Pt on Cu-modified CeO$_2$ via redox-coupled atomic layer deposition for CO oxidation. *Nat. Commun.* **2020**, *11*, 4240. [CrossRef]
81. Wei, H.; Liu, X.; Wang, A.; Zhang, L.; Qiao, B.; Yang, X.; Huang, Y.; Miao, S.; Liu, J.; Zhang, T. FeO$_x$-supported platinum single-atom and pseudo-single-atom catalysts for chemoselective hydrogenation of functionalized nitroarenes. *Nat. Commun.* **2014**, *5*, 5634. [CrossRef] [PubMed]
82. Sun, X.; Chen, C.; Liu, S.; Hong, S.; Zhu, Q.; Qian, Q.; Han, B.; Zhang, J.; Zheng, L. Aqueous CO$_2$ Reduction with High Efficiency Using alpha-Co(OH)$_2$-Supported Atomic Ir Electrocatalysts. *Angew. Chem. Int. Ed.* **2019**, *58*, 4669–4673. [CrossRef] [PubMed]
83. Xu, L.; Chen, X.; Jing, H.; Wang, L.; Wei, J.; Han, Y. Design and performance of Ag nanoparticle-modified graphene/SnAgCu lead-free solders. *Mater. Sci. Eng. A* **2016**, *667*, 87–96. [CrossRef]
84. Yan, D.; Chen, J.; Jia, H. Temperature-Induced Structure Reconstruction to Prepare a Thermally Stable Single-Atom Platinum Catalyst. *Angew. Chem. Int. Ed.* **2020**, *59*, 13562–13567. [CrossRef]
85. Zhao, C.; Wang, Y.; Li, Z.; Chen, W.; Xu, Q.; He, D.; Xi, D.; Zhang, Q.; Yuan, T.; Qu, Y.; et al. Solid-Diffusion Synthesis of Single-Atom Catalysts Directly from Bulk Metal for Efficient CO$_2$ Reduction. *Joule* **2019**, *3*, 584–594. [CrossRef]
86. Wen, X.; Zhang, Q.; Guan, J. Applications of metal-organic framework-derived materials in fuel cells and metal-air batteries. *Coord. Chem. Rev.* **2020**, *409*, 213214. [CrossRef]
87. Han, A.; Chen, W.; Zhang, S.; Zhang, M.; Han, Y.; Zhang, J.; Ji, S.; Zheng, L.; Wang, Y.; Gu, L.; et al. A Polymer Encapsulation Strategy to Synthesize Porous Nitrogen-Doped Carbon-Nanosphere-Supported Metal Isolated-Single-Atomic-Site Catalysts. *Adv. Mater.* **2018**, *30*, e1706508. [CrossRef]

88. Wang, J.; Huang, Z.; Liu, W.; Chang, C.; Tang, H.; Li, Z.; Chen, W.; Jia, C.; Yao, T.; Wei, S.; et al. Design of N-Coordinated Dual-Metal Sites: A Stable and Active Pt-Free Catalyst for Acidic Oxygen Reduction Reaction. *J. Am. Chem. Soc.* **2017**, *139*, 17281–17284. [CrossRef]
89. Liu, J.J. Advanced Electron Microscopy of Metal-Support Interactions in Supported Metal Catalysts. *ChemCatChem* **2011**, *3*, 934–948. [CrossRef]
90. Wang, X.; Pi, M.; Zhang, D.; Li, H.; Feng, J.; Chen, S.; Li, J. Insight into the Superior Electrocatalytic Performance of a Ternary Nickel Iron Poly-Phosphide Nanosheet Array: An X-ray Absorption Study. *ACS Appl. Mater. Interfaces* **2019**, *11*, 14059–14065. [CrossRef]
91. Cheng, N.; Zhang, L.; Doyle-Davis, K.; Sun, X. Single-Atom Catalysts: From Design to Application. *Electrochem. Energy Rev.* **2019**, *2*, 539–573. [CrossRef]
92. Zhu, Y.; Wang, J.; Chu, H.; Chu, Y.; Chen, H.M. In Situ/Operando Studies for Designing Next-Generation Electrocatalysts. *ACS Energy Lett.* **2020**, *5*, 1281–1291. [CrossRef]
93. Hui, Y.; Liu, Y.F.; Liu, X.L.; Wang, X.K.; Tian, H.; Waterhouse, G.I.; Kruger, P.E.; Telfer, S.G.; Ma, S.Q. Large-scale synthesis of N-doped carbon capsules supporting atomically dispersed iron for efficient oxygen reduction reaction electrocatalysis. *eScience* **2022**, *2*, 227–234.
94. Rao, P.; Wu, D.X.; Wang, T.J.; Li, J.; Deng, P.L.; Chen, Q.; Shen, Y.J.; Chen, Y.; Tian, X.L. Single atomic cobalt electrocatalyst for efficient oxygen reduction reaction. *eScience* **2022**, *2*, 399–404. [CrossRef]
95. Zheng, C.Y.; Zhang, X.; Zhou, Z.; Hu, Z.P. A first-principles study on the electrochemical reaction activity of 3d transition metal single-atom catalysts in nitrogen-doped graphene: Trends and hints. *eScience* **2022**, *2*, 219–226. [CrossRef]
96. Thangadurai, V.; Narayanan, S.; Pinzaru, D. Garnet-type solid-state fast Li ion conductors for Li batteries: Critical review. *Chem. Soc. Rev.* **2014**, *43*, 4714–4727. [CrossRef] [PubMed]
97. Lu, Y.; Meng, X.; Alonso, J.A.; Fernandez-Diaz, M.T.; Sun, C. Effects of Fluorine Doping on Structural and Electrochemical Properties of $Li_{6.25}Ga_{0.25}La_3Zr_2O_{12}$ as Electrolytes for Solid-State Lithium Batteries. *ACS Appl. Mater. Interfaces* **2019**, *11*, 2042–2049. [CrossRef]
98. Thangadurai, V.; Pinzaru, D.; Narayanan, S.; Baral, A.K. Fast Solid-State Li Ion Conducting Garnet-Type Structure Metal Oxides for Energy Storage. *J. Phys. Chem. Lett.* **2015**, *6*, 292–299. [CrossRef]
99. Wang, Y.; Song, S.; Xu, C.; Hu, N.; Molenda, J.; Lu, L. Development of solid-state electrolytes for sodium-ion battery-A short review. *Nano Mater. Sci.* **2019**, *1*, 91–100. [CrossRef]
100. Lu, Y.; Li, L.; Zhang, Q.; Niu, Z.; Chen, J. Electrolyte and Interface Engineering for Solid-State Sodium Batteries. *Joule* **2018**, *2*, 1747–1770. [CrossRef]
101. Bachman, J.; Muy, S.; Grimaud, A.; Chang, H.; Pour, N.; Lux, S.; Paschos, O.; Maglia, F.; Lupart, S.; Lamp, P.; et al. Inorganic Solid-State Electrolytes for Lithium Batteries: Mechanisms and Properties Governing Ion Conduction. *Chem. Rev.* **2016**, *116*, 140–162. [CrossRef] [PubMed]
102. Famprikis, T.; Canepa, P.; Dawson, J.A.; Islam, M.S.; Masquelier, C. Fundamentals of inorganic solid-state electrolytes for batteries. *Nat. Mater.* **2019**, *18*, 1278–1291. [CrossRef] [PubMed]
103. Zhang, Z.; Zhang, Q.; Shi, J.; Chu, Y.S.; Yu, X.; Xu, K.; Ge, M.; Yan, H.; Li, W.; Gu, L.; et al. A Self-Forming Composite Electrolyte for Solid-State Sodium Battery with Ultralong Cycle Life. *Adv. Energy Mater.* **2017**, *7*, 1601196. [CrossRef]
104. Yue, L.; Ma, J.; Zhang, J.; Zhao, J.; Dong, S.; Liu, Z.; Cui, G.; Chen, L. All solid-state polymer electrolytes for high-performance lithium ion batteries. *Energy Storage Mater.* **2016**, *5*, 139–164. [CrossRef]
105. Song, S.; Duong, H.M.; Korsunsky, A.M.; Hu, N.; Lu, L. A Na(+) Superionic Conductor for Room-Temperature Sodium Batteries. *Sci. Rep.* **2016**, *6*, 32330. [CrossRef]
106. Zhu, Z.; Chu, I.-H.; Deng, Z.; Ong, S.P. Role of Na^+ Interstitials and Dopants in Enhancing the Na^+ Conductivity of the Cubic Na_3PS_4 Superionic Conductor. *Chem. Mater.* **2015**, *27*, 8318–8325. [CrossRef]
107. Park, H.; Jung, K.; Nezafati, M.; Kim, C.S.; Kang, B. Sodium Ion Diffusion in Nasicon ($Na_3Zr_2Si_2PO_{12}$) Solid Electrolytes: Effects of Excess Sodium. *ACS Appl. Mater. Interfaces* **2016**, *8*, 27814–27824. [CrossRef]
108. Ma, Q.; Guin, M.; Naqash, S.; Tsai, C.-L.; Tietz, F.; Guillon, O. Scandium-Substituted $Na_3Zr_2(SiO_4)_2(PO_4)$ Prepared by a Solution-Assisted Solid-State Reaction Method as Sodium-Ion Conductors. *Chem. Mater.* **2016**, *28*, 4821–4828. [CrossRef]
109. Ruan, Y.; Song, S.; Liu, J.; Liu, P.; Cheng, B.; Song, X.; Battaglia, V. Improved structural stability and ionic conductivity of $Na_3Zr_2Si_2PO_{12}$ solid electrolyte by rare earth metal substitutions. *Ceram. Int.* **2017**, *43*, 7810–7815. [CrossRef]
110. Goodenough, J.B.; Hong, H.; Kafalas, J. Fast Na^+-ion transport in skeleton structures. *Mater. Res. Bull.* **1976**, *11*, 203–220. [CrossRef]
111. Hong, Y. Crystal structures and crystal chemistry in the system $Na_{1+x}Zr_2Si_xP_{3-x}O_1$. *Mater. Res. Bull.* **1976**, *11*, 173–182. [CrossRef]
112. Im, E.; Ryu, J.H.; Baek, K.; Moon, G.D.; Kang, S.J. "Water-in-salt" and NASICON Electrolyte-Based Na-CO_2 Battery. *Energy Storage Mater.* **2021**, *37*, 424–432. [CrossRef]
113. Lu, L.; Sun, C.; Hao, J.; Wang, Z.; Mayer, S.F.; Fernández-Díaz, M.T.; Alonso, J.A.; Zou, B. A High-Performance Solid-State Na-CO_2 Battery with Poly(Vinyliden Fluoride-co-Hexafluoropropylene)-$Na_{3.2}Zr_{1.9}Mg_{0.1}Si_2PO_{12}$ Electrolyte. *Energy Environ. Mater.* **2022**, 1–9.
114. Jung, J.I.; Kim, D.; Kim, H.; Jo, Y.N.; Park, J.S.; Kim, Y. Progressive Assessment on the Decomposition Reaction of Na Superionic Conducting Ceramics. *ACS Appl. Mater. Interfaces* **2017**, *9*, 304–310. [CrossRef] [PubMed]

115. Sun, C.; Liu, J.; Gong, Y.; Wilkinson, D.P.; Zhang, J. Recent advances in all-solid-state rechargeable lithium batteries. *Nano Energy* **2017**, *33*, 363–386. [CrossRef]
116. Lu, L.; Lu, Y.; Alonso, J.A.; Lopez, C.A.; Fernandez-Diaz, M.T.; Zou, B.; Sun, C. A Monolithic Solid-State Sodium-Sulfur Battery with Al-Doped $Na_{3.4}Zr_2(Si_{0.8}P_{0.2}O_4)_3$ Electrolyte. *ACS Appl. Mater. Interfaces* **2021**, *13*, 42927–42934. [CrossRef]
117. Shen, L.; Yang, J.; Liu, G.; Avdeev, M.; Yao, X. High ionic conductivity and dendrite-resistant NASICON solid electrolyte for all-solid-state sodium batteries. *Mater. Today Energy* **2021**, *20*, 100691. [CrossRef]
118. Ihlefeld, J.F.; Gurniak, E.; Jones, B.H.; Wheeler, D.R.; Rodriguez, M.A.; McDaniel, A.H.; Dunn, B. Scaling Effects in Sodium Zirconium Silicate Phosphate ($Na_{1+x}Zr_2Si_xP_{3-x}O_{12}$) Ion-Conducting Thin Films. *J. Am. Ceram. Soc.* **2016**, *99*, 2729–2736. [CrossRef]
119. Yao, Y.; Kummer, J. Ion exchange properties of and rates of ionic diffusion in beta-alumina. *J. Inorg. Nucl. Chem.* **1967**, *29*, 2453–2475.
120. Hueso, K.B.; Palomares, V.; Armand, M.; Rojo, T. Challenges and perspectives on high and intermediate-temperature sodium batteries. *Nano Res.* **2017**, *10*, 4082–4114. [CrossRef]
121. Lu, X.; Xia, G.; Lemmon, J.P.; Yang, Z. Advanced Materials for sodium-beta alumina batteries: Status, challenges and perspectives. *J. Power Source* **2010**, *195*, 2431–2442. [CrossRef]
122. Baffier, N.; Badot, J.; Colomban, P. Conductivity of ion rich β and β″ alumina: Sodium and potassium compounds. *Mater. Res. Bull.* **1981**, *16*, 259–265. [CrossRef]
123. Wu, Y.; Zhuo, L.; Ming, J.; Yu, Y.; Zhao, F. Coating of Al_2O_3 on layered $Li(Mn_{1/3}Ni_{1/3}Co_{1/3})O_2$ using CO_2 as green precipitant and their improved electrochemical performance for lithium ion batteries. *J. Energy Chem.* **2013**, *22*, 468–476. [CrossRef]
124. Ghadbeigi, L.; Szendrei, A.; Moreno, P.; Sparks, T.D.; Virkar, A.V. Synthesis of iron-doped Na-β″-alumina+yttria-stabilized zirconia composite electrolytes by a vapor phase process. *Solid State Ion.* **2016**, *290*, 77–82. [CrossRef]
125. Virkar, A.V.; Gordon, R.S. Fracture Properties of Polycrystalline Lithia-Stabilized β″ Alumina. *J. Am. Ceram. Soc.* **1977**, *60*, 58–61. [CrossRef]
126. Wang, Y.; Wang, Z.X.; Wu, D.X.; Niu, Q.H.; Lu, P.S.; Ma, T.H.; Su, Y.B.; Chen, L.Q.; Li, H.; Wu, F. Stable Ni-rich layered oxide cathode for sulfide-based all-solid-state lithium battery. *eScience* **2022**, *2*, 537–545. [CrossRef]
127. Hayashi, A.; Noi, K.; Sakuda, A.; Tatsumisago, M. Superionic glass-ceramic electrolytes for room-temperature rechargeable sodium batteries. *Nat. Commun.* **2012**, *3*, 856. [CrossRef]
128. Shang, S.L.; Yu, Z.X.; Wang, Y.; Wang, D.H.; Liu, Z.K. Origin of Outstanding Phase and Moisture Stability in a $Na_3P_{1-x}As_xS_4$ Superionic Conductor. *ACS Appl. Mater. Interfaces* **2017**, *9*, 16261–16269. [CrossRef]
129. Tian, Y.; Shi, T.; Richards, W.D.; Li, J.; Kim, J.C.; Bo, S.-H.; Ceder, G. Compatibility issues between electrodes and electrolytes in solid-state batteries. *Energy Environ. Sci.* **2017**, *10*, 1150–1166. [CrossRef]
130. Zhang, L.; Yang, K.; Mi, J.; Lu, L.; Zhao, L.; Wang, L.; Li, Y.; Zeng, H. Na_3PSe_4: A Novel Chalcogenide Solid Electrolyte with High Ionic Conductivity. *Adv. Energy Mater.* **2015**, *5*, 1501294. [CrossRef]
131. Nishimura, S.; Tanibata, N.; Hayashi, A.; Tatsumisago, M.; Yamada, A. The crystal structure and sodium disorder of high-temperature polymorph β-Na_3PS_4. *J. Mater. Chem. A* **2017**, *5*, 25025–25030. [CrossRef]
132. Jansen, M.; Henseler, U. Synthesis, Structure Determination, and Ionic Conductivity of Sodium Tetrathiophosphate. *J. Solid State Chem.* **1992**, *99*, 110–119. [CrossRef]
133. Yu, Z.; Shang, S.L.; Seo, J.H.; Wang, D.; Luo, X.; Huang, Q.; Chen, S.; Lu, J.; Li, X.; Liu, Z.K.; et al. Exceptionally High Ionic Conductivity in $Na_3P_{0.62}As_{0.38}S_4$ with Improved Moisture Stability for Solid-State Sodium-Ion Batteries. *Adv. Mater.* **2017**, *29*, 1605561. [CrossRef] [PubMed]
134. Zhang, L.; Zhang, D.; Yang, K.; Yan, X.; Wang, L.; Mi, J.; Xu, B.; Li, Y. Vacancy-Contained Tetragonal Na_3SbS_4 Superionic Conductor. *Adv. Sci.* **2016**, *3*, 1600089. [CrossRef]
135. Wang, Y.; Yang, N.; Shuai, Y.; Chen, K. Low-Cost Raw Materials Synthesized $Na_{2.9375}PS_{3.9375}Cl_{0.0625}$ Solid Electrolyte. *Energy Technol.* **2020**, *8*, 2000606. [CrossRef]
136. de Klerk, N.J.J.; Wagemaker, M. Diffusion Mechanism of the Sodium-Ion Solid Electrolyte Na_3PS_4 and Potential Improvements of Halogen Doping. *Chem. Mater.* **2016**, *28*, 3122–3130. [CrossRef]
137. Krauskopf, T.; Pompe, C.; Kraft, M.A.; Zeier, W.G. Influence of Lattice Dynamics on Na^+ Transport in the Solid Electrolyte $Na_3PS_{4-x}Se_x$. *Chem. Mater.* **2017**, *29*, 8859–8869. [CrossRef]
138. Wang, H.; Chen, Y.; Hood, Z.D.; Sahu, G.; Pandian, A.S.; Keum, J.K.; An, K.; Liang, C. An Air-Stable Na_3SbS_4 Superionic Conductor Prepared by a Rapid and Economic Synthetic Procedure. *Angew. Chem. Int. Ed. Engl.* **2016**, *55*, 8551–8555. [CrossRef]
139. Zhang, Z.; Zhang, J.; Sun, Y.; Jia, H.; Peng, L.; Zhang, Y.; Xie, J. $Li_{4-x}Sb_xSn_{1-x}S_4$ solid solutions for air-stable solid electrolytes. *J. Energy Chem.* **2020**, *41*, 171–176. [CrossRef]
140. Oguchi, H.; Matsuo, M.; Kuromoto, S.; Kuwano, H.; Orimo, S. Sodium-ion conduction in complex hydrides $NaAlH_4$ and Na_3AlH_6. *J. Appl. Phys.* **2012**, *111*, 036102. [CrossRef]
141. Matsuo, M.; Oguchi, H.; Sato, T.; Takamura, H.; Tsuchida, E.; Ikeshoji, T.; Orimo, S.-I. Sodium and magnesium ionic conduction in complex hydrides. *J. Alloys Compd.* **2013**, *580*, S98–S101. [CrossRef]
142. Udovic, T.J.; Matsuo, M.; Unemoto, A.; Verdal, N.; Stavila, V.; Skripov, A.V.; Rush, J.J.; Takamura, H.; Orimo, S. Sodium superionic conduction in $Na_2B_{12}H_{12}$. *Chem. Commun.* **2014**, *50*, 3750–3752. [CrossRef] [PubMed]

143. Verdal, N.; Udovic, T.J.; Stavila, V.; Tang, W.S.; Rush, J.J.; Skripov, A.V. Anion Reorientations in the Superionic Conducting Phase of Na$_2$B$_{12}$H$_{12}$. *J. Phys. Chem. C* **2014**, *118*, 17483–17489. [CrossRef]
144. Tang, W.S.; Yoshida, K.; Soloninin, A.V.; Skoryunov, R.V.; Babanova, O.A.; Skripov, A.V.; Dimitrievska, M.; Stavila, V.; Orimo, S.-I.; Udovic, T.J. Stabilizing Superionic-Conducting Structures via Mixed-Anion Solid Solutions of Monocarba-closo-borate Salts. *ACS Energy Lett.* **2016**, *1*, 659–664. [CrossRef]
145. Tang, W.S.; Matsuo, M.; Wu, H.; Stavila, V.; Zhou, W.; Talin, A.A.; Soloninin, A.V.; Skoryunov, R.V.; Babanova, O.A.; Skripov, A.V.; et al. Liquid-Like Ionic Conduction in Solid Lithium and Sodium Monocarba-closo-Decaborates Near or at Room Temperature. *Adv. Energy Mater.* **2016**, *6*, 1502237. [CrossRef]
146. Duchêne, L.; Kühnel, R.S.; Rentsch, D.; Remhof, A.; Hagemann, H.; Battaglia, C. A highly stable sodium solid-state electrolyte based on a dodeca/deca-borate equimolar mixture. *Chem. Commun.* **2017**, *53*, 4195–4198. [CrossRef]
147. Duchêne, L.; Kühnel, R.S.; Stilp, E.; Cuervo Reyes, E.; Remhof, A.; Hagemann, H.; Battaglia, C. A stable 3 V all-solid-state sodium–ion battery based on a closo-borate electrolyte. *Energy Environ. Sci.* **2017**, *10*, 2609–2615. [CrossRef]
148. Udovic, T.J.; Matsuo, M.; Tang, W.S.; Wu, H.; Stavila, V.; Soloninin, A.V.; Skoryunov, R.V.; Babanova, O.A.; Skripov, A.V.; Rush, J.J.; et al. Exceptional superionic conductivity in disordered sodium decahydro-closo-decaborate. *Adv. Mater.* **2014**, *26*, 7622–7626. [CrossRef]
149. Sun, Y.; Wang, Y.; Liang, X.; Xia, Y.; Peng, L.; Jia, H.; Li, H.; Bai, L.; Feng, J.; Jiang, H.; et al. Rotational Cluster Anion Enabling Superionic Conductivity in Sodium-Rich Antiperovskite Na$_3$OBH$_4$. *J. Am. Chem. Soc.* **2019**, *141*, 5640–5644. [CrossRef]
150. Dimitrievska, M.; Shea, P.; Kweon, K.E.; Bercx, M.; Varley, J.B.; Tang, W.S.; Skripov, A.V.; Stavila, V.; Udovic, T.J.; Wood, B.C. Carbon Incorporation and Anion Dynamics as Synergistic Drivers for Ultrafast Diffusion in Superionic LiCB$_{11}$H$_{12}$ and NaCB$_{11}$H$_{12}$. *Adv. Energy Mater.* **2018**, *8*, 1703422. [CrossRef]
151. Tang, W.S.; Matsuo, M.; Wu, H.; Stavila, V.; Unemoto, A.; Orimo, S.-I.; Udovic, T.J. Stabilizing lithium and sodium fast-ion conduction in solid polyhedral-borate salts at device-relevant temperatures. *Energy Storage Mater.* **2016**, *4*, 79–83. [CrossRef]
152. Fenton, D.E.; Parker, J.M.; Wright, P.V. Complexes of alkali metal ions with poly(ethylene oxide). *Polymer* **1973**, *14*, 589. [CrossRef]
153. Feuillade, G.; Perche, P. Ion-conductive macromolecular gels and membranes for solid lithium cells. *J. Appl. Electrochem.* **1975**, *5*, 63–69. [CrossRef]
154. Chen, R.; Qu, W.; Guo, X.; Li, L.; Wu, F. The pursuit of solid-state electrolytes for lithium batteries: From comprehensive insight to emerging horizons. *Mater. Horiz.* **2016**, *3*, 487–516. [CrossRef]
155. Xue, Z.; He, D.; Xie, X. Poly(ethylene oxide)-based electrolytes for lithium-ion batteries. *J. Mater. Chem. A* **2015**, *3*, 19218–19253. [CrossRef]
156. Ratner, M.A.; Johansson, P.; Shriver, D.F. Polymer Electrolytes: Ionic Transport Mechanisms and Relaxation Coupling. *MRS Bull.* **2000**, *3*, 31–37. [CrossRef]
157. Zhang, Q.; Liu, K.; Ding, F.; Liu, X. Recent advances in solid polymer electrolytes for lithium batteries. *Nano Res.* **2017**, *10*, 4139–4174. [CrossRef]
158. Hou, H.D.; Xu, Q.K.; Pang, Y.K.; Li, L.; Wang, J.L.; Zhang, C.; Sun, C.W. Efficient Storing Energy Harvested by Triboelectric Nanogenerators Using a Safe and Durable All-solid-state Sodium-ion Battery. *Adv. Sci.* **2017**, *4*, 1700072. [CrossRef]
159. Hu, X.; Li, Z.; Zhao, Y.; Sun, J.; Zhao, Q.; Wang, J.; Tao, Z.; Chen, J. Quasi-solid state rechargeable Na-CO$_2$ batteries with reduced graphene oxide Na anodes. *Sci. Adv.* **2017**, *3*, e1602396. [CrossRef]
160. Nagasubramanian, G.; Stefano, S.D. 12-Crown-4 Ether-Assisted Enhancement of Ionic Conductivity and Interfacial Kinetics in Polyethylene Oxide Electrolytes. *J. Electrochem. Soc.* **1990**, *12*, 3830. [CrossRef]
161. Song, J.; Wang, Y.; Wan, C. Review of gel-type polymer electrolytes for lithium-ion batteries. *J. Power Source* **1999**, *2*, 183–197. [CrossRef]
162. Yu, Q.; Lu, Q.; Qi, X.; Zhao, S.; He, Y.-B.; Liu, L.; Li, J.; Zhou, D.; Hu, Y.-S.; Yang, Q.-H.; et al. Liquid electrolyte immobilized in compact polymer matrix for stable sodium metal anodes. *Energy Storage Mater.* **2019**, *23*, 610–616. [CrossRef]
163. Zhou, D.; He, Y.-B.; Cai, Q.; Qin, X.; Li, B.; Du, H.; Yang, Q.-H.; Kang, F. Investigation of cyano resin-based gel polymer electrolyte: In situ gelation mechanism and electrode-electrolyte interfacial fabrication in lithium-ion battery. *J. Mater. Chem. A* **2014**, *2*, 20059–20066. [CrossRef]
164. Zhang, W.; Yi, Q.; Li, S.; Sun, C. An ion-conductive Li$_7$La$_3$Zr$_2$O$_{12}$-based composite membrane for dendrite-free lithium metal batteries. *J. Power Source* **2020**, *450*, 227710. [CrossRef]
165. Skaarup, S.; West, K.; Zachau-Christiansen, B. Mixed phase solid electrolytes. *Solid State Ion.* **1988**, *28–30*, 975–978. [CrossRef]
166. Wieczorek, W. Modifications of crystalline structure of peo polymer electrolytes with ceramic additives. *Solid State Ion.* **1989**, *3–4*, 255–257. [CrossRef]
167. Liu, L.; Qi, X.; Yin, S.; Zhang, Q.; Liu, X.; Suo, L.; Li, H.; Chen, L.; Hu, Y.-S. In Situ Formation of a Stable Interface in Solid-State Batteries. *ACS Energy Lett.* **2019**, *4*, 1650–1657. [CrossRef]
168. Ni'mah, Y.L.; Cheng, M.-Y.; Cheng, J.H.; Rick, J.; Hwang, B.-J. Solid-state polymer nanocomposite electrolyte of TiO$_2$/PEO/NaClO$_4$ for sodium ion batteries. *J. Power Source* **2015**, *278*, 375–381. [CrossRef]
169. Thakur, A.K.; Upadhyaya, H.M.; Hashmi, S.A. Polyethylene oxide based sodium ion conducting composite polymer electrolytes dispersed with Na$_2$SiO$_3$. *Indian J. Pure Appl. Phys.* **1999**, *37*, 302–305.
170. Zhang, Z.; Zhang, Q.; Ren, C.; Luo, F.; Ma, Q.; Hu, Y.-S.; Zhou, Z.; Li, H.; Huang, X.; Chen, L. A ceramic/polymer composite solid electrolyte for sodium batteries. *J. Mater. Chem. A* **2016**, *4*, 15823–15828. [CrossRef]

171. Zhang, Q.Q.; Su, X.; Lu, Y.X.; Hu, Y.S. A Composite Solid-state Polymer Electrolyte for Solid-state Sodium Batteries. *J. Chin. Ceram. Soc.* **2020**, *48*, 939–946.
172. Colò, F.; Bella, F.; Nair, J.R.; Destro, M.; Gerbaldi, C. Cellulose-based novel hybrid polymer electrolytes for green and efficient Na-ion batteries. *Electrochim. Acta* **2015**, *174*, 185–190. [CrossRef]
173. Gao, H.; Guo, B.; Song, J.; Park, K.; Goodenough, J.B. A Composite Gel-Polymer/Glass-Fiber Electrolyte for Sodium-Ion Batteries. *Adv. Energy Mater.* **2015**, *5*, 1402235. [CrossRef]
174. Villaluenga, I.; Bogle, X.; Greenbaum, S.; Gil de Muro, I.; Rojo, T.; Armand, M. Cation only conduction in new polymer-SiO$_2$ nanohybrids: Na$^+$ electrolytes. *J. Mater. Chem. A* **2013**, *1*, 8348–8352. [CrossRef]
175. Lu, Y.; Alonso, J.A.; Yi, Q.; Lu, L.; Wang, Z.L.; Sun, C.W. A High-performance Monolithic Solid-State Sodium Battery with Ca^{2+} Doped Na$_3$Zr$_2$Si$_2$PO$_{12}$ Electrolyte. *Adv. Energy Mater.* **2019**, *9*, 1901205. [CrossRef]
176. Zhang, X.; Sun, C.W. Recent Advances in Dendrite-free Lithium Metal Anode for High-Performance Batteries. *Phys. Chem. Chem. Phys.* **2022**, *24*, 19996–20011. [CrossRef] [PubMed]
177. Yang, G.P.; Li, N.W.; Sun, C.W. High-performance sodium metal batteries with sodium-bismuth alloy anode. *ACS Appl. Energy Mater.* **2020**, *3*, 12607–12612. [CrossRef]
178. Lu, Y.; Lu, L.; Qiu, G.R.; Sun, C.W. Flexible Quasi-Solid-State Sodium Battery for Storing Pulse Electricity Harvested by Triboelectric Nanogenerators. *ACS Appl. Mater. Interfaces* **2020**, *12*, 39342–39351. [CrossRef]
179. Zhou, W.; Li, Y.; Xin, S.; Goodenough, J.B. Rechargeable Sodium All-Solid-State Battery. *ACS Cent. Sci.* **2017**, *3*, 52–57. [CrossRef]
180. Yi, Q.; Lu, Y.; Sun, X.R.; Zhang, H.; Yu, H.L.; Sun, C.W. Fluorinated ether based electrolyte with low flammability enabling sodium-metal batteries with exceptional cycling stability. *ACS Appl. Mater. Interfaces* **2019**, *11*, 46965–46972. [CrossRef]
181. Chen, D.; Zhu, M.; Kang, P.; Zhu, T.; Yuan, H.; Lan, J.; Yang, X.; Sui, G. Self-Enhancing Gel Polymer Electrolyte by In Situ Construction for Enabling Safe Lithium Metal Battery. *Adv. Sci.* **2022**, *9*, e2103663. [CrossRef] [PubMed]
182. Wang, J.N.; Ma, Q.Y.; Sun, S.Y.; Yang, K.; Cai, Q.; Olsson, E.; Chen, X.; Wang, Z.; Abdelkader, A.M.; Li, Y.S.; et al. Highly aligned lithiophilic electrospun nanofiber membrane for the multiscale suppression of Li dendrite growth. *eScience* **2022**, *2*, 655–665. [CrossRef]
183. Yi, Q.; Zhang, W.Q.; Li, S.Q.; Sun, C.W. A durable sodium battery with a flexible Na$_3$Zr$_2$Si$_2$PO$_{12}$-PVDF-HFP composite electrolyte and sodium/carbon cloth anode. *ACS Appl. Mater. Interfaces* **2018**, *10*, 35039–35046. [CrossRef] [PubMed]
184. Zhao, Q.; Stalin, S.; Zhao, C.Z.; Archer, L.A. Designing solid-state electrolytes for safe, energy-dense batteries. *Nat. Rev. Mater.* **2020**, *5*, 229–252. [CrossRef]
185. Feng, L.L.; Yu, T.S.; Cheng, D.H.; Sun, C.W. Progress of Sodium Battery Failure Research. *Sci. Sin. Chim.* **2020**, *50*, 1801–1815.
186. Eng, S.Y.S.; Soni, C.B.; Lum, Y.; Khoo, E.; Yao, Z.P.; Vineeth, S.K.; Kumar, V.; Lu, J.; Johnson, C.S.; Wolverton, C.; et al. Theory-guided experimental design in battery materials research. *Sci. Adv.* **2022**, *8*, eabm2422. [CrossRef]
187. Liu, Q.; Tang, Y.; Sun, H.; Yang, T.; Sun, Y.; Du, C.; Jia, P.; Ye, H.; Chen, J.; Peng, Q.; et al. In Situ Electrochemical Study of Na-O$_2$/CO$_2$ Batteries in an Environmental Transmission Electron Microscope. *ACS Nano* **2020**, *14*, 13232–13245. [CrossRef]
188. Yang, M.; Bi, R.; Wang, J.; Yu, R.; Wang, D. Decoding lithium batteries through advanced in situ characterization techniques. *Int. J. Miner. Metall. Mater.* **2022**, *29*, 965–989. [CrossRef]
189. Dixit, M.B.; Park, J.-S.; Kenesei, P.; Almer, J.; Hatzell, K.B. Status and prospect of in situ and operando characterization of solid-state batteries. *Energy Environ. Sci.* **2021**, *14*, 4672–4711. [CrossRef]
190. Hou, D.; Xia, D.; Gabriel, E.; Russell, J.A.; Graff, K.; Ren, Y.; Sun, C.-J.; Lin, F.; Liu, Y.; Xiong, H. Spatial and Temporal Analysis of Sodium-Ion Batteries. *ACS Energy Letters* **2021**, *6*, 4023–4054. [CrossRef]
191. Liu, D.; Shadike, Z.; Lin, R.; Qian, K.; Li, H.; Li, K.; Wang, S.; Yu, Q.; Liu, M.; Ganapathy, S.; et al. Review of Recent Development of In Situ/Operando Characterization Techniques for Lithium Battery Research. *Adv. Mater.* **2019**, *31*, e1806620. [CrossRef] [PubMed]
192. Gao, F.; Tian, X.D.; Lin, J.S.; Dong, J.C.; Lin, X.M.; Li, J.F. In situ Raman, FTIR, and XRD spectroscopic studies in fuel cells and rechargeable batteries. *Nano Res.* **2022**. [CrossRef]
193. Bak, S.-M.; Shadike, Z.; Lin, R.; Yu, X.; Yang, X.-Q. In situ/operando synchrotron-based X-ray techniques for lithium-ion battery research. *NPG Asia Mater.* **2018**, *10*, 563–580. [CrossRef]
194. Bragg, W. The structure of some crystals as indicated by their diffraction of X-rays. *Proc. R. Soc. Lond. Ser. A Contain. Pap. A Math. Phys. Character* **1997**, *89*, 248–277.
195. Ameh, E.S. A review of basic crystallography and x-ray diffraction applications. *Int. J. Adv. Manuf. Technol.* **2019**, *105*, 3289–3302. [CrossRef]
196. Xia, M.; Liu, T.; Peng, N.; Zheng, R.; Cheng, X.; Zhu, H.; Yu, H.; Shui, M.; Shu, J. Lab-Scale In Situ X-ray Diffraction Technique for Different Battery Systems: Designs, Applications, and Perspectives. *Small Methods* **2019**, *3*, 1900119. [CrossRef]
197. Yu, W.; Yang, W.; Liu, R.; Qin, L.; Lei, Y.; Liu, L.; Zhai, S.; Li, B.; Kang, F. A soluble phenolic mediator contributing to enhanced discharge capacity and low charge overpotential for lithium-oxygen batteries. *Electrochem. Commun.* **2017**, *79*, 68–72. [CrossRef]
198. Sheng, C.; Yu, F.; Wu, Y.; Peng, Z.; Chen, Y. Disproportionation of Sodium Superoxide in Metal-air Batteries. *Angew. Chem. Int. Ed. Engl.* **2018**, *57*, 9906–9910. [CrossRef]
199. Ganapathy, S.; Adams, B.D.; Stenou, G.; Anastasaki, M.S.; Goubitz, K.; Miao, X.F.; Nazar, L.F.; Wagemaker, M. Nature of Li$_2$O$_2$ oxidation in a Li-O$_2$ battery revealed by operando X-ray diffraction. *J. Am. Chem. Soc.* **2014**, *136*, 16335–16344. [CrossRef]

200. Lin, F.; Liu, Y.; Yu, X.; Cheng, L.; Singer, A.; Shpyrko, O.G.; Xin, H.L.; Tamura, N.; Tian, C.; Weng, T.C.; et al. Synchrotron X-ray Analytical Techniques for Studying Materials Electrochemistry in Rechargeable Batteries. *Chem. Rev.* **2017**, *117*, 13123–13186. [CrossRef]
201. Sun, C.W.; López, C.A.; Alonso, J.A. Elucidating the Diffusion Pathway of Protons in Ammonium Polyphosphate: A potential Electrolyte for Intermediate Temperature Fuel Cells. *J. Mater. Chem. A* **2017**, *5*, 7839–7844. [CrossRef]
202. Sun, C.W.; Chen, L.L.; Shi, S.Q.; Reeb, B.; López, C.A.; Alonso, J.A.; Stimming, U. Visualization of the Diffusion Pathway of Protons in the Ammonium Polyphosphate for Intermediate Temperature Fuel Cells. *Inorg. Chem.* **2018**, *57*, 676–680. [CrossRef] [PubMed]
203. Li, S.Q.; Meng, X.Y.; Alonso, J.A.; Fernández-Díaz, M.T.; Sun, C.W.; Wang, Z.L. Structural and Electrochemical Properties of $LiMn_{0.6}Fe_{0.4}PO_4$ as a Cathode Material for Flexible Lithium-ion Batteries and Self-charging Power Pack. *Nano Energy* **2018**, *52*, 510–516. [CrossRef]
204. Lu, Y.; López, C.A.; Wang, J.; Alonso, J.A.; Sun, C.W. Insight into the Structure and Functional Application of Mg doped $Na_{0.5}Bi_{0.5}TiO_3$ electrolyte for Solid Oxide Fuel Cells. *J. Alloys Compd.* **2018**, *752*, 213–219. [CrossRef]
205. Gong, Y.D.; Sun, C.W.; Huang, Q.A.; Alonso, J.A.; Fernández-Díaz, M.T.; Chen, L.Q. Dynamic octahedral breathing in oxygen-deficient $Ba_{0.9}Co_{0.7}Fe_{0.2}Nb_{0.1}O_{3-\delta}$ perovskite performing as a cathode for intermediate-temperature solid oxide fuel cells. *Inorg. Chem.* **2016**, *55*, 3091–3097. [CrossRef]
206. Yang, W.; Zhang, H.R.; Sun, C.W.; Liu, L.L.; Alonso, J.A.; Fernández-Díaz, M.T.; Chen, L.Q. Insight into the structure and functional application of $Sr_{0.95}Ce_{0.05}CoO_{3-\delta}$ cathode for solid oxide fuel cells. *Inorg. Chem.* **2015**, *54*, 3477–3485. [CrossRef]
207. Ma, Z.H.; Wang, Y.S.; Sun, C.W.; Alonso, J.A.; Fernández-Díaz, M.T.; Chen, L.Q. Experimental visualization of the diffusion pathway of sodium ions in the $Na_3[Ti_2P_2O_{10}F]$ anode for sodium-ion battery. *Sci. Rep.* **2014**, *4*, 7231. [CrossRef]
208. Wang, J.; Sun, C.W.; Gong, Y.D.; Zhang, H.R.; Alonso, J.A.; Fernández-Díaz, M.T.; Wang, Z.L.; Goodenough, J.B. Imaging of the diffusion pathway of Al^{3+} ion in NASICON-type $(Al_{0.2}Zr_{0.8})_{20/19}Nb(PO_4)_3$ electrolyte for high-temperature solid-state Al batteries. *Chin. Phys. B* **2018**, *27*, 128201. [CrossRef]
209. Rong, X.; Liu, J.; Hu, E.; Liu, Y.; Wang, Y.; Wu, J.; Yu, X.; Page, K.; Hu, Y.-S.; Yang, W.; et al. Structure-Induced Reversible Anionic Redox Activity in Na Layered Oxide Cathode. *Joule* **2018**, *2*, 125–140. [CrossRef]
210. Castellanos, M.M.; McAuley, A.; Curtis, J.E. Investigating Structure and Dynamics of Proteins in Amorphous Phases Using Neutron Scattering. *Comput. Struct. Biotechnol. J.* **2017**, *15*, 117–130. [CrossRef]
211. Matthey, J. In the Lab: Synchrotron Radiation Based Solid Phase Characterisation of Industrial Catalysts and Materials. *Johnson Matthey Technol. Rev.* **2016**, *60*, 158–160.
212. Wang, X.; Tan, S.; Yang, X.-Q.; Hu, E. Pair distribution function analysis: Fundamentals and application to battery materials. *Chin. Phys. B* **2020**, *29*, 028802. [CrossRef]
213. Castillo-Blas, C.; Moreno, J.M.; Romero-Muniz, I.; Platero-Prats, A.E. Applications of pair distribution function analyses to the emerging field of non-ideal Metal-organic framework materials. *Nanoscale* **2020**, *12*, 15577–15587. [CrossRef] [PubMed]
214. Yadav, N.; Folastre, N.; Bolomont, M.; Jamali, A.; Morcrette, M.; Davoisne, C. Understanding the Battery Degradation Mechanism in All-solid-state Batteries via In-situ SEM. *Microsc. Microanal.* **2021**, *27*, 105–106. [CrossRef]
215. Han, S.; Cai, C.; Yang, F.; Zhu, Y.; Sun, Q.; Zhu, Y.G.; Li, H.; Wang, H.; Shao-Horn, Y.; Sun, X.; et al. Interrogation of the Reaction Mechanism in a $Na-O_2$ Battery Using In Situ Transmission Electron Microscopy. *ACS Nano* **2020**, *14*, 3669–3677. [CrossRef]
216. Zhang, L.; Tang, Y.; Liu, Q.; Yang, T.; Du, C.; Jia, P.; Wang, Z.; Tang, Y.; Li, Y.; Shen, T.; et al. Probing the charging and discharging behavior of $K-CO_2$ nanobatteries in an aberration corrected environmental transmission electron microscope. *Nano Energy* **2018**, *53*, 544–549. [CrossRef]
217. Liang, Z.; Zou, Q.; Wang, Y.; Lu, Y.C. Recent Progress in Applying In Situ/Operando Characterization Techniques to Probe the Solid/Liquid/Gas Interfaces of $Li-O_2$ Batteries. *Small Methods* **2017**, *1*, 1700150. [CrossRef]
218. Cortes, H.A.; Corti, H.R. In-situ characterization of discharge products of lithium-oxygen battery using Flow Electrochemical Atomic Force Microscopy. *Ultramicroscopy* **2021**, *230*, 113369. [CrossRef]
219. Wang, Y.; Wang, W.; Xie, J.; Wang, C.H.; Yang, Y.W.; Lu, Y.-C. Electrochemical reduction of CO_2 in ionic liquid: Mechanistic study of $Li–CO_2$ batteries via in situ ambient pressure X-ray photoelectron spectroscopy. *Nano Energy* **2021**, *83*, 105830. [CrossRef]
220. Wang, M.; Árnadóttir, L.; Xu, Z.J.; Feng, Z. In Situ X-ray Absorption Spectroscopy Studies of Nanoscale Electrocatalysts. *Nano-Micro Lett.* **2019**, *11*, 47. [CrossRef]
221. Kimura, Y.; Fakkao, M.; Nakamura, T.; Okumura, T.; Ishiguro, N.; Sekizawa, O.; Nitta, K.; Uruga, T.; Tada, M.; Uchimoto, Y.; et al. Influence of Active Material Loading on Electrochemical Reactions in Composite Solid-State Battery Electrodes Revealed by Operando 3D CT-XANES Imaging. *ACS Appl. Energy Mater.* **2020**, *3*, 7782–7793. [CrossRef]
222. Gao, R.; Zhou, D.; Ning, D.; Zhang, W.; Huang, L.; Sun, F.; Schuck, G.; Schumacher, G.; Hu, Z.; Liu, X. Probing the Self-Boosting Catalysis of $LiCoO_2$ in $Li-O_2$ Battery with Multiple In Situ/Operando Techniques. *Adv. Funct. Mater.* **2020**, *30*, 2002223. [CrossRef]
223. Ding, S.; Yu, X.; Ma, Z.-F.; Yuan, X. A review of rechargeable aprotic lithium–oxygen batteries based on theoretical and computational investigations. *J. Mater. Chem. A* **2021**, *9*, 8160–8194. [CrossRef]
224. Deng, Q.; Yang, Y.; Qu, S.; Wang, W.; Zhang, Y.; Ma, X.; Yan, W.; Zhang, Y. Electron structure and reaction pathway regulation on porous cobalt-doped CeO_2/graphene aerogel: A free-standing cathode for flexible and advanced $Li-CO_2$ batteries. *Energy Storage Mater.* **2021**, *42*, 484–492. [CrossRef]

225. Jiang, H.R.; Tan, P.; Liu, M.; Zeng, Y.K.; Zhao, T.S. Unraveling the Positive Roles of Point Defects on Carbon Surfaces in Nonaqueous Lithium-Oxygen Batteries. *J. Phys. Chem. C* **2016**, *120*, 18394–18402. [CrossRef]
226. Yuan, M.W.; Sun, Z.M.; Yang, H.; Wang, D.; Liu, Q.M.; Nan, C.Y.; Li, H.F.; Sun, G.B.; Chen, S.W. Self-Catalyzed Rechargeable Lithium-Air Battery by in-situ Metal Ion Doping of Discharge Products: A Combined Theoretical and Experimental Study. *Energy Environ. Mater.* **2021**, 1–9. [CrossRef]
227. Liu, Y.K.; Li, J.; Shen, Q.Y.; Zhang, J.; He, P.G.; Qu, X.H.; Liu, Y.C. Advanced characterizations and measurements for sodium-ion batteries with NASICON-type cathode materials. *eScience* **2022**, *2*, 10–31. [CrossRef]
228. Jain, A.; Hautier, G.; Moore, C.J.; Ping Ong, S.; Fischer, C.C.; Mueller, T.; Persson, K.A.; Ceder, G. A high-throughput infrastructure for density functional theory calculations. *Comp. Mater. Sci.* **2011**, *50*, 2295–2310. [CrossRef]
229. Wang, H.S.; Ji, Y.J.; Li, Y.Y. Simulation and design of energy materials accelerated by machine learning. *WIREs Comput. Mol. Sci.* **2019**, *10*, e1421. [CrossRef]

Disclaimer/Publisher's Note: The statements, opinions and data contained in all publications are solely those of the individual author(s) and contributor(s) and not of MDPI and/or the editor(s). MDPI and/or the editor(s) disclaim responsibility for any injury to people or property resulting from any ideas, methods, instructions or products referred to in the content.

Article

Grain Boundary Characterization and Potential Percolation of the Solid Electrolyte LLZO

Shuo Fu [1,2], Yulia Arinicheva [2,3], Claas Hüter [1], Martin Finsterbusch [2] and Robert Spatschek [1,4,*]

1. IEK-2, Forschungszentrum Jülich, 52425 Jülich, Germany
2. IEK-1, Forschungszentrum Jülich, 52425 Jülich, Germany
3. Department of Safety, Chemistry and Biomedical Laboratory Sciences, Faculty of Engineering and Science, Western Norway University of Applied Sciences (HVL), 5020 Bergen, Norway
4. JARA-ENERGY, 52425 Jülich, Germany
* Correspondence: r.spatschek@fz-juelich.de

Abstract: The influence of different processing routes and grain size distributions on the character of the grain boundaries in $Li_7La_3Zr_2O_{12}$ (LLZO) and the potential influence on failure through formation of percolating lithium metal networks in the solid electrolyte are investigated. Therefore, high quality hot-pressed $Li_7La_3Zr_2O_{12}$ pellets are synthesised with two different grain size distributions. Based on the electron backscatter diffraction measurements, the grain boundary network including the grain boundary distribution and its connectivity via triple junctions are analysed concerning potential Li plating along certain susceptible grain boundary clusters in the hot-pressed LLZO pellets. Additionally, the study investigates the possibility to interpret short-circuiting caused by Li metal plating or penetration in all-solid-state batteries through percolation mechanisms in the solid electrolyte microstructure, in analogy to grain boundary failure processes in metallic systems.

Keywords: all-solid-state batteries; Li plating; grain boundary network; triple junctions

Citation: Fu, S.; Arinicheva, Y.; Hüter, C.; Finsterbusch, M.; Spatschek, R. Grain Boundary Characterization and Potential Percolation of the Solid Electrolyte LLZO. Batteries 2023, 9, 222. https://doi.org/10.3390/batteries9040222

Academic Editors: Chunwen Sun, Siqi Shi and Yongjie Zhao

Received: 3 March 2023
Revised: 31 March 2023
Accepted: 6 April 2023
Published: 8 April 2023

Copyright: © 2023 by the authors. Licensee MDPI, Basel, Switzerland. This article is an open access article distributed under the terms and conditions of the Creative Commons Attribution (CC BY) license (https://creativecommons.org/licenses/by/4.0/).

1. Introduction

All-solid-state batteries (ASSBs) belong to the most promising next generation electrochemical energy storage systems [1–3]. A solid electrolyte, one of the key components enabling rechargeable ASSBs, allows safety concerns of the conventional lithium ion batteries to be overcome, substituting flammable organic electrolytes, and offers the potential for a significant improvement of energy density and battery life when metallic lithium is used as an anode [3,4]. Li-ion conducting garnets, in particular $Li_7La_3Zr_2O_{12}$ (LLZO), satisfy a number of the technological requirements for the application as solid electrolytes in ASSBs, such as high ionic and negligible electronic conductivities, a wide voltage window, as well as chemical and electrochemical compatibility with metallic lithium [5].

However, short-circuiting caused by Li dendrite formation within LLZO solid electrolytes during battery cycling has still been a challenge according to recent publications. Recently, it was shown that LLZO garnets can fail during operation by the development of microcracks due to mechanical stressing, followed by Li intrusion [6]. This effect, however, does not exclude other failure mechanisms, which may be related to the intrinsic microstructure of polycrystalline battery components, commonly found in industrial applications. Ren et al. [7] directly observed that lithium dendrites grow along grain boundaries and through interconnected pores. Aside from these effects, internal lithium plating was observed in isolated pores, which are considered as a trap of electrons and can reduce Li^+ to metallic Li [8,9]. Cheng et al. [10] proved that Li preferentially propagates intergranularly along the grain boundaries in LLZO. Motoyama et al. [4] demonstrated using in situ SEM that Li plating tends to occur at grain boundaries and triple junctions. In order to enable large-scale applications such as electric vehicles, dendrite formation and propagation should be prevented during cycling over a wide range of current densities [11] and operating temperatures [12]. Key factors for dendrite formation or propagation in LLZO are relative density or porosity [7], interface properties (contact between the electrolyte and Li, roughness of the electrolyte surface, defects) [4,13], microstructure (i.e., grain size and boundary character) [11,14,15], etc. These aspects cover a large fraction of accessible characteristics of grain boundaries in the context of Li dendrite penetration, but they do not yet consider the influence of the connectivity of the grain boundary network. Such conclusions would require a detailed analysis of the grain boundary structures, which has not been performed so far. Due to the complexity of this topic, we pursue in this using an experimental analysis of the grain boundary and trijunction characteristics, and on top of this a critical inspection of percolation models, to shed light on the question, whether such an approach, which is well established in other materials science disciplines, may also be useful for all-solid-state battery materials.

In general, percolation is the formation of a conducting path through a network. The random occupation of sites or bonds in connected lattices or networks is referred to as site percolation and bond percolation, respectively. Percolation works in an all-or-nothing mode and it is a threshold phenomenon, which means that if the threshold value is achieved, the percolation is certain to happen. The description of grain-boundary-related failure in terms of percolation theory has been established with investigations on austenitic steels [16]. The prediction of the percolation threshold of an intergranular failure for 3D grain boundary networks at 23% of active bonds is supported by experimental findings in Ref. [17]. With the derivation of the influence of crystallographic constraints (e.g., the Σ-product rule) on percolation in 2D and 3D grain boundary networks, the substantial discrepancy of the percolation threshold between constrained and non-constrained networks could be explained. Aside from the mentioned applications of percolation theory to grain boundary failure driven by corrosion, Perrior et al. [18] found that the vacancy-mediated cation diffusion in disordered pyrochlore is enhanced once a percolation network is established. A similar effect is also observed by Lee et al. [19] for Li diffusion in lithium transition metal oxides, where the diffusion is facile along the percolating network of channels with excess content of Li. However, these observations only provide indications of preferred cation diffusion in percolating networks, whereas the interpretation of the Li plating along grain boundaries and triple junctions in solid electrolytes in the spirit of a percolation analysis remains questionable. Depending on the answer to this question, it may be conceivable that grain boundary degradation phenomena such as Li plating or dendrite propagation in a polycrystalline solid electrolyte could be evaluated using percolation theory as a step towards grain boundary engineering to control the fraction and distribution of specific grain boundaries, which are resistant to intergranular percolation phenomena, in analogy to the intergranular corrosion in austenitic stainless steels [20,21]. To test such a hypothesis, it is necessary to first overcome the lack of careful investigations into the grain and grain boundary structure of LLZO, which is essential for the understanding of the mechanical and electrochemical performance of the solid electrolyte. As the direct test of such a potential picture is difficult from an experimental perspective alone, indirect and theoretically supported arguments are needed to uphold or counter this hypothesis.

Consequently, the present work focuses on the joint experimental and theoretical investigation of the influence of microstructure, and in particular of the grain boundary connectivity, on Li dendrite formation and propagation using percolation theory. The first and main goal of the paper is a detailed analysis of the microstructure and the grain boundaries of LLZO using two different synthesis routes. The second aspect is dedicated to the question of whether established percolation models for intergranular corrosion in metallic systems, for which a deep understanding of the different types of grain boundaries and their resistance to corrosion has been achieved as explained above, can similarly be applied to failure mechanisms in LLZO-based all-solid-state batteries.

2. Materials and Methods

2.1. Experimental Investigations

2.1.1. Synthesis and Fabrication of LLZO Samples

The different grain sizes of LLZO pellets are obtained from two different synthesis routes of the precursor powders, a conventional solid-state reaction (SSR) [22] and a solution-assisted solid-state reaction (SASSR). Then, both precursor powders are sintered using hot-pressing in order to achieve a high density, i.e., close to theoretical density of a defect free and perfect crystal. The optimal sintering temperature for each precursor powder was determined from the onset temperature for shrinkage from densification curves obtained by dilatometric analysis.

For the solid-state reaction, the starting reagents listed here are mixed in stoichiometric amounts: $LiOH \cdot H_2O$ (98%, Merck, Darmstadt, Germany, with 20 mol% excess for the compensation of lithium loss during the next calcination and sintering steps), La_2O_3 (99.9%, Merck, Darmstadt, Germany, dried at 900 °C for 10 h), ZrO_2 (99.5%, Treibacher, Treibach, Austria), Ta_2O_5 (99.5%, Inframat Corp., Manchester, USA) and 5 mol% $\alpha\text{-}Al_2O_3$ (99.9%, Inframat Corp., Manchester, NH, USA, as a sintering additive). The reaction mixture is homogenized in a motor grinder (Retsch RM 200) for 30 min at a rotational speed of 100 rpm. Next, the resulting powder is pressed into pellets and tempered in an Al_2O_3 crucible at 850 °C for 20 h with a heating and cooling rate of 5 K/min in air. The pellets are subsequently ground and pressed into pellets and calcined at 1000 °C for 20 h.

Alternatively, for the solution-assisted solid-state reaction, the starting reagents $LiNO_3$ (99%, water-free, Alfa Aeser, Ward Hill, MA, USA, with 20 mol% excess), $ZrO(NO_3)_2 \cdot 6H_2O$ (99%, Sigma-Aldrich, St. Louis, MO, USA) and $La(NO_3)_3 \cdot 6H_2O$ (99%, Alfa Aeser, Ward Hill, USA) are dissolved in distilled water. Then, $C_{10}H_{25}O_5Ta$ (99.9%, Strem Chemicals, Newburyport, MA, USA) is slowly added dropwise to the metal salt solution upon continuous stirring on a magnetic stirrer. The reaction mixture is dried at 80 °C overnight and calcined at 400 °C for 3 h to burn out the organic residues, and then at 750 °C for 4 h in an Al_2O_3 crucible.

After both synthesis routes of the precursor powders, the high density of the pellets is achieved by a hot-pressing (HP) technique. The calcined pellets are thoroughly ground in an agate mortar and the resulting powders are pre-pressed at 100 MPa into pellets with a diameter of 13 mm using a uniaxial press (Paul-Otto Weber). For the powder from the solid-state synthesis, the hot-pressing of the pellets is conducted at 1150 °C and 50 MPa under a flowing N_2 atmosphere for 3 h. Similarly, for the powder from the solution-assisted solid-state synthesis, the hot-pressing is conducted at 1075 °C and 50 MPa under the same flowing N_2 atmosphere for 3 h. The hot-pressed pellets are then cut into $\sim$0.65 mm thick slices by using a diamond saw under ethanol.

2.1.2. Sample Characterization

The purity of the phases and the crystal structure of the hot-pressed LLZO samples are characterized by X-ray diffraction (XRD) using the Bruker D4 Endeavor diffractometer with Cu-K_α radiation in a 2θ range from 10° to 80° with a step size of 0.02° at room temperature.

The relative densities of the HP pellets are determined via Archimedes's method using water as a liquid medium.

Inductively coupled plasma optical emission spectroscopy (ICP-OES) (Thermo Elemental, IRIS Intrepid) is used to determine the elemental composition of the hot-pressed LLZO samples. For this, 50 mg LLZO samples are dissolved in a solution of 2 g ammonium sulphate and 4 mL H_2SO_4, until the powder is completely dissolved. The obtained solution is diluted to 50 mL by using distilled water for the ICP-OES analysis. The experimental inaccuracy of ICP-OES analyses is about 3% of the measured concentration.

For the microstructural investigations, the pellets are mechanically ground using SiC sandpapers up to 4000 grit and mirror-polished by a water-free diamond suspension up to 0.5 µm under a 10 N force. In order to remove the surface contamination of the LiOH and Li_2CO_3 layer and to reduce the surface roughness, thermal etching under flowing

Ar for 2 h at 800 °C and the plasma-etching in the glove box (with Ar atmosphere) are conducted before polishing. For the scanning electron microscopy (SEM) and electron backscatter diffraction (EBSD) measurements, the pellets are transferred into the chamber immediately after the final polishing step is finished, in order to minimize the surface contamination, e.g., by carbonate formation, from the exposure to the moist air. The SEM of the precursor powders and HP pellets and the EBSD of the HP pellets are performed on the scanning electron microscope JEOL JSM-7000F (2006) with a combined EDX/EBSD-System EDAX Pegasus.

The EBSD measurements are conducted at an accelerating voltage of 30 kV with a lateral resolution of 1.2 nm and the HP pellets are tilted at 70° toward the EBSD detector. A working distance of 20 mm and a step size of 1 µm are used for EBSD mapping. The scanning area of the HP pellets is large enough (500 µm × 1000 µm), so that a sufficient statistical accuracy can be achieved with a huge number of investigated grains and grain boundaries. The grain boundary (GB) and triple junction (TJ) analysis of LLZO pellets is conducted from the collected EBSD data using the open source toolbox MTEX [23] in Matlab. The grain boundary misorientation angle θ is calculated from the rotation angle with respect to the crystal symmetry, which can be obtained from the Euler angles (ϕ_1, Φ, ϕ_2) of the adjacent grains.

2.2. Percolation Models

As reported in Ref. [20], percolation theory can predict intergranular corrosion degradation preferentially occurring along the grain boundaries in steels. The basic idea includes a binary classification of the grain boundaries present in the material, i.e., coincidence site lattices (CSLs) and random grain boundaries. The CSLs are grain boundaries for which the adjacent periodic crystal lattices have common atomic positions in order to obtain the most stable energetic state [24,25]. They are characterised by the quantity Σ, which is the density of the coincident sites in crystal lattice [26,27]. For example, $\Sigma 5$ means that every fifth lattice site is a coincidence site. This distinction reflects whether an individual grain boundary is expected to block percolation of ions ("resistant grain boundary" for CSLs) or support it ("susceptible grain boundaries" for random grain boundaries). In the analysis of grain boundary microstructures in polycrystalline materials, grain boundaries with a misorientation angle $\theta \leq 15°$ are considered as low angle grain boundaries (LAGBs) based on the dislocation structure [26]. LAGBs consist of isolated dislocations and are not participating in the percolating process [20,21]. High angle grain boundaries ($\theta > 15°$) with low-Σ values are defined as CSL boundaries (CSLs) and Brandon's criterion [25], $\Delta \theta = 15° / \Sigma^{0.5}$, is used to calculate the deviations from CSLs in the cubic structure. Other grain boundaries with high-Σ value or grain boundaries that cannot be described by CSLs, i.e., the misorientation angle of the grain boundary is far beyond the deviations from Brandon's criterion, are considered as random grain boundaries. This classification proposes a distinction between grain boundaries, which may be susceptible to Li plating, and resistant grain boundaries. According to the raised hypothesis, fatal failure could occur in such a network if a percolating path of susceptible grain boundaries may form inside the solid electrolyte, connecting anode and cathode.

In addition to the first picture, where grain boundaries themselves are considered as the limiting elements to form a conducting network, the focus of the second picture is on the triple junctions, where three adjacent grain boundaries come together. The connection between two conducting grain boundaries has to cross the triple junction, and the triple junction itself can be considered as a conducting or blocking element. Hence, according to this picture, triple junctions may play an important role in the connectivity of a grain boundary network. The triple junctions in the microstructure can be distinguished by the number of CSLs connected, namely the "resistant triple junctions" with 3-CSL or 2-CSL connected, and the "susceptible triple junctions" with 1-CSL or 0-CSL connected [28,29], which are schematically shown in Figure 1.

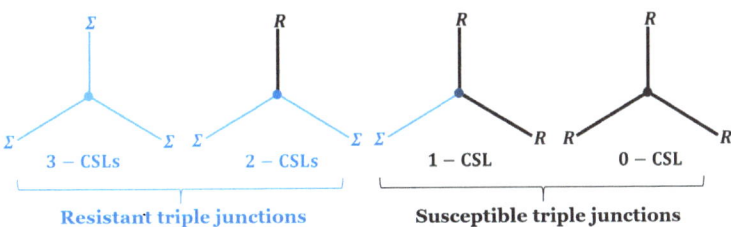

Figure 1. Schematic sketch of two different types of triple junctions in terms of percolation probability: the triple junctions with 3-CSL and 2-CSL are resistant triple junctions, which are expected to block Li percolation; the triple junctions with 1-CSL and 0-CSL are susceptible triple junctions, which are potentially active for the Li plating process. Here, thin blue lines denote CSLs, while bold black lines denote random grain boundaries.

Only the triple junctions with at least two random grain boundaries are considered to be susceptible to percolation, which are potentially the active entities in the process of Li plating and percolation along grain boundaries in the solid electrolyte.

In a two-dimensional calculation of the homogeneous bond percolation for the hexagonal or honeycomb lattice, which is assumed to approximately represent polycrystalline microstructures considering the topological rather than the geometric nature appropriately, the theoretical percolation threshold (p_c) of the susceptible grain boundaries/triple junctions is about 65% (fraction of resistant grain boundaries/triple junctions is 35%) [30,31]. This percolation threshold is confirmed by both analytical models [32] and experimental observations [33], since the honeycomb lattice can reflect the connectivity of the polycrystalline structure. Based on the aforementioned distinction of grain boundaries and triple junctions, we can investigate whether such a percolation model is suitable for the lithium plating failure across grain boundary networks in the sense of homogeneous bond percolation.

3. Results and Discussion

3.1. Microstuctural Analysis

Dense LLZO samples (99% of the theoretical density of a perfect crystal) with larger grains (LG) and smaller grains (SG) are obtained by hot-pressing of LLZO precursor powders with different particle morphologies (see Figure 2 and Table 1), synthesized by the conventional solid-state method and solution-assisted solid-state synthesis method, respectively.

Table 1. Summary of synthesis route information leading to different grain sizes in the samples.

Sample	Powder Synthesis	Surface Roughness	Grain Size
SG	SASSR	147 ± 10 nm	7.25 ± 4.29 µm
LG	SSR	202 ± 10 nm	10.30 ± 6.07 µm

The actual composition for the hot-pressed LLZO obtained from both synthesis routes is $Li_{6.3}Al_{0.01}La_3Zr_{1.6}Ta_{0.4}O_{12}$, according to the ICP-OES analysis. XRD analysis confirms the formation of the pure cubic garnet phase (space group $Ia\bar{3}d$) of the SG and LG samples. The SEM micrographs of the surfaces of the hot-pressed LLZO samples and corresponding XRD patterns are presented in Figure 3.

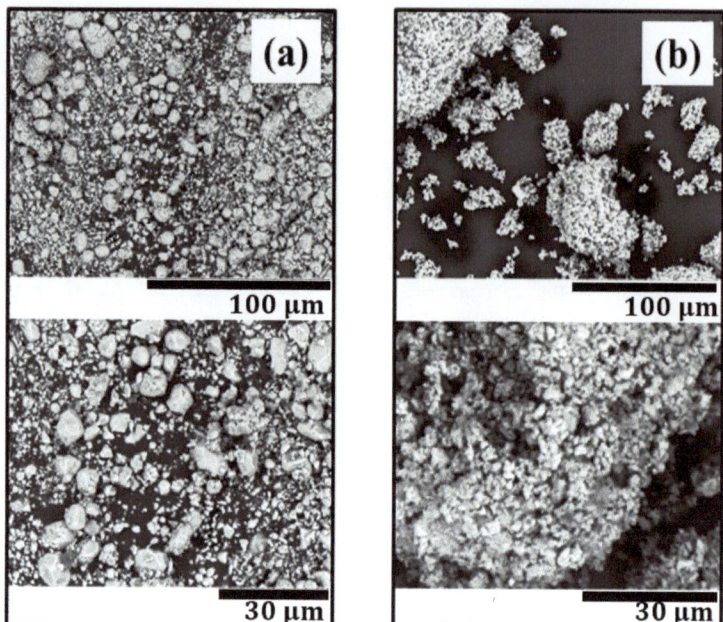

Figure 2. SEM images of precursor powders synthesized from: (**a**) conventional solid-state reaction (SSR) and (**b**) solution-assisted solid-state reaction (SASSR).

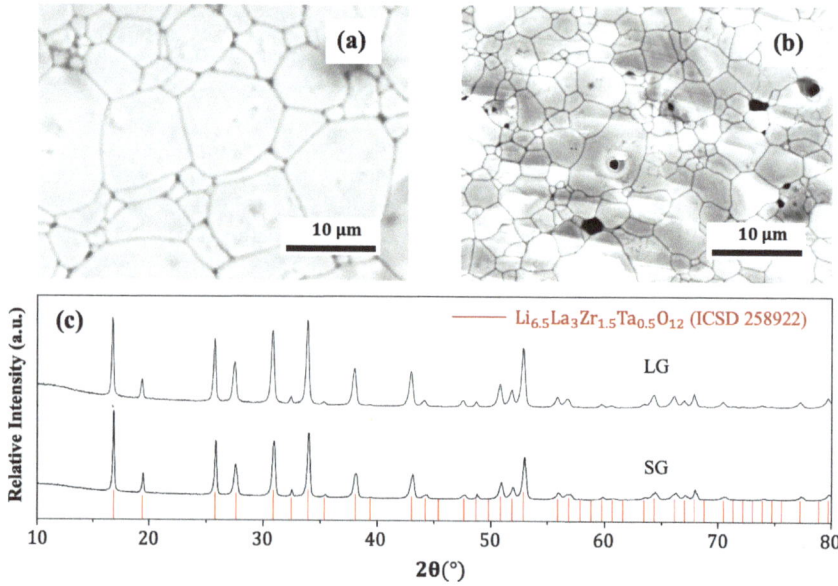

Figure 3. SEM images of the pellets with (**a**) larger grains (LG) and (**b**) smaller grains (SG). The visible surface pores are removed by polishing before further processing. The corresponding (**c**) XRD patterns of the pellets are compared with the cubic phase of $Li_{6.5}La_3Zr_{1.5}Ta_{0.5}O_{12}$, calculated from the results of Awaka et al. [34].

3.2. Grain Boundary and Triple Junction Classification

From the EBSD grain orientation maps of the hot-pressed pellets (Figure 4), the grain size (largest grain diameter) statistics is evaluated and shown in Figure 5.

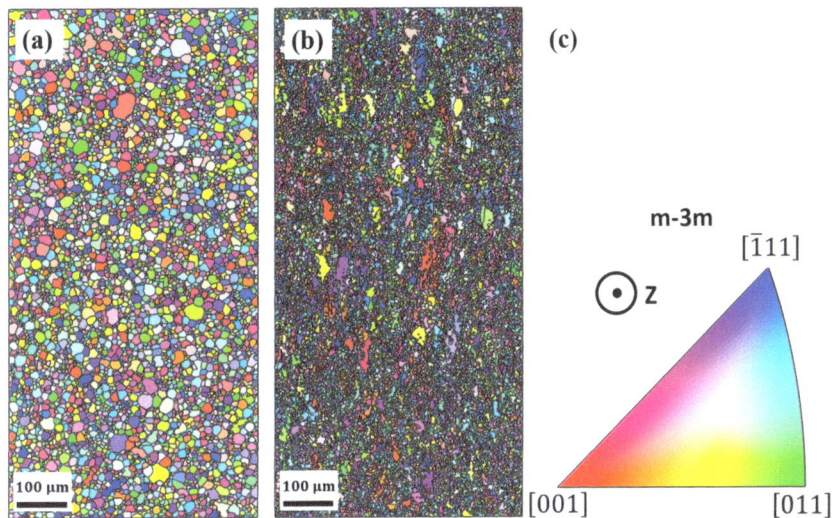

Figure 4. EBSD mean grain orientation maps with grain boundaries in solid black line of the hot-pressed pellets (**a**) LG, (**b**) SG and (**c**) the inverse pole figure [001] to indicate the cystallographic orientations.

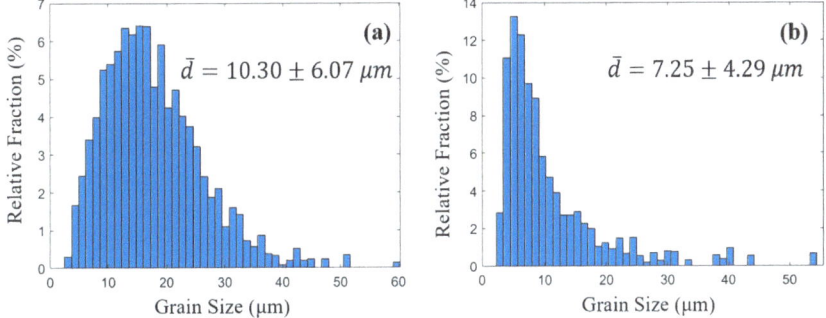

Figure 5. Grain size distribution histograms obtained from EBSD analysis of the hot-pressed pellets (**a**) LG, (**b**) SG.

The mean grain size of each hot-pressed pellet and its corresponding standard deviation are summarised in Table 1. For the sample of small grains (SG), the grain size is more centrally distributed than that of the sample of large grains (LG).

The grain boundary misorientation is also investigated from EBSD data and compared with the Mackenzie distribution [35] to determine its randomness (Figure 6).

A wide range of misorientation angles from 15° to 65° is observed in both LG and SG pellets. Apparently, no low angle grain boundaries (LAGBs), which have a misorientation angle below 15°, are detected, and therefore the calculated length fraction of LAGBs is confirmed to be 0%. Hence, all grain boundaries in the pellets are either coincidence site lattices (CSLs) or random grain boundaries. The fraction of grain boundaries with relatively lower misorientation angle (from 15° to 35°) in SG samples is larger than the theoretical Mackenzie distribution, indicating that the grain boundaries in SG pellets prefer to have a lower lattice mismatch. In contrast, in LG samples, the distribution of grain boundary misorientation is in coincidence with the Mackenzie distribution, indicating a non-preferential orientation of the grains.

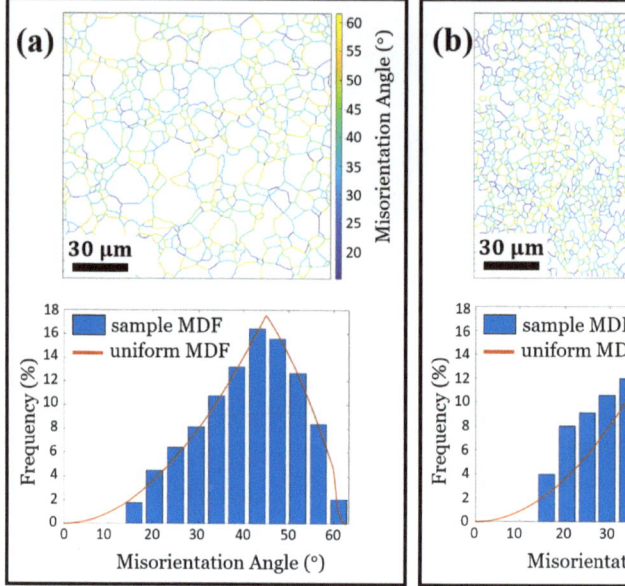

Figure 6. Grain boundary misorientation angle maps and distribution histograms compared with the misorientation density function (MDF) according to the Mackenzie distribution [35] in the selected area of the pellets: (**a**) LG, (**b**) SG.

Based on the crystal symmetry and the orientation from the EBSD measurements, we calculate the misorientation angles ($\theta°$) for CSLs (up to Σ49) from their Euler angles and evaluate a specific deviation angle ($\Delta\theta°$) for each CSL according to Brandon's criterion [25], $\Delta\theta = 15°/\Sigma^{0.5}$, which is summarised in Table 2.

The grain boundary fraction of each type is given by the length fraction, i.e., the total grain boundary length of one type normalised to the total length of all the grain boundaries. The fraction of different types of triple junctions is given by its number fraction. The evaluation for the grain boundary distribution is conducted through the entire EBSD scanning area of 500 µm × 1000 µm. The calculated fraction of CSLs and random grain boundaries are summarised in Table 3.

Table 2. List of CSLs ($\Sigma \leq 49$) with Brandon's criterion in cubic LLZO (CSL type Σ, Euler angles (ϕ_1, Φ, ϕ_2) in degree, misorientation angle $\theta°$ and its deviation $\Delta\theta°$).

Σ	ϕ_1	Φ	ϕ_2	$\theta°$	$\Delta\theta°$
3	63.43	48.19	333.44	60	8.66
5	53.13	0	0	36.87	6.71
7	56.31	31	326.31	38.21	5.67
9	104.04	27.27	284.04	38.94	5
11	108.44	35.10	288.44	50.48	4.52
13a	22.62	0	0	22.62	4.16
13b	53.13	22.62	323.13	27.80	4.16
15	111.80	21.04	291.80	48.19	3.87
17a	28.07	0	0	28.07	3.64
17b	85.24	45.10	318.37	61.93	3.64
19a	99.46	18.67	279.46	26.53	3.44
19b	59.04	37.86	329.04	46.83	3.44
21a	51.34	17.75	321.34	21.79	3.27

Table 2. Cont.

Σ	ϕ_1	Φ	ϕ_2	$\theta°$	$\Delta\theta°$
21b	116.57	25.21	296.57	44.42	3.27
23	116.57	16.96	296.57	40.46	3.13
25a	16.26	0	0	16.26	3
25b	90	36.87	306.87	51.68	3
27a	105.95	15.64	285.95	35.43	2.89
27b	101.31	22.19	281.31	31.59	2.89
29	79.38	34.15	312.51	46.40	2.79
31	113.20	29.43	293.20	52.20	2.69
33a	97.13	14.14	277.13	20.05	2.61
33b	119.75	14.14	299.75	33.56	2.61
33c	68.20	40.75	338.20	58.99	2.61
35	50.04	19.46	329.04	34.05	2.54
37a	18.98	0	0	18.98	2.46
37b	110.56	13.35	290.56	43.14	2.47
39a	54.46	26.18	324.46	32.20	2.40
39b	83.99	29.33	317.12	50.13	2.40
39c	85.60	41.96	302.47	50.13	2.40
41a	12.68	0	0	12.68	2.34
41b	108.44	17.97	288.44	40.88	2.34
43a	102.53	12.38	282.53	27.91	2.29
43b	76.87	45.76	324.25	60.77	2.29
45a	75.96	27.27	309.09	36.87	2.24
45b	81.87	38.94	315	53.13	2.24
47	108.44	23.81	288.44	43.66	2.19
49a	113.96	11.60	293.96	43.57	2.14
49b	116.57	33.20	296.57	49.23	2.14

Table 3. Summary of grain boundary characteristics for small grain (SG) and large grain (LG) samples of HP-LLZO.

Property	SG	LG
$\bar{d}$ (µm)	7.25	10.30
Grain Boundary Distribution		
CSL%	12.29	13.31
Σ3%	1.42	1.78
Σ9%	1.00	0.91
Σ27%	0.71	0.65
Triple Junction Distribution		
f_{0CSL}%	62.65	61.55
f_{1CSL}%	32.05	31.70
f_{2CSL}%	5.02	6.20
f_{3CSL}%	0.27	0.54
$f = f_{2CSL}/1 - f_{3CSL}$%	5.04	6.23

For reasons which will be discussed in the following section, we use here a rather high limiting value of Σ49 for the distinction between CSL and random grain boundaries. We can conclude that both SG and LG samples have a very low fraction (< 15%) of CSL grain boundaries and a high fraction of random grain boundaries. The fraction of the former in LG sample (13.31%) is slightly larger than that in SG sample (12.29%), but this difference is not significant.

In general, the fraction of CSLs in a microstructure is dependent on the stacking-fault energy of the material. Metallic polycrystals such as stainless steels have a low stacking-fault energy, resulting in a high fraction of CSLs [36]. Twinning often occurs when there are not enough slip systems to accommodate deformation or when the material has a very low stacking-fault energy. A fundamental effect of the twin formation on the CSL distributions is via the crystallographic constraints related to the CSL framework, i.e., the so-called Σ-product rule, also named as coincidence index combination rule [27,36,37]. It describes that the integer product or quotient of the Σ values of any two CSLs connected at one triple junction is equal to the Σ value of the third CSL, i.e., $\Sigma x \times \Sigma y = \Sigma xy$ or $\Sigma x/y$. This product rule is strictly valid only for triple junctions with three non-random grain boundaries and for the material with a cubic symmetry [27]. One typical example is the $\Sigma 3^n$ (n = 1, 2, 3) product effect, which indicates an interaction between Σ3-, Σ9- and Σ27-CSLs. If the fraction of Σ3-CSLs is increased through deformation and an annealing process, the corresponding fraction of $\Sigma 3^n$ type boundary increases and thus the microstructure is enhanced with increased resistant triple junction fraction [20,33].

In the further analysis of the grain boundary distribution in our LLZO pellets, the twin-characterized Σ3 grain boundary is observed to have a small fraction, that is 1.42% in SG and 1.78% in LG, corresponding to the low fraction of the CSLs. The total fraction of $\Sigma 3^n$-CSLs, which are dominant in a twin-limited microstructure [33], is 3.13% for SG and 3.34% for LG. As listed in Table 3, these fractions are much smaller than the total CSL fraction in each pellet and even far smaller than the $\Sigma 3^n$-CSL fraction in the twinned metallic microstructure (> 20%) [20,33,38]. Furthermore, the fraction of the triple junctions with 3-CSLs in both pellets is $f_{3CSL} < 0.55\%$, which means that the possibility for a triple junction to have 3-CSLs connected is very small. Thus, we can conclude that the garnet type hot-pressed ceramic LLZO is not a twin-limited microstructure and the product rule for $\Sigma 3^n$-CSLs is not applicable in the investigated samples. Aside from the absence of slip systems in LLZO, which describe the dominant energy scale for plasticity-driven recrystalisation in metallic systems, electrochemical aspects might also contribute to the found discrepancy between LLZO and metallic grain boundary resilience. Specifically, we note that recent investigations on grain boundary diffusivites have exhibited few cases of Li diffusivities lower in GBs than in the bulk for twin GBs and comparable to bulk diffusivity for certain types of screw GBs in LLZO, see Ref. [39,40]. The absence of a preference for low-energy states under long time tempering increases the number of fundamentally different characteristics between metallic and ceramic grain boundaries concerning transport properties [20,21,33,38].

Finally, Figure 7 presents the distributions of CSLs, random grain boundaries and different triple junction structures in the LLZO pellets.

The results are summarized in Table 3. We can see clearly the dominance of random grain boundaries and triple junctions with 0-CSL (>60%) and 1-CSL (>30%).

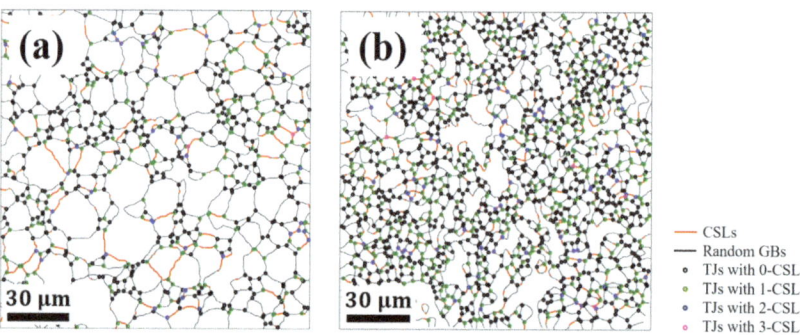

Figure 7. CSL grain boundaries ($\Sigma \leq 49$), random grain boundaries and different triple junction structures (with different numbers of connected CSLs) in the selected area of the pellets: (**a**) LG and (**b**) SG.

3.3. Percolation Hypothesis

In this section we return to the raised question, whether established grain boundary and triple junction percolation models, e.g., for corrosion or embrittlement in metallic systems, could also be useful for failure in LLZO-based all-solid-state batteries. This question can now be addressed based on the extended analysis of the preceding section.

There is extended experimental evidence from the percolation analysis in the corrosion tests of stainless steel that the low-Σ CSLs, especially these with $\Sigma \leq 29$, have a higher resistance to percolating degradation than the random grain boundaries [20,21,33,38]. In contrast, the above results for LLZO show a clear dominance of random grain boundaries. To ensure a specific choice of a cutoff criterion between CSL grain boundaries that are considered as resistant, and random grain boundaries that may be susceptible to Li plating, is not affecting the result, we shift the transition between these two categories even up to $\Sigma 49$. The percolating paths in the microstructure are broken according to the conjecture if the resistant grain boundary fraction is larger than 65% (or susceptible grain boundary fraction is smaller than 35%) from the homogeneous bond percolation theory of the honeycomb lattice [20,30,31,33]. Obviously, with a random grain boundary fraction of about 87%, which is far above 35%, this criterion is fulfilled under all circumstances, hence percolation should always occur. We can therefore conclude that a distinction of grain boundaries according to CSLs and random grain boundaries does not lead to a applicable criterion to predict percolation-based failure of solid electrolytes. Aside from the honeycomb structure, we arrive at the same conclusion for a random network of either CSLs or random grain boundaries [20,41], with a slightly higher tendency for percolation in SG pellets. A consideration on the basis of resistant triple junction also leads to a similar picture. As suggested by Kumar [33], the fraction of resistant triple junctions is defined as $f = f_{2CSL}/1 - f_{3CSL}$ and is listed in Table 3. Here it is shown that the resistant triple junction fraction f and the CSL boundary fraction in SG samples are smaller than those in LG pellets, which is consistent with other publications [20,21]. From our results above, the fraction of resistant triple junctions of the pellets f (5.04% in SG, 6.23% in LG pellets) would be far smaller than the threshold for percolation suppression, contrary to the experimental findings that the majority of samples does not suffer from short-circuiting at the first battery cycles. However, LLZO-based all-solid-state batteries can still fail at post cycling stage due to intergranular Li penetration [4,42,43].

We can therefore conclude this section by the statement that a percolation criterion for Li plating based on the classification between resistant and susceptible grain boundaries similar to metallic systems does not seem to be appropriate. Nevertheless, the distinction of grain boundaries may still have an influence on the Li plating kinetics and could therefore affect the long term stability of solid electrolysis. This topic may be the subject of future investigations.

4. Conclusions

In this paper, the microstructure, and in particular the grain boundary and triple junction characteristics of LLZO hot-pressed pellets, which were synthesized using two different processing routes, were analyzed. Based on these results, an investigation of the transferability of a percolation-based failure model from metallic systems to ceramic solid electrolyte systems was performed. The key results are summarised as follows:

- The synthesis methods of the precursor powder and hot-pressed pellets are successful in producing large amounts of pure cubic phase of Ta:LLZO with very high relative density, the same composition and surface treatment, but with different grain size distributions, namely small grain (SG) and large grain (LG) samples.
- The fraction of CSL grain boundaries are slightly larger in the LG sample than those in the SG sample.
- The fraction of random grain boundaries is significantly higher than in typical metallic materials.
- Our investigations indicate that percolation-based failure criteria, which are successfully used, e.g., for corrosion of austenitic steels, cannot be applied for the prediction of Li plating along grain boundaries in LLZO as solid electrolytes. This conclusion

does however not exclude that different grain boundary characteristics can influence the delayed failure after long operation times.

Overall, the results suggest that the solid electrolyte LLZO is rather robust against Li penetration from the perspective of grain boundary percolation, and at least we do not find strong model-based arguments for such a failure mode. Therefore, the early manufacturing steps of the solid electrolyte, which are decisive for the grain structure, and which were proposed as origin for grain boundary percolation, are probably not competing with fracture induced failure during later fabrication and operation [6]. In turn, tuning of the grain structure may therefore be a suitable way to improve the mechanical stability of the solid electrolyte against fracturing or even to enable self-healing features.

Author Contributions: Conceptualization, S.F., C.H. and R.S.; methodology, S.F. and Y.A.; software, S.F.; validation, S.F., C.H. and R.S.; formal analysis, S.F.; investigation, S.F. and Y.A.; resources, M.F.; data curation, S.F.; writing—original draft preparation, S.F.; writing—review and editing, C.H. and R.S.; visualization, S.F.; supervision, R.S.; project administration, M.F. and R.S.; funding acquisition, M.F. and R.S. All authors have read and agreed to the published version of the manuscript.

Funding: This research was funded by the German Ministry of Research and Education (BMBF) in the framework of the project MEET Hi-End III. The authors gratefully acknowledge the computing time granted by the JARA Vergabegremium and provided on the JARA Partition part of the supercomputer JURECA at Forschungszentrum Jülich [44]. Open access was funded by the Deutsche Forschungsgemeinschaft (DFG, German Research Foundation)—491111487.

Data Availability Statement: The data presented in this study are available on request.

Acknowledgments: The authors would also like to especially thank to Alexander Schwedt, Fadli Rohman for the support of EBSD measurements and Stefan Neumeier for the support of hot-pressing.

Conflicts of Interest: There is no conflict to declare.

Abbreviations

The following abbreviations are used in this manuscript:

ASSB	All-Solid-State Batteries
LLZO	$Li_7La_3Zr_2O_{12}$
EBSD	Electron Back Scatter Diffraction
CSL	Coincidence Site Lattice
LAGB	Low Angle Grain Boundary
GB	Grain Boundary
HP	Hot-Pressing
XRD	X-Ray Diffraction
ICP-OES	Inductively Coupled Plasma Optical Emission Spectroscopy
SEM	Scanning Emission Microscopy
MDF	Misorientation Density Function
TJ	Triple Junction
LG	Large Grain
SG	Small Grain
SSR	Solid-State Reaction
SASSR	Solution-Assisted Solid-State Reaction

References

1. Goodenough, J.B. Rechargeable batteries: challenges old and new. *J. Solid State Electrochem.* **2012**, *16*, 2019–2029. [CrossRef]
2. Murugan, R.; Thangadurai, V.; Weppner, W. Fast lithium ion conduction in garnet-type $Li_7La_3Zr_2O_{12}$. *Angew. Chem. Int. Ed.* **2007**, *46*, 7778–7781. [CrossRef] [PubMed]
3. Arinicheva, Y.; Wolff, M.; Lobe, S.; Dellen, C.; Fattakhova-Rohlfing, D.; Guillon, O.; Böhm, D.; Zoller, F.; Schmuch, R.; Li, J.; et al. Ceramics for electrochemical storage. In *Advanced Ceramics for Energy Conversion and Storage*; Elsevier: Amsterdam, The Netherlands, 2020; pp. 549–709.

4. Motoyama, M.; Tanaka, Y.; Yamamoto, T.; Tsuchimine, N.; Kobayashi, S.; Iriyama, Y. The Active Interface of Ta-Doped $Li_7La_3Zr_2O_{12}$ for Li Plating/Stripping Revealed by Acid Aqueous Etching. *ACS Appl. Energy Mater.* **2019**, *2*, 6720–6731. [CrossRef]
5. Manthiram, A.; Yu, X.; Wang, S. Lithium battery chemistries enabled by solid-state electrolytes. *Nat. Rev. Mater.* **2017**, *2*, 1–16. [CrossRef]
6. McConohy, G.; Xu, X.; Cui, T.; Barks, E.; Wang, S.; Kaeli, E.; Melamed, C.; Gu, X.W.; Chueh, W.C. Mechanical regulation of lithium intrusion probability in garnet solid electrolytes. *Nat. Energy* **2023**, *8*, 241–250. [CrossRef]
7. Ren, Y.; Shen, Y.; Lin, Y.; Nan, C.W. Direct observation of lithium dendrites inside garnet-type lithium-ion solid electrolyte. *Electrochem. Commun.* **2015**, *57*, 27–30. [CrossRef]
8. Han, F.; Westover, A.S.; Yue, J.; Fan, X.; Wang, F.; Chi, M.; Leonard, D.N.; Dudney, N.J.; Wang, H.; Wang, C. High electronic conductivity as the origin of lithium dendrite formation within solid electrolytes. *Nat. Energy* **2019**, *4*, 187–196. [CrossRef]
9. Song, Y.; Yang, L.; Zhao, W.; Wang, Z.; Zhao, Y.; Wang, Z.; Zhao, Q.; Liu, H.; Pan, F. Revealing the short-circuiting mechanism of garnet-based solid-state electrolyte. *Adv. Energy Mater.* **2019**, *9*, 1900671. [CrossRef]
10. Cheng, E.J.; Sharafi, A.; Sakamoto, J. Intergranular Li metal propagation through polycrystalline $Li_{6.25}Al_{0.25}La_3Zr_2O_{12}$ ceramic electrolyte. *Electrochim. Acta* **2017**, *223*, 85–91. [CrossRef]
11. Sharafi, A.; Haslam, C.; Kerns, R.D.; Wolfenstine, J.; Sakamoto, J.S. Controlling and correlating the effect of grain size with the mechanical and electrochemical properties of $Li_7La_3Zr_2O_{12}$ solid-state electrolyte. *J. Mater. Chem.* **2017**, *5*, 21491–21504. [CrossRef]
12. Yonemoto, F.; Nishimura, A.; Motoyama, M.; Tsuchimine, N.; Kobayashi, S.; Iriyama, Y. Temperature effects on cycling stability of Li plating/stripping on Ta-doped $Li_7La_3Zr_2O_{12}$. *J. Power Sources* **2017**, *343*, 207–215. [CrossRef]
13. Tsai, C.L.; Roddatis, V.; Chandran, C.V.; Ma, Q.; Uhlenbruck, S.; Bram, M.; Heitjans, P.; Guillon, O. $Li_7La_3Zr_2O_{12}$ Interface Modification for Li Dendrite Prevention. *ACS Appl. Mater. Interfaces* **2016**, *8*, 10617–10626. [CrossRef] [PubMed]
14. Ren, Y.; Shen, Y.; Lin, Y.; Nan, C.W. Microstructure Manipulation for Enhancing the Resistance of Garnet-Type Solid Electrolytes to "Short Circuit" by Li Metal Anodes. *ACS Appl. Mater. Interfaces* **2019**, *116*, 5928–5937. [CrossRef]
15. Cheng, L.; Chen, W.; Kunz, M.; Persson, K.A.; Tamura, N.; Chen, G.; Doeff, M.M. Effect of surface microstructure on electrochemical performance of garnet solid electrolytes. *ACS Appl. Mater. Interfaces* **2015**, *7*, 2073–2081. [CrossRef]
16. Wells, D.B.; Stewart, J.; Herbert, A.W.; Scott, P.M.; Williams, D.E. The Use of Percolation Theory to Predict the Probability of Failure of Sensitized, Austenitic Stainless Steels by Intergranular Stress Corrosion Cracking. *Corrosion* **1989**, *45*, 649–660. [CrossRef]
17. Gaudett, M.A.; Scully, J.R. Applicability of bond percolation theory to intergranular stress-corrosion cracking of sensitized AlSI 304 stainless steel. *Metall. Trans.* **1994**, *25*, 775–787. [CrossRef]
18. Perriot, R.; Uberuaga, B.P.; Zamora, R.J.; Perez, D.; Voter, A.F. Evidence for percolation diffusion of cations and reordering in disordered pyrochlore from accelerated molecular dynamics. *Nat. Commun.* **2017**, *8*, 618. [CrossRef] [PubMed]
19. Lee, J.; Urban, A.; Li, X.; Su, D.; Hautier, G.; Ceder, G. Unlocking the Potential of Cation-Disordered Oxides for Rechargeable Lithium Batteries. *Science* **2014**, *343*, 519–522. [CrossRef] [PubMed]
20. Tsurekawa, S.; Nakamichi, S.; Watanabe, T. Correlation of grain boundary connectivity with grain boundary character distribution in austenitic stainless steel. *Acta Mater.* **2006**, *54*, 3617–3626. [CrossRef]
21. Kobayashi, S.; Kobayashi, R.; Watanabe, T. Control of grain boundary connectivity based on fractal analysis for improvement of intergranular corrosion resistance in SUS316L austenitic stainless steel. *Acta Mater.* **2016**, *102*, 397–405. [CrossRef]
22. Tsai, C.L.; Dashjav, E.; Hammer, E.M.; Finsterbusch, M.; Tietz, F.; Uhlenbruck, S.; Buchkremer, H.P. High conductivity of mixed phase Al-substituted $Li_7La_3Zr_2O_{12}$. *J. Electroceramics* **2015**, *35*, 25–32. [CrossRef]
23. Mainprice, D.; Hielscher, R.; Schaeben, H. Calculating anisotropic physical properties from texture data using the MTEX open-source package. *Geol. Soc. Lond. Spec. Publ.* **2011**, *360*, 175–192. [CrossRef]
24. Kronberg, M.; Wilson, F. Secondary recrystallization in copper. *JOM* **1949**, *1*, 501–514. [CrossRef]
25. Brandon, D. The structure of high-angle grain boundaries. *Acta Metall.* **1966**, *14*, 1479–1484. [CrossRef]
26. Gottstein, G.; Shvindlerman, L. *Grain Boundary Migration in Metals: Thermodynamics, Kinetics, Applications*; Engineering & Technology, Physical Sciences, CRC Press: Boca Raton, FL, USA, 2009.
27. Priester, L. *Grain Boundaries: From Theory to Engineering*; Springer Series in Materials Science, Springer: Dordrecht, The Netherlands, 2013.
28. Fortier, P.; Aust, K.; Miller, W. Effects of symmetry, texture and topology on triple junction character distribution in polycrystalline materials. *Acta Metall. Mater.* **1995**, *43*, 339–349. [CrossRef]
29. Fortier, P.; Miller, W.; Aust, K. Triple junction and grain boundary character distributions in metallic materials. *Acta Mater.* **1997**, *45*, 3459–3467. [CrossRef]
30. Sykes, M.F.; Essam, J.W. Exact Critical Percolation Probabilities for Site and Bond Problems in Two Dimensions. *J. Math. Phys.* **1964**, *5*, 1117–1127. [CrossRef]
31. Stauffer, D.; Aharony, A. *Introduction to Percolation Theory*, 2nd ed.; Taylor & Francis: London, UK, 1992.
32. Frary, M.; Schuh*, C.A. Connectivity and percolation behaviour of grain boundary networks in three dimensions. *Philos. Mag.* **2005**, *85*, 1123–1143. [CrossRef]

33. Kumar, M.; King, W.E.; Schwartz, A.J. Modifications to the microstructural topology in fcc materials through thermomechanical processing. *Acta Mater.* **2000**, *48*, 2081–2091. [CrossRef]
34. Awaka, J.; Takashima, A.; Kataoka, K.; Kijima, N.; Idemoto, Y.; Akimoto, J. Crystal Structure of Fast Lithium-ion-conducting Cubic $Li_7La_3Zr_2O_{12}$. *Chem. Lett.* **2011**, *40*, 60–62. [CrossRef]
35. Mackenzie, J.K. Second paper on statistics associated with the random disorientation of cubes. *Biometrika* **1958**, *45*, 229–240. [CrossRef]
36. Don, J.; Majumdar, S. Creep cavitation and grain boundary structure in type 304 stainless steel. *Acta Metall.* **1986**, *34*, 961–967. [CrossRef]
37. Palumbo, G.; Aust, K.T.; Erb, U.; King, P.J.; Brennenstuhl, A.M.; Lichtenberger, P.C. On annealing twins and CSL distributions in F.C.C. polycrystals. *Phys. Status Solidi* **1992**, *131*, 425–428. [CrossRef]
38. Schuh, C.A.; Kumar, M.; King, W.E. Analysis of grain boundary networks and their evolution during grain boundary engineering. *Acta Mater.* **2003**, *51*, 687–700. [CrossRef]
39. Yu, S.; Siegel, D. Grain boundary contributions to Li-ion transport in the solid electrolyte $Li_7La_3Zr_2O_{12}$ (LLZO). *Chem. Mater.* **2017**, *29*, 9639–9647. [CrossRef]
40. Zhu, Y. Atomistic Modeling of Solid Interfaces in All-Solid-State Li-ion Batteries. Ph.D. Thesis, University of Maryland, College Park, MD, USA, 2018.
41. Ren, J.; Zhang, L.; Siegmund, S. How Inhomogeneous Site Percolation Works on Bethe Lattices: Theory and Application. *Sci. Rep.* **2016**, *6*, 22420. [CrossRef]
42. Aguesse, F.; Manalastas, W.; Buannic, L.; Lopez del Amo, J.M.; Singh, G.; Llordés, A.; Kilner, J. Investigating the dendritic growth during full cell cycling of garnet electrolyte in direct contact with Li metal. *ACS Appl. Mater. Interfaces* **2017**, *9*, 3808–3816. [CrossRef]
43. Liu, X.; Garcia-Mendez, R.; Lupini, A.R.; Cheng, Y.; Hood, Z.D.; Han, F.; Sharafi, A.; Idrobo, J.C.; Dudney, N.J.; Wang, C.; et al. Local electronic structure variation resulting in Li 'filament' formation within solid electrolytes. *Nat. Mater.* **2021**, *20*, 1485–1490. [CrossRef]
44. Krause, D.; Thörnig, P. JURECA: Modular supercomputer at Jülich Supercomputing Centre. *J. Large-Scale Res. Facil.* **2018**, *4*, 1–9. [CrossRef]

Disclaimer/Publisher's Note: The statements, opinions and data contained in all publications are solely those of the individual author(s) and contributor(s) and not of MDPI and/or the editor(s). MDPI and/or the editor(s) disclaim responsibility for any injury to people or property resulting from any ideas, methods, instructions or products referred to in the content.

Article

Engineered Grain Boundary Enables the Room Temperature Solid-State Sodium Metal Batteries

Yang Li [1,2], Zheng Sun [1], Haibo Jin [1] and Yongjie Zhao [1,2,*]

[1] Beijing Key Laboratory of Construction Tailorable Advanced Functional Materials and Green Applications, School of Materials Science and Engineering, Beijing Institute of Technology, Beijing 100081, China
[2] Yangtze Delta Region Academy, Beijing Institute of Technology, Jiaxing 314000, China
* Correspondence: zhaoyj14@bit.edu.cn

Abstract: The NASICON-type (Sodium Super Ionic Conductor) $Na_3Zr_2Si_2PO_{12}$ solid electrolyte is one of the most promising electrolytes for solid-state sodium metal batteries. When preparing $Na_3Zr_2Si_2PO_{12}$ ceramic using a traditional high-temperature solid-state reaction, the high-densification temperature would result in the volatilization of certain elements and the consequent generation of impurity phase, worsening the functional and mechanical performance of the NASICON electrolyte. We rationally introduced the sintering additive B_2O_3 to the NASICON matrix and systemically investigated the influence of B_2O_3 on the crystal structure, microstructure, electrical performance, and electrochemical performance of the NASICON electrolytes. The results reveal that B_2O_3 can effectively reduce the densification sintering temperature and promote the performance of the $Na_3Zr_2Si_2PO_{12}$ electrolyte. The $Na_3Zr_2Si_2PO_{12}$-2%B_2O_3-1150 °C achieves the highest ionic conductivity of 4.7×10^{-4} S cm^{-1} (at 25 °C) with an activation energy of 0.33 eV. Furthermore, the grain boundary phase formed during the sintering process could improve the mechanical behavior of the grain boundary and inhibit the propagation of metallic sodium dendrite within the NASICON electrolyte. The assembled Na/$Na_3Zr_2Si_2PO_{12}$-2%B_2O_3/$Na_3V_{1.5}Cr_{0.5}(PO_4)_3$ cell reveals the initial discharge capacity of 98.5 mAh g^{-1} with an initial Coulombic efficiency of 84.14% and shows a capacity retention of 70.3% at 30 mA g^{-1} over 200 cycles.

Keywords: NASICON; B_2O_3; microstructure; grain boundary

Citation: Li, Y.; Sun, Z.; Jin, H.; Zhao, Y. Engineered Grain Boundary Enables the Room Temperature Solid-State Sodium Metal Batteries. *Batteries* **2023**, *9*, 252. https://doi.org/10.3390/batteries9050252

Academic Editor: Hirotoshi Yamada

Received: 6 March 2023
Revised: 23 April 2023
Accepted: 24 April 2023
Published: 27 April 2023

Copyright: © 2023 by the authors. Licensee MDPI, Basel, Switzerland. This article is an open access article distributed under the terms and conditions of the Creative Commons Attribution (CC BY) license (https://creativecommons.org/licenses/by/4.0/).

1. Introduction

At present, fossil energy is being consumed at an excessive rate; thus, there is an urgent need to expand research into other types of new energy sources. In the meantime, new types of energy storage systems are also required [1–5]. Among these, electrochemical energy storage has attracted popular research interest owing to its high-energy conversion efficiency and ease of maintenance [6–8]. Currently, the consumer market is dominated by commercial lithium-ion batteries, but the limited reserves of lithium resources hinder their further development and utilization. The sodium-ion battery is deemed one of the promising candidates for the lithium-ion battery system because of its wide availability and homogeneous distribution on Earth [9–11]. However, sodium-ion batteries with organic liquid electrolytes have issues of easy leakage, flammability, and the growth of sodium dendrites. Solid-state batteries with solid-state electrolytes rather than traditional organic electrolytes can well address the above-mentioned issues [12–14]. The solid-state electrolyte plays a crucial role in the performance of the solid-state battery. In general, solid-state electrolytes are required to have favorable room-temperature ionic conductivity, high stability in the air, and excellent mechanical properties [15,16].

The NASICON electrolyte with a stoichiometry of $Na_{1+x}Zr_2Si_xP_{3-x}O_{12}$ ($0 \leq x \leq 3$) is obtained by partially replacing P with Si in $NaZr_2P_3O_{12}$ [17,18]. The NASICON electrolyte has the highest room-temperature ionic conductivity of about 10^{-4} S cm^{-1} when $x = 2$

($Na_3Zr_2Si_2PO_{12}$). Recently, various emerging methods of preparing $Na_3Zr_2Si_2PO_{12}$ ceramic electrolyte with high performance have appeared, such as spark plasma sintering (SPS), microwave-assisted sintering, hot pressing (HP), and the cold sintering process (CSP), but these preparation methods always need special equipment and are costly [19–22]. Usually, the high-temperature solid-state reaction is utilized to prepare $Na_3Zr_2Si_2PO_{12}$ ceramic. However, the high-sintering temperature would bring about the volatilization of sodium and phosphorus elements, producing impurity phase ZrO_2 at the grain boundary [23–25]. Moreover, abnormal grain growth is also accompanied by the deteriorated mechanical property of NASICON electrolytes. Therefore, low-cost and low-sintering temperatures are imperative for the large-scale synthesis of NASICON electrolytes. The sintering additive-assisted solid-state reaction is a well-known approach for ceramic sintering. At a certain temperature, the sintering additives would change into a liquid phase at the grain boundary. Furthermore, on account of the surface tension, the liquid phase would spontaneously fill the voids between the grains. Accordingly, this liquid phase plays an important role in facilitating mass migration during the sintering process and effectively lowers the densification sintering temperature. After sintering, the liquid phase would present as an amorphous phase lying at the grain boundary, resulting in dense microstructure and enhanced mechanical behavior.

In this work, $Na_3Zr_2Si_2PO_{12}$-xB_2O_3 ceramic electrolytes were synthesized through the solid-state reaction. The B_2O_3 would change into a liquid phase during the high-temperature sintering process, consequently promoting the densification sintering process. The charge transfer capability of $Na_3Zr_2Si_2PO_{12}$-xB_2O_3 is improved owing to the as-obtained dense microstructure and reduced grain boundary resistance. In the meantime, the interface compatibility between the ceramic electrolyte and metallic Na anode has also been greatly boosted. As a result, the assembled solid-state full battery operates well at room temperature and reveals excellent electrochemical performance, highlighting the effectiveness of grain boundary engineering in a solid-state electrolyte toward a rechargeable solid-state metal battery.

2. Experimental Section

2.1. Synthesis of $Na_3Zr_2Si_2PO_{12}$-B_2O_3 Ceramics

$Na_3Zr_2Si_2PO_{12}$-xB_2O_3 (abbreviated as NZSP-xB_2O_3, x = 0, 1, 2, and 4 wt.%) ceramics were synthesized through the conventional solid-state reaction. Na_2CO_3 (99.5%), ZrO_2 (99.99%), SiO_2 (99.5%), and $NH_4H_2PO_4$ (99.0%) were utilized as raw materials without further treatment. In order to compensate for the volatilization of sodium element during the sintering process, an excessive amount of Na_2CO_3 (10 wt.%) was added. Anhydrous ethanol was used as the milling medium when the weighted raw materials in accordance with as-designed compositions were thoroughly ball-milled. Thereafter, the mixtures were dried in an electric oven at 60 °C for 12 h. The dried powder was pre-calcinated in a muffle at 1000 °C/10 h (with a heating rate of 5 °C min^{-1}). Afterwards, the as-obtained products with different amounts of B_2O_3 (0, 1, 2, and 4 wt.%) were ball-milled again (same conditions as previous) and pressed into pellets with a diameter of 10 mm under 10 MPa and then sintered at different temperatures (1050, 1100, 1150, and 1250 °C) for 10 h. After sintering, the as-obtained ceramic pellets had a thickness of about 1 mm and a diameter of 8–9 mm. The as-sintered ceramics were carefully polished with sand paper prior to use.

2.2. Microstructure and Performance Characterization

The phase structure of the samples was identified with an X-ray diffractometer with Cu Kα radiation (λ = 1.5418 Å, Bruker D8 Advance). A field emission scanning electron microscope (Hitachi Regulus8230) equipped with an energy dispersive spectrometer was adopted for microstructural and composition analysis. The ionic conductivity was assessed using AC impedance analysis (Chenhua 660E). The measured frequency range was 1 MHz~0.1 Hz. The equivalent circuit was fitted in the Nyquist plot using the Z-view software. The total ionic conductivity (σ_t) can be calculated using $\sigma = \frac{L}{S_1 R}$, wherein S_1 is the area of the ceramic pellet, R is the total resistance of the ceramic pellet, and L is the thickness of the ceramic pellet. The interfacial area specific resistance (ASR) is computed using ASR = $R_{\text{interface}} \times S_2$, wherein $R_{\text{interface}}$ is the interfacial resistance between the metallic sodium anode and the solid electrolyte, and S_2 is the contact area of the sodium metal anode with the solid electrolyte. The galvanostatic charge/discharge measurement was conducted on a battery testing system (Land multi-channel battery test instrument, Wuhan, China).

2.3. Synthesis of the Composite Cathode

The cathode active material $Na_3V_{1.5}Cr_{0.5}(PO_4)_3$ was prepared according to our previous research [26,27]. $Na_3V_{1.5}Cr_{0.5}(PO_4)_3$ (the carbon content of about 3 wt.%), plastic-crystal electrolyte, polyvinylidene fluoride (PVDF), and carbon black were mixed with a weight ratio of 60:25:10:5 to create the composite cathode. Herein, the plastic-crystal electrolyte was prepared by heating a 20:1 molar mixture of succinonitrile and $NaClO_4$ to achieve a clear solution at 65 °C, then allowing it naturally cool to 25 °C. This composite cathode was dispersed in N-methyl-2-pyrrolidone and stirred for 12 h. The as-achieved slurry was carefully coated onto aluminum foil and dried at 40 °C for another 12 h under vacuum. The loading mass of active material was about 2–3 mg cm^{-2}.

2.4. Assemble and Disassemble Cells

To test the electrochemical performance of the cells, CR2032 coin cells with ceramics as the electrolyte and sodium metal (with an area of about 0.3 cm^2) as the anode were assembled. The ceramic electrolyte was sandwiched between sodium foil and the composite cathode. Before assembling the solid-state sodium metal battery, the as-sintered ceramics were carefully polished with sand paper. The Na metal was pressed onto one side of the ceramic electrolyte by hand; the as-synthesized cathode was placed onto the other side. Next, the CR2032 cell was sealed at 10 MPa.

The packaged cell case was opened and the sodium metal on the solid electrolyte surface was removed. To observe the combination of metallic sodium anode with the ceramic electrolyte, we broke the ceramic electrolyte along its diameter to obtain a complete cross-section. The whole process of assembling and disassembling was completed in a glove box filled with Ar and the transfer for XPS analysis was also finished in a homemade container with the protection of Ar.

3. Results and Discussion

The crystal structure of the as-obtained ceramics was identified using XRD analysis. Figure 1a shows the XRD patterns of $Na_3Zr_2Si_2PO_{12}$-xB_2O_3 sintered at 1150 °C (abbreviated as NZSP-xB$_2$O$_3$—1150 °C). All diffraction peaks show a good match to the monoclinic phase $Na_3Zr_2Si_2PO_{12}$ (PDF#: 84-1200), suggesting that the introduction of B_2O_3 could not vary the main phase structure of the NASICON matrix. In addition, no obvious diffraction peaks associated with B_2O_3 could be identified, and it is possible that B_2O_3 exists as an amorphous phase at the grain boundary after sintering. In addition, the ZrO_2 phase can be clearly observed, which is attributed to the sodium and phosphorus elements volatilizing during the high-temperature sintering process [23–25].

The sectional SEM images of NZSP-xB$_2$O$_3$—1150 °C are presented in Figure 1b and the SEM analysis results of the other samples are summarized in Figure S1. The result indicates that bare NZSP has clear grain boundaries together with a certain number of pores. In addition, an intergranular fracture morphology and an abnormal grain growth are observed for the bare NZSP, implying a weak grain–grain bonding strength. In contrast, the NZSP-xB$_2$O$_3$ samples present bigger grain sizes, fewer pores, and blurred grain boundaries covered by the amorphous phase, demonstrating a much denser microstructure. The reason for this is that the liquid phase would form at the grain boundaries when the temperature reaches the melting point of B$_2$O$_3$ (450 °C) and fill the pores between these grains, resulting in a closer grain contact. With the increase in B$_2$O$_3$ content, the fracture behavior of ceramics changes from an intergranular to a transcrystalline fracture, indicating that the grain bonding strength was observably enhanced [28]. Moreover, by observing the microstructure of the ceramic electrolytes obtained at different sintering temperatures, it can be concluded that the B$_2$O$_3$-modified NZSP ceramics can obtain a denser microstructure at lower temperatures, suggesting that the liquid phase formed during the sintering process can effectively facilitate the migration of mass and reduce the densification temperature of the NASICON matrix. The energy dispersive spectroscopy (EDS) results of the surface for NZSP-2%B$_2$O$_3$—1150 °C (Figure 1c) and the results with an enlarged scale (Figures S2 and S3) show the uniform distribution of elements over the sample. Furthermore, element B is uniformly dispersed at the grain boundary. The shrinkage variation of NZSP-xB$_2$O$_3$ is exhibited in Figure S4. The B$_2$O$_3$-modified NZSP sintered at low temperatures also can shrink considerably compared with bare NZSP. Again, this reconfirms that the liquid phase can promote the densification sintering, which is consistent with the SEM analysis results.

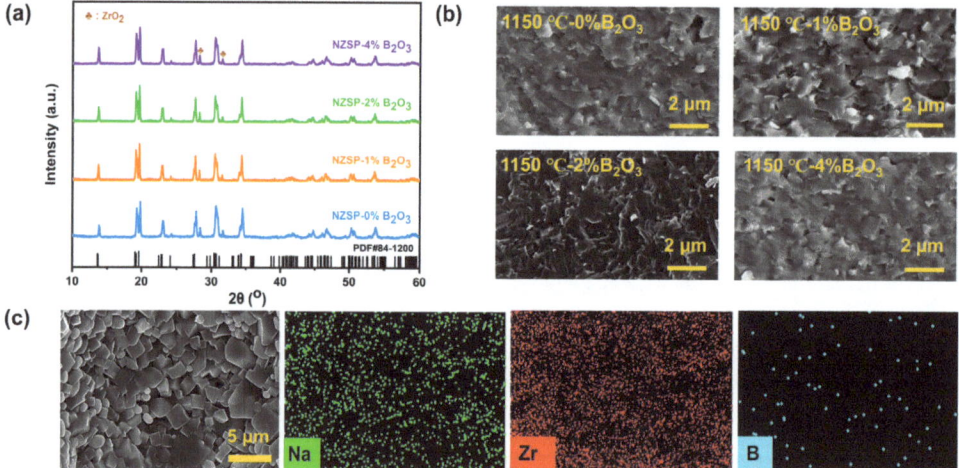

Figure 1. (**a**) XRD patterns of NZSP-xB$_2$O$_3$—1150 °C. (**b**) Sectional SEM images. (**c**) EDS mapping results for the surface of NZSP-2%B$_2$O$_3$—1150 °C.

Figure 2 exhibits the Rietveld refinement results of the samples and Table S1 displays detailed information about the crystal structure. Generally, all of these samples possess a typical NASICON structure with a space group of C2/c, implying that the B_2O_3 as a sintering additive does not change the structure of the NASICON matrix. As listed in Table S1, the lattice parameters and cell volume changed after B_2O_3, suggesting that the composition of phases changed after B_2O_3 addition. This can be explained by the present matrix phase possibly being a deviation from $Na_3Zr_2Si_2PO_{12}$, that is, for example, Na-rich NZSP ($Na_{3+x}Zr_{2-y}Si_2PO_{12}$). Then, additional Na may react with B_2O_3 to form a Na-B-O compound that may act as a good sintering additive located at the grain boundary, which is consistent with the results of the SEM-EDS.

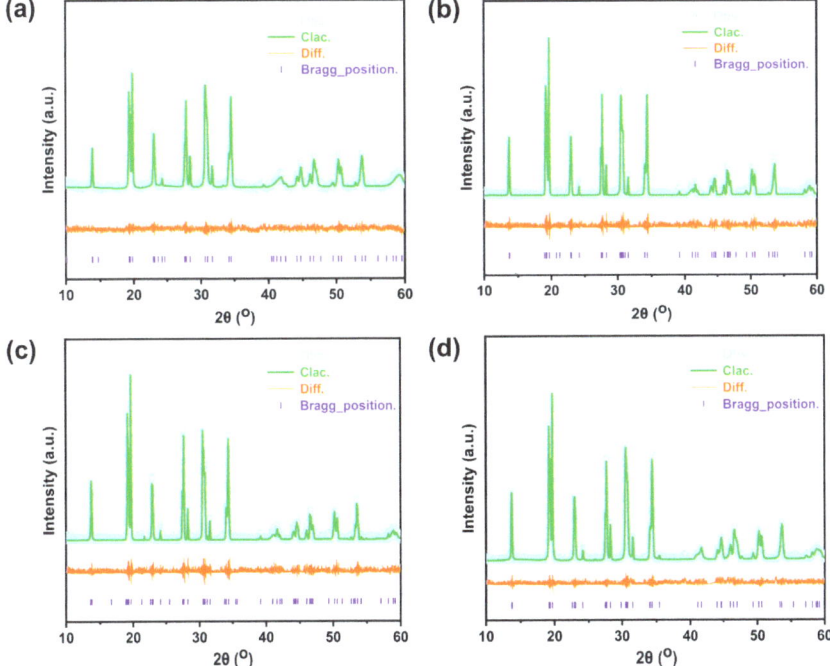

Figure 2. Rietveld refinement plots of (**a**) NZSP—1250 °C, (**b**) NZSP-1%B_2O_3—1150 °C, (**c**) NZSP-2%B_2O_3—1150 °C, and (**d**) NZSP-4%B_2O_3—1150 °C.

Ionic conductivity is an important parameter to assess the capability of ion transport in the electrolyte. Electrochemical impedance spectra (EIS) analysis was performed with these ceramic samples. The EIS plots are collected in Figure 3a, Figures S5 and S6. Typically, the impedance spectra of NZSP-xB_2O_3 consist of a sloping line in a low-frequency region and a semicircle in a high-frequency range [29,30]. Furthermore, the room temperature ionic conductivity was calculated according to the resistance obtained from the impedance spectra and corresponding results are collected in Figure 3b. The ionic conductivity of NZSP-xB_2O_3—1050 °C and NZSP-xB_2O_3—1100 °C increases monotonously with increasing B_2O_3 content, but NZSP-xB_2O_3—1150 °C increases first and then decreases. NZSP-4%B_2O_3—1050 °C, NZSP-4%B_2O_3—1100 °C, and NZSP-2%B_2O_3—1150 °C achieve ionic conductivity of 4.4×10^{-4}, 4.6×10^{-4}, and 4.7×10^{-4} S cm^{-1}. In other words, NZSP-xB_2O_3 sintered at lower sintering temperatures can achieve the same or even better room temperature ionic conductivity than bare NZSP sintered at 1250 °C (3.4×10^{-4} S cm^{-1}). This can be ascribed to the dense microstructure and the increased average grain size [31–34]. However, when too much content of B_2O_3 is added, the ionic conductivity decreases. Figure 3c,d, Figures S7 and S8 show temperature-dependent Nyquist plots and Arrhenius plots of conductivities for these ceramic electrolytes. Furthermore, the activation energy (E_a) was computed and the E_a of NZSP-4%B$_2$O$_3$—1050 °C, NZSP-4%B$_2$O$_3$—1100 °C, NZSP-2%B$_2$O$_3$—1150 °C, and NZSP-1250 °C is 0.37, 0.37, 0.33, and 0.36 eV, respectively.

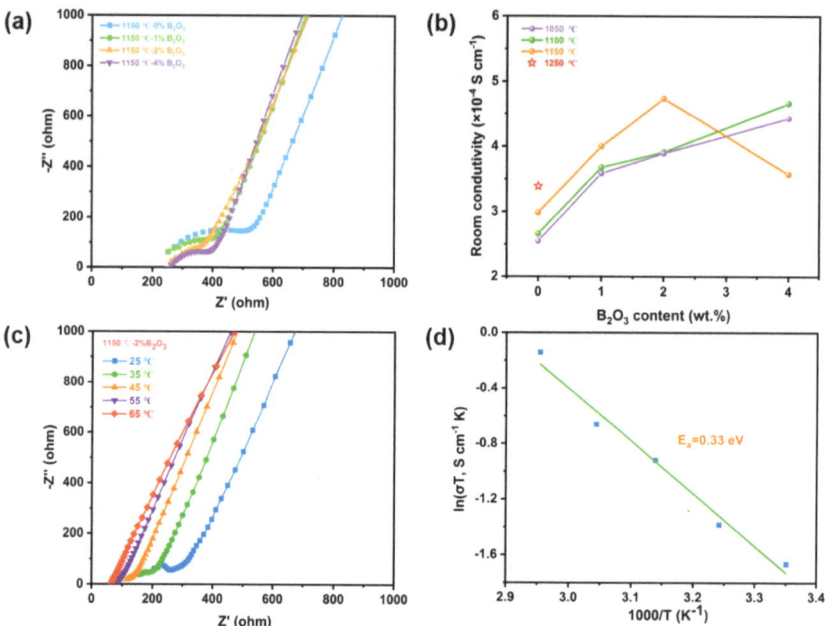

Figure 3. (**a**) Nyquist plots of the impendence spectroscopy of NZSP-xB$_2$O$_3$—1150 °C. (**b**) The room temperature ionic conductivity of NZSP-xB$_2$O$_3$ obtained at different sintering temperatures. (**c**) The temperature-dependent Nyquist plots and (**d**) Arrhenius temperature-dependent total ionic conductivity plot of NZSP-2%B$_2$O$_3$—1150 °C.

Symmetrical sodium metal cells were assembled by using NZSP—1250 °C and NZSP-2%B$_2$O$_3$—1150 °C as the solid electrolytes to investigate the interfacial electrochemical behavior. Figure 4a shows the EIS plots of as-assembled symmetric batteries and the inset is the interfacial area specific resistance (ASR). Figure S9 shows the result of the fitted circuit simulation and the total impedance consists of electrolyte impedance and interface impedance. The impedance of the Na/NZSP-2%B$_2$O$_3$/Na symmetric battery is smaller than that of the Na/NZSP/Na symmetric battery. In addition, the symmetric battery with bare NZSP electrolyte has an ASR of 148 ohm cm^2. Moreover, Na/NZSP-2%B$_2$O$_3$/Na presents an ASR of 67 ohm cm^2, indicating that the engineered grain boundary can observably promote Na$^+$ migration at the interface and improve the interfacial compatibility of metallic sodium and solid electrolyte. The total impedance of Na/NZSP-2%B$_2$O$_3$/Na symmetrical battery gets significantly reduced with increasing the temperature (Figure 4b). According to the corresponding Arrhenius plot (Figure 4c), the activation energy of sodium ion transfer at the Na/NZSP-2%B$_2$O$_3$ interface is about 0.33 eV, such a low value indicates that ion migration at the interfaces is dramatically facilitated [35]. Galvanostatic charge/discharge measurements were utilized to assess the cycling stability of as-assembled cells. As shown (Figure 4d), at room temperature the Na/NZSP-2%B$_2$O$_3$/Na battery presents a stable cycling performance at 0.05 mA cm^{-2} up to 1000 h with no significant variation in the polarization response, indicating the effective inhibition of metallic sodium dendrite growth. Moreover, the cycling performance of the symmetrical cell under variable current densities of 0.1–0.3 mA cm^{-2} at 60 °C was also assessed (Figure 4e,f). As observed, the polarization voltage hysteresis linearly increases with increasing the current density. The favorable cycling stability and rate performance are attributed to improved interfacial compatibility and effective inhibition of sodium dendrite growth.

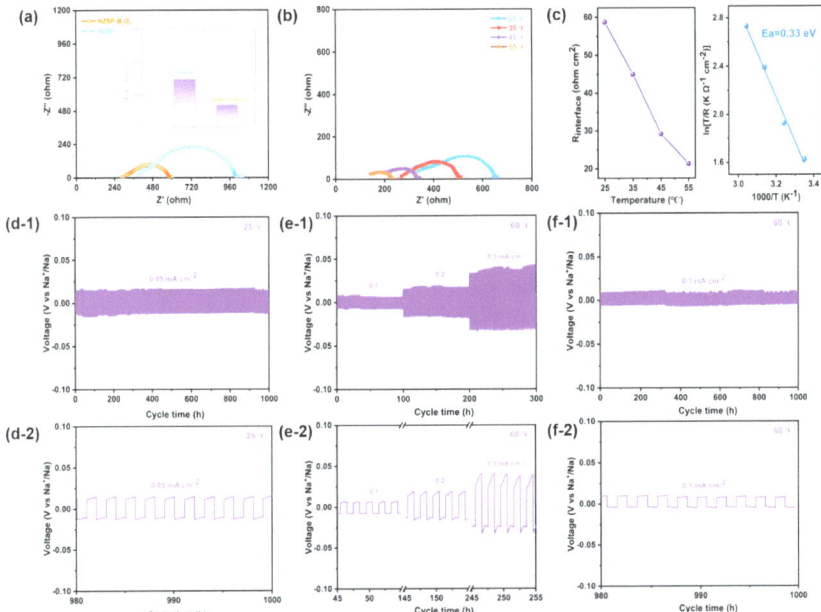

Figure 4. (**a**) Room temperature EIS profiles of symmetric sodium metal batteries based on NZSP—1250 °C and NZSP-2%B$_2$O$_3$—1150 °C ceramic electrolytes. The inset is the comparison of the ASR of the interface between metallic sodium anode and solid electrolyte. (**b**) Temperature-dependence EIS profiles of symmetrical sodium metal batteries with NZSP-2%B$_2$O$_3$—1150 °C ceramic electrolyte. (**c**) The fitted overall areal specific resistance of the Na/NZSP-2%B$_2$O$_3$/Na cell against the temperature and the corresponding Arrhenius plot. (**d-1**) Voltage profiles of Na/NZSP-2%B$_2$O$_3$/Na cell at 0.05 mA cm^{-2} and 25 °C. (**e-1,f-1**) Voltage profiles of Na/NZSP-2%B$_2$O$_3$/Na cell at 0.1–0.3 mA cm^{-2} and 60 °C. (**d-2,e-2,f-2**) are corresponding locally enlarged images.

We dissembled the cycled symmetrical cell to gain more information about the interphase between the metallic sodium anode and NZSP-2%B$_2$O$_3$—1150 °C electrolyte. Figure 5a shows XPS analysis results of the section and surface of the ceramic electrolytes after electrochemical cycling. Obviously, the integral area and relative intensity of the Na 1s spectrum become larger owing to the increased surface roughness and coverage of interfacial products. In the meantime, the peak of Si 2p shifts toward the lower-binding energy indicating that the reduction in Si^{4+} would occur as the sodium metal contacts with the ceramic electrolyte, and the Zr 3d, P 2p, and B 1s are virtually unaffected. The overlay SEM-EDS mapping results of the cross-section of cycled symmetrical cells are provided in Figure 5b. After cycling, the ceramic electrolyte still has a dense microstructure, illustrating favorable mechanical properties; sodium metal is uniformly deposited onto the ceramic electrolyte surface, and no visible "dead sodium" can be observed. Furthermore, the sodium metal and ceramic electrolyte are tightly combined together and there is no appreciable gap between them, suggesting a high integrity of the interface. It should be noted that the evident demarcations within the region of Na metal are induced during the sample preparation process.

The nucleation and growth of metal dendrites seriously affect the performance of solid-state metal batteries. In addition, metal dendrites are mainly formed at the interface between the solid electrode and the electrolyte, and the grain boundary of the solid-state electrolyte [36,37]. Usually, the grain boundary in polycrystalline ceramic electrolytes has low-ionic conductivity, poor mechanical strength, and high-electronic conductivity compared to the grain bulk [28,38]. On the one hand, the high-electronic conductivity of the grain boundary would drive the reduction in Na^+, leading to the formation and growth of sodium dendrite at the grain boundary. Furthermore, the voids between the grains would provide an additional barrier to the Na^+ transport, resulting in an increase in the grain boundary resistance for ionic transportation [28]. With the addition of B_2O_3 to the NASICON matrix, the amorphous phase formed after the sintering process can reduce the electronic conductivity of the NASICON grain boundary, hindering the formation of metallic sodium dendrite. In addition, the bonding strength of the grain boundaries is also increased and the penetration path of the sodium dendrites is blocked.

The nucleation and growth of dendrites at the interface are mainly ascribed to the inhomogeneous distribution of Na^+ flux at the interface, and the loose microstructure of ceramic electrolyte would further degrade this circumstance [39,40]. However, using the B_2O_3-modified NASICON electrolyte can enhance the interface contact of the metallic sodium anode and ceramic electrolyte, alleviating the edging effect and homogenizing the distribution of Na^+ (Figure 5c).

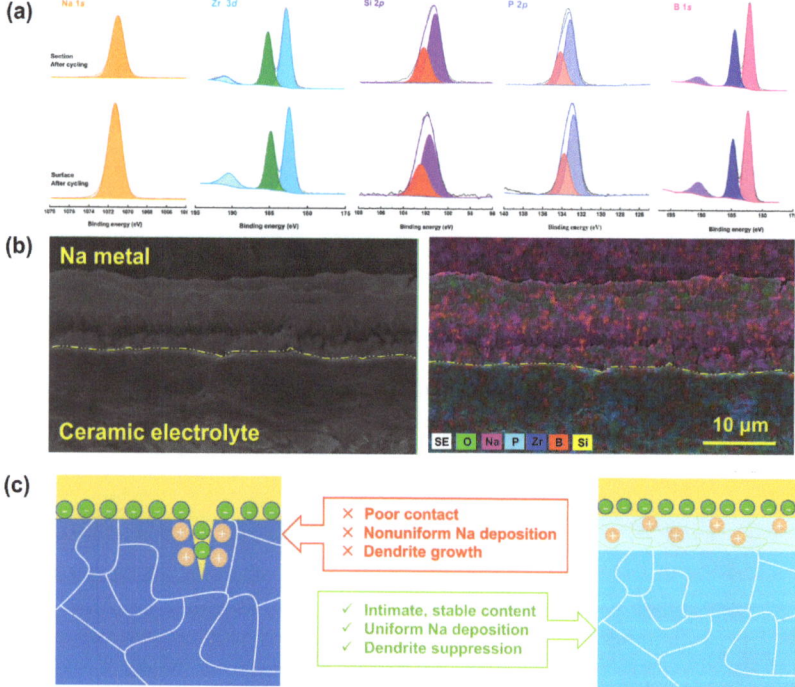

Figure 5. (a) XPS spectra of the surface and section of cycled NZSP-2%B_2O_3—1150 °C ceramic electrolyte. (b) Cross−sectional SEM image and EDS mapping results of NZSP-2%B_2O_3—1150 °C/Na interface disassembled from cycled symmetric sodium metal cell. (c) The demonstration of interface contact between the ceramic electrolyte and the Na metal anode; the left is for Na/NZSP—1250 °C and the right is for Na/NZSP-2%B_2O_3—1150 °C.

The solid-state sodium battery is assembled with $Na_3V_{1.5}Cr_{0.5}(PO_4)_3$ as the cathode active material and NZSP-2%B_2O_3—1150 °C as the ceramic electrolyte. Figure 6a shows the schematic diagram of the solid-state sodium battery and the room temperature EIS plot fitted according to the equivalent circuit displayed in the illustration. Figure 6b shows the initial charge/discharge profiles of the as-assembled solid-state battery at 30 mA g^{-1}

and the initial discharge capacity of 98.5 mAh g^{-1} achieves a Coulombic efficiency of 84.14%. The long-term cycling performance of the solid-state battery with NZSP—1250 °C as the ceramic electrolyte is demonstrated in Figure 6c. The capacity of the solid-state battery decays rapidly with a capacity retention of 48% after 200 cycles. However, the solid-state battery with NZSP-2%B$_2$O$_3$—1150 °C electrolyte has a capacity retention of 70.3% after 200 cycles at 30 mA g^{-1} (Figure 6c), which is mainly ascribed to the enhanced ionic conductivity of the ceramic electrolyte and the improved interfacial compatibility of metallic Na and NASICON electrolytes. As shown in Figure 6d, the solid-state sodium metal battery can offer reversible capacities of 106.8, 78.4, and 70.6 mAh g^{-1} at 30, 100, and 200 mAh g^{-1}, respectively. The specific capacity can be fully restored to its initial state following a complete rate trial, highlighting the advantage of an all-solid-state sodium metal battery based on a stable Na/NZSP-2%B$_2$O$_3$—1150 °C interface.

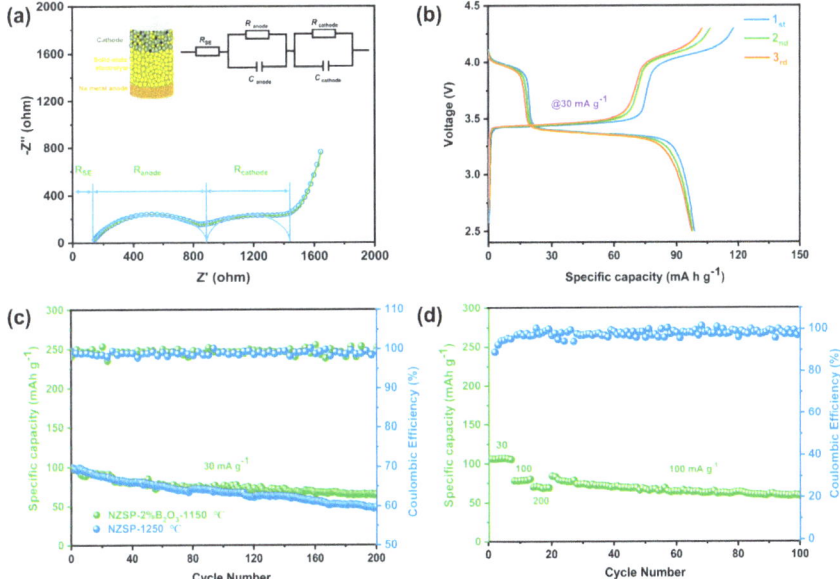

Figure 6. Electrochemical performance of the solid-state sodium metal batteries based on NZSP-2%B$_2$O$_3$—1150 °C ceramic electrolytes. (**a**) The impedance profile of solid-state sodium metal battery. (**b**) The initial charge/discharge curves of the solid-state sodium metal batteries at a current density of 30 mA g^{-1}. (**c**) Cycling performance of the solid-state sodium metal batteries based on NZSP—1250 °C and NZSP-2%B$_2$O$_3$—1150 °C ceramic electrolytes. (**d**) Rate performance of the solid-state sodium metal batteries based on NZSP-2%B$_2$O$_3$—1150 °C ceramic electrolytes.

4. Conclusions

In summary, NZSP-xB$_2$O$_3$ ceramics were prepared using a sintering additive-assisted solid-state reaction. With a low-melting point of 450 °C, B$_2$O$_3$ would change into a liquid phase at a certain temperature during the sintering process and fill the voids between the grains, thus, accelerating the process of densification sintering and then achieving a dense microstructure. After sintering, the liquid phase exists at the grain boundary as an amorphous phase, reducing the electronic conductivity of the grain boundary and impeding the formation of metallic sodium dendrites. Furthermore, the engineered grain boundary is also beneficial to promote the interface contact between the ceramic electrolyte and the metallic sodium anode, effectively lowering the ASR from 148 to 67 ohm cm^2. Eventually, the B$_2$O$_3$-modified NASICON-type electrolyte-based all-solid-state sodium batteries were constructed and the desirable electrochemical performance was realized.

Supplementary Materials: The following supporting information can be downloaded at: https://www.mdpi.com/article/10.3390/batteries9050252/s1, Figure S1. Sectional SEM images of the NZSP-xB$_2$O$_3$—1050 °C and NZSP-xB$_2$O$_3$—1100 °C ceramics. Figure S2. EDS mapping results with an enlarged scale for NZSP-2%B$_2$O$_3$—1150 °C ceramic. Figure S3. Energy spectrum element analysis of grain boundaries for NZSP-2%B2O3—1150 °C ceramic. Figure S4. Shrinkage variation of the NZSP-xB$_2$O$_3$ ceramics pellets sintered at different temperatures. Figure S5. Nyquist plots of the room temperature EIS of (**a**) NZSP-xB$_2$O$_3$—1050 °C and (**b**) NZSP-xB$_2$O$_3$—1100 °C ceramic pellets. Figure S6. Nyquist plots of NZSP—1250 °C. Figure S7. The temperature-dependent Nyquist plots of (**a**) NZSP-4%B$_2$O$_3$—1050 °C, (**c**) NZSP-4%B$_2$O$_3$—1100 °C, and (**e**) NZSP—1250 °C. The Arrhenius temperature-dependent total conductivity plots of (**b**) NZSP-4%B$_2$O$_3$—1050 °C, (**d**) NZSP-4%B$_2$O$_3$—1100 °C, and (**f**) NZSP—1250 °C. Figure S8. Nyquist plot of NZSP-2%B$_2$O$_3$—1150 °C at room temperature, simulation results based on the equivalent circuit, and illustration of analysis of R_b and R_{gb}. Figure S9. Room temperature Nyquist plots of symmetric sodium metal batteries based on NZSP—1250 °C and NZSP-2%B$_2$O$_3$—1150 °C ceramic electrolytes. The insets are the equivalent circuit and illustration of analysis of R_{SE} and R_t. Table S1. Cell Parameters and volume of NZSP-xB$_2$O$_3$.

Author Contributions: Conceptualization, Y.Z.; Formal analysis, Y.L. and Z.S.; Funding acquisition, Y.Z.; Investigation, H.J. and Y.Z.; Methodology, Y.L. and Z.S.; Project administration, H.J. and Y.Z.; Software, Z.S.; Validation, Y.L.; Writing—original draft, Y.L.; Writing—review & editing, Y.Z. All authors have read and agreed to the published version of the manuscript.

Funding: Y.Z. acknowledges funding support from the National Natural Science Foundation of China (No. 52072033), the State Key Laboratory of New Ceramic and Fine Processing Tsinghua University (No. 202211), the Open Research Fund of Songshan Lake Materials Laboratory (No. 2022SLABFN24), and the Graduate Interdisciplinary Innovation Project of Yangtze Delta Region Academy of Beijing Institute of Technology (Jiaxing), No. GIIP2022-014.

Data Availability Statement: The data presented in this study are available upon request from the corresponding author.

Conflicts of Interest: The authors declare no conflict of interest.

References

1. Dehghani-Sanij, A.R.; Tharumalingam, E.; Dusseault, M.B.; Fraser, R. Study of energy storage systems and environmental challenges of batteries. *Renew. Sust. Energ. Rev.* **2019**, *104*, 192–208. [CrossRef]
2. Olabi, A.G. Renewable energy and energy storage systems. *Energy* **2017**, *136*, 1–6. [CrossRef]
3. Ould Amrouche, S.; Rekioua, D.; Rekioua, T.; Bacha, S. Overview of energy storage in renewable energy systems. *Int. J. Hydrog. Energy* **2016**, *41*, 20914–20927. [CrossRef]
4. Guney, M.S.; Tepe, Y. Classification and assessment of energy storage systems. *Renew. Sust. Energ. Rev.* **2017**, *75*, 1187–1197. [CrossRef]
5. Olabi, A.G.; Onumaegbu, C.; Wilberforce, T.; Ramadan, M.; Abdelkareem, M.A.; Al-Alami, A.H. Critical review of energy storage systems. *Energy* **2021**, *214*, 118987. [CrossRef]
6. Chen, K.F.; Xue, D.F. Materials chemistry toward electrochemical energy storage. *J. Mater. Chem. A* **2016**, *4*, 7522–7537. [CrossRef]
7. Ragupathy, P.; Bhat, S.D.; Kalaiselvi, N. Electrochemical energy storage and conversion: An overview. *WIREs Energy Environ.* **2022**, *12*, e464. [CrossRef]
8. Yang, Y.Q.; Bremner, S.; Menictas, C.; Kay, M. Battery energy storage system size determination in renewable energy systems: A review. *Renew. Sust. Energ. Rev.* **2018**, *91*, 109–125. [CrossRef]
9. Vaalma, C.; Buchholz, D.; Weil, M.; Passerini, S. A cost and resource analysis of sodium-ion batteries. *Nat. Rev. Mater.* **2018**, *3*, 18013. [CrossRef]
10. Hwang, J.-Y.; Myung, S.-T.; Sun, Y.-K. Sodium-ion batteries: Present and future. *Chem. Soc. Rev.* **2017**, *46*, 3529–3614. [CrossRef]
11. Hou, D.W.; Xia, D.W.; Gabriel, E.; Russell, J.A.; Graff, K.; Ren, Y.; Sun, C.-J.; Lin, F.; Liu, Y.Z.; Xiong, H. Spatial and temporal analysis of sodium-ion batteries. *ACS Energy Lett.* **2021**, *6*, 4023–4054. [CrossRef] [PubMed]
12. Wang, L.G.; Li, J.; Lu, G.L.; Li, W.Y.; Tao, Q.Q.; Shi, C.H.; Huile, J.; Chen, G.; Wang, S. Fundamentals of electrolytes for solid-state batteries: Challenges and perspectives. *Front. Mater.* **2020**, *7*, 111. [CrossRef]
13. Yu, X.Q.; Chen, R.S.; Gan, L.Y.; Li, H.; Chen, L.Q. Battery safety: From lithium-ion to solid-state batteries. *Engineering* **2022**, *21*, 9–14. [CrossRef]
14. Zhao, Q.; Stalin, S.; Zhao, C.-Z.; Archer, L.A. Designing solid-state electrolytes for safe, energy-dense batteries. *Nat. Rev. Mater.* **2020**, *5*, 229–252. [CrossRef]

15. Murgia, F.; Brighi, M.; Piveteau, L.; Avalos, C.E.; Gulino, V.; Nierstenhöfer, M.C.; Ngene, P.; Jongh, P.; CČerny, R. Enhanced room-temperature ionic conductivity of NaCB$_{11}$H$_{12}$ via high-energy mechanical milling. *ACS Appl. Mater Inter.* **2021**, *13*, 61346–61356. [CrossRef]
16. Till, P.; Agne, M.T.; Kraft, M.A.; Courty, M.; Famprikis, T.; Ghidiu, M.; Krauskopf, T.; Masquelier, C.; Zeier, W.G. Two-dimensional substitution series Na$_3$P$_{1-x}$Sb$_x$S$_{4-y}$Se$_y$: Beyond static description of structural bottlenecks for Na$^+$ Transport. *Chem. Mater.* **2022**, *34*, 2410–2421. [CrossRef]
17. Liu, L.M.; Liang, D.S.; Zhou, X.L.; Liu, Y.J.; Su, J.W.; Xu, Y.; Peng, J.S. Enhancing Na-ion conducting capacity of NASICON ceramic electrolyte Na$_{3.4}$Zr$_2$Si$_{2.4}$P$_{0.6}$O$_{12}$ by NaF sintering aid. *J. Mater. Sci.* **2022**, *57*, 11774–11782. [CrossRef]
18. Oh, J.A.S.; He, L.C.; Plewa, A.; Morita, M.; Zhao, Y.; Sakamoto, T.; Song, X.; Zhai, W.; Zeng, K.Y.; Lu, L. Composite NASICON (Na$_3$Zr$_2$Si$_2$PO$_{12}$) solid-state electrolyte with enhanced Na$^+$ ionic conductivity: Effect of liquid phase sintering. *Acs App. Mater. Interfaces* **2019**, *11*, 40125–40133. [CrossRef]
19. Grady, Z.; Fan, Z.M.; Ndayishimiye, A.; Randall, C.A. Design and sintering of all-solid-state composite cathodes with tunable mixed conduction properties via the cold sintering process. *ACS Appl. Mater. Interfaces* **2021**, *13*, 48071–48087. [CrossRef]
20. Leng, H.Y.; Huang, J.J.; Nie, J.Y.; Luo, J. Cold sintering and ionic conductivities of Na$_{3.256}$Mg$_{0.128}$Zr$_{1.872}$Si$_2$PO$_{12}$ solid electrolytes. *J. Power Sources* **2018**, *391*, 170–179. [CrossRef]
21. Li, Y.; Li, M.; Sun, Z.; Ni, Q.; Jin, H.B.; Zhao, Y.J. Recent advance on NASICON electrolyte in solid-state sodium metal batteries. *Energy Storage Mater.* **2023**, *56*, 582–599. [CrossRef]
22. Zhang, X.; Wang, J.X.; Wen, J.W.; Wang, Y.; Li, N.; Wang, J.; Fan, L.J. Improvement of ionic conductivity and densification of Na$_3$Zr$_2$Si$_2$PO$_{12}$ solid electrolyte rapidly prepared by microwave sintering. *Ceram. Int.* **2022**, *48*, 18999–19005. [CrossRef]
23. Fuentes, R.O.; Figueiredo, F.M.; Marques, F.M.B.; Franco, J.I. Processing and electrical properties of NASICON prepared from yttria-doped zirconia precursors. *J. Eur. Ceram. Soc.* **2001**, *21*, 737–743. [CrossRef]
24. Krok, F. Influence of sintering conditions on chemical composition of NASICON. *Solid State Ion.* **1987**, *24*, 21–28. [CrossRef]
25. Li, Y.; Sun, Z.; Sun, C.; Jin, H.B.; Zhao, Y.J. Exploring the origin of ZrO$_2$ phase in Na$_3$Zr$_2$Si$_2$PO$_{12}$ ceramic electrolyte. *Ceram. Int.* **2023**, *49*, 3094–3098. [CrossRef]
26. Zhao, Y.J.; Gao, X.W.; Gao, H.C.; Jin, H.B.; Goodenough, J.B. Three electron reversible redox reaction in sodium vanadium chromium phosphate as a high-energy-density cathode for sodium-ion batteries. *Adv. Funct. Mater.* **2020**, *30*, 1908680. [CrossRef]
27. Sun, C.; Zhao, Y.J.; Ni, Q.; Sun, Z.; Yuan, X.Y.; Li, J.B.; Jin, H.B. Reversible multielectron redox in NASICON cathode with high energy density for low-temperature sodium-ion batteries. *Energy Storage Mater.* **2022**, *49*, 291–298. [CrossRef]
28. Zheng, C.J.; Lu, Y.; Su, J.M.; Song, Z.; Xiu, T.P.; Jin, J.; Badding, M.E.; Wen, Z.Y. Grain boundary engineering enabled high-performance garnet-type electrolyte for lithium dendrite free lithium metal batteries. *Small Methods* **2022**, *6*, e2200667. [CrossRef]
29. Thokchom, J.S.; Kumar, B. The effects of crystallization parameters on the ionic conductivity of a lithium aluminum germanium phosphate glass-ceramic. *J. Power Sources* **2010**, *195*, 2870–2876. [CrossRef]
30. Zhao, X.C.; Luo, Y.S.; Zhao, X.J. Effect of TeO$_2$ sintering aid on the microstructure and electrical properties of Li$_{1.3}$Al$_{0.3}$Ti$_{1.7}$(PO$_4$)$_3$ solid electrolyte. *J. Alloy. Compd.* **2022**, *927*, 167019. [CrossRef]
31. He, S.N.; Xu, Y.L.; Chen, Y.J.; Ma, X.N. Enhanced ionic conductivity of an F$^-$-assisted Na$_3$Zr$_2$Si$_2$PO$_{12}$ solid electrolyte for solid-state sodium batteries. *J. Mater. Chem. A* **2020**, *8*, 12594–12602. [CrossRef]
32. Naqash, S.; Ma, Q.; Tietz, F.; Guillon, O. Na$_3$Zr$_2$(SiO$_4$)$_2$(PO$_4$) prepared by a solution-assisted solid state reaction. *Solid State Ion.* **2017**, *302*, 83–91. [CrossRef]
33. Park, H.; Jung, K.; Nezafati, M.; Kim, C.-S.; Kang, B. Sodium ion diffusion in Nasicon (Na$_3$Zr$_2$Si$_2$PO$_{12}$) solid electrolytes: Effects of excess sodium. *ACS Appl. Mater. Interfaces* **2016**, *8*, 27814–27824. [CrossRef] [PubMed]
34. Shao, Y.J.; Zhong, G.M.; Lu, Y.X.; Liu, L.L.; Zhao, C.L.; Zhang, Q.Q.; Hu, Y.-S.; Yang, Y.; Chen, L.Q. A novel NASICON-based glass-ceramic composite electrolyte with enhanced Na-ion conductivity. *Energy Storage Mater.* **2019**, *23*, 514–521. [CrossRef]
35. Zhao, Y.J.; Wang, C.Z.; Dai, Y.J.; Jin, H.B. Homogeneous Na$^+$ transfer dynamic at Na/Na$_3$Zr$_2$Si$_2$PO$_{12}$ interface for all solid-state sodium metal batteries. *Nano Energy* **2021**, *88*, 106293. [CrossRef]
36. Aslam, M.K.; Niu, Y.B.; Hussain, T.; Tabassum, H. How to avoid dendrite formation in metal batteries: Innovative strategies for dendrite suppression. *Nano Energy* **2021**, *86*, 106142. [CrossRef]
37. Gao, Z.H.; Yang, J.Y.; Li, G.C.; Ferber, T.; Feng, J.R.; Li, Y.Y.; Fu, H.Y.; Jaegermann, W.; Monroe, C.W.; Huang, Y.H. TiO$_2$ as second phase in Na$_3$Zr$_2$Si$_2$PO$_{12}$ to suppress dendrite growth in sodium metal solid-state batteries. *Adv. Energy Mater.* **2022**, *12*, 2103607. [CrossRef]
38. Wang, X.X.; Chen, J.J.; Mao, Z.Y.; Wang, D.J. Effective resistance to dendrite growth of NASICON solid electrolyte with lower electronic conductivity. *Chem. Eng. J.* **2022**, *427*, 130899. [CrossRef]
39. Oh, J.A.S.; He, L.C.; Chua, B.; Zeng, K.Y.; Lu, L. Inorganic sodium solid-state electrolyte and interface with sodium metal for room-temperature metal solid-state batteries. *Energy Storage Mater.* **2021**, *34*, 28–44. [CrossRef]
40. Ruan, Y.L.; Guo, F.; Liu, J.J.; Song, S.D.; Jiang, N.Y.; Cheng, B.W. Optimization of Na$_3$Zr$_2$Si$_2$PO$_{12}$ ceramic electrolyte and interface for high performance solid-state sodium battery. *Ceram. Inter.* **2019**, *45*, 1770–1776. [CrossRef]

Disclaimer/Publisher's Note: The statements, opinions and data contained in all publications are solely those of the individual author(s) and contributor(s) and not of MDPI and/or the editor(s). MDPI and/or the editor(s) disclaim responsibility for any injury to people or property resulting from any ideas, methods, instructions or products referred to in the content.

Review

Recent Advances in Ionic Liquids—MOF Hybrid Electrolytes for Solid-State Electrolyte of Lithium Battery

Ruifan Lin, Yingmin Jin *, Yumeng Li, Xuebai Zhang and Yueping Xiong *

MIIT Key Laboratory of Critical Materials Technology for New Energy Conversion and Storage, School of Chemistry and Chemical Engineering, Harbin Institute of Technology, Harbin 150001, China; 1181400512@stu.hit.edu.cn (R.L.); 22b925039@stu.hit.edu.cn (Y.L.); 20b925081@stu.hit.edu.cn (X.Z.)
* Correspondence: jymjinyingmin@163.com (Y.J.); ypxiong@hit.edu (Y.X.); Tel.: +86-18245029032 (Y.J.); +86-13904658015 (Y.X.)

Abstract: Li-ion batteries are currently considered promising energy storage devices for the future. However, the use of liquid electrolytes poses certain challenges, including lithium dendrite penetration and flammable liquid leakage. Encouragingly, solid electrolytes endowed with high stability and safety appear to be a potential solution to these problems. Among them, ionic liquids (ILs) packed in metal organic frameworks (MOFs), known as ILs@MOFs, have emerged as a hybrid solid-state material that possesses high conductivity, low flammability, and strong mechanical stability. ILs@MOFs plays a crucial role in forming a continuous interfacial conduction network, as well as providing internal ion conduction pathways through the ionic liquid. Hence, ILs@MOFs can not only act as a suitable ionic conduct main body, but also be used as an active filler in composite polymer electrolytes (CPEs) to meet the demand for higher conductivity and lower cost. This review focuses on the characteristic properties and the ion transport mechanism behind ILs@MOFs, highlighting the main problems of its applications. Moreover, this review presents an introduction of the advantages and applications of Ils@MOFs as fillers and the improvement directions are also discussed. In the conclusion, the challenges and suggestions for the future improvement of ILs@MOFs hybrid electrolytes are also prospected. Overall, this review demonstrates the application potential of ILs@MOFs as a hybrid electrolyte material in energy storage systems.

Keywords: ionic liquids; metal organic frameworks; solid-state electrolyte; Li battery

Citation: Lin, R.; Jin, Y.; Li, Y.; Zhang, X.; Xiong, Y. Recent Advances in Ionic Liquids—MOF Hybrid Electrolytes for Solid-State Electrolyte of Lithium Battery. *Batteries* 2023, 9, 314. https://doi.org/10.3390/batteries9060314

Academic Editor: Claudio Gerbaldi

Received: 15 April 2023
Revised: 24 May 2023
Accepted: 4 June 2023
Published: 6 June 2023

Copyright: © 2023 by the authors. Licensee MDPI, Basel, Switzerland. This article is an open access article distributed under the terms and conditions of the Creative Commons Attribution (CC BY) license (https://creativecommons.org/licenses/by/4.0/).

1. Introduction

The increasing demand of clean energy calls for the progression of advanced energy storage systems, which helps to regulate the unstable energy output using renewable energy [1]. Nowadays, electrochemical energy storage, such as Li-ion batteries, is considered to be one of the most promising future energy storage techniques [2]. The rapid development of Li-ion batteries has drawn much attention from researchers due to their distinct advantages, such as high theoretical energy density, stable energy output and low memory effect. However, the highly flammable electrolytes, complex temperature management and limited practical capacity still restrict the further development of Li-ion batteries. In comparison, lithium batteries which utilize an Li-metal anode show significant superiority in high energy density due to their ultra-high theoretical capacity of 3860 mAh/g and ultra-low electrode potential of −3.04 V (vs. SHE), which reveals promising prospects in alleviating the "range anxiety" of electrical vehicles [3]. Unfortunately, serious challenges remain to be solved before the practical application of lithium batteries, including the infinite volume change of Li metal and the generation of Li dendrites, and even "dead lithium" caused by the pulverization of the Li anode [4].

The emergence of solid-state electrolytes with a comparatively higher safety and longer life span offers a potential solution to the challenges faced in Li metal batteries. Solid-state electrolytes with a high shear modulus can provide sufficient mechanical strength to suppress the uneven Li deposition. In addition, solid-state lithium batteries employing solid electrolytes with high thermal stability prevent the potential thermal runaways, which greatly improves the safety of high energy-density devices. Solid-state electrolyte can be mainly classified into inorganic ceramic electrolyte, solid polymer electrolyte and a combination of the two. As a solid electrolyte with promising prospects for practical application, solid polymer electrolytes which exhibit the advantages of shape versatility, low weight, flexibility and low processing costs have become the research focus [5]. However, on the one hand, the high crystallinity of polymer chains at room temperature results in an undesired ionic conductivity. On the other hand, the low electrochemical stability of polymers can cause interfacial side reactions, resulting in an increased interfacial resistance and structural degradation of electrode materials. Moreover, due to volume effects, poor contact between the electrode and electrolyte can also occur during long-term cycling. These challenges need to be addressed to improve the performance of polymer-based electrolytes in lithium-ion batteries [6,7]. To solve the problem, adding inorganic fillers into solid polymer electrolyte and forming composite polymer electrolyte is regarded as the ultimate approach to construct solid-state electrolyte with advanced comprehensive properties. However, the nano effect of inorganic fillers leads to two challenges: firstly, the unsatisfied dispersity of filler, and secondly, the low upper limit of filler addition. Therefore, the properties of composite polymer electrolyte have not yet met the established standards for practical operation of batteries. For example, the low migration number of lithium ions of composite polymer electrolytes can easily form a lithium-ion concentration gradient on the electrode surface, which accelerates dendrite growth. Moreover, the conduction of Li-ions can easily be impeded by overly added fillers, leading to discontinuous Li^+ transmission pathways, which attenuates the high C-rate performance of batteries. Furthermore, during the Li^+ deposition process, the uneven electric field on the electrode surface resulting from the inhomogeneous distribution of inorganic fillers can also accelerate electrode decay. Hence, in view of composite polymer systems, the structure and the chemical composition of fillers has a significant impact on the conductive property of the polymer electrolyte chain, and can therefore enhance comprehensive performance of polymer electrolyte [8]. Therefore, developing advanced fillers for high-performance solid electrolyte is considered an urgent requirement for Li-metal solid-state battery manufacturing.

Notably, in view of the advantage of high conductivity, stable structure and high chemical compatibility, other types of newly developed solid electrolytes have become substantial alternatives for ceramic electrolytes and composite polymer electrolytes [9–11]. Most recently, ionic liquids (ILs)@metal organic frameworks (MOFs) have emerged as a promising candidate material for potential utility because of their high ion conductivity, abundant metal sites, large specific surface area and modulable ability [12]. Unlike conventional carbonate solvents, ionic liquids are a class of molten salts that exist entirely in ionic form at room temperature. Generally speaking, the cations of ionic liquids are derivatives of 1-methylimidazole and anions are conjugate bases of inorganic acids [13]. Equipped with unique properties such as nonflammability, low vapor pressure and electrochemical stability [14], ionic liquids have been widely used in Li-ion batteries to replace carbonate solvents or participate in the formation of functional SEI [15]. Apart from that, ionic liquids can also be used as stabilizing agents in solid-state batteries to improve interface stability [16]. However, the direct use of ionic liquids as liquid electrolytes still cannot avoid the problem of Li dendrites caused by low Li^+ transference number (t_{Li+}) and possible liquid leakage. The low transference number originates from the free anion and cation migration. Although the nonflammable ILs avoid electrolyte combustion, the dendrite remains can penetrate the separator and cause a short circuit under abuse conditions. Meanwhile, the high cost of the ionic liquids system impedes the application progress of lithium-ion batteries. Hence, some researchers use MOFs to confine ILs, which achieves high performance solid-state electrolytes with a low ILs dosage, contributing to the transition metal ion or clusters and organic ligands on MOFs [13]. In the 1990s, Robison and Hoskins reported the first successfully synthesized MOF [N(CH3)4][CuIZnII(CN)4] [17], the stable porous structure, diverse combination of metal and organic units and the tunable electrochemical property of which attracted the interest of researchers [18–21]. Usually, simple and inexpensive methods such as the microwave-assisted heating method, hydrothermal method and solvent self-assembly methods have been proposed to synthesize MOFs [22–24]. Through a self-assembling procedure in a solution experiment, the chemical bonds formed between organic ligands and metal ions can bring the unique properties which cannot be achieved in other skeletal compounds [25]. Compared with ILs hybrid electrolyte and SPEs in Figure 1, using MOFs to confine ILs can not only avoid leakage of ILs and provide mechanical suppor, but also facilitate the ion conductivity of MOFs material. Consequently, with tunable porosity, rich Lewis acidic active sites and a modular nature, ILs@MOFs have been regarded as promising materials for applications as electrolyte of solid-state lithium batteries.

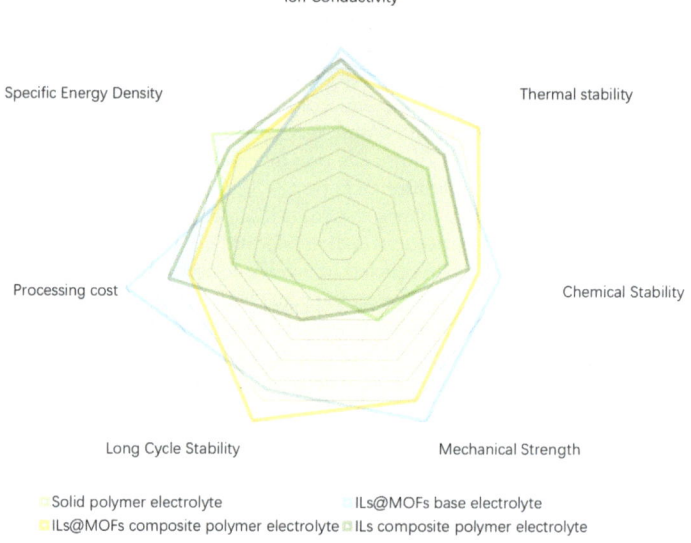

Figure 1. The performance comparison diagram of different organic electrolyte: SPE, ILs@MOFs-based electrolyte, ILs@MOFs CPE and ILs CPE.

In this review, the application of ILs@MOFs materials as the main body of solid electrolytes in lithium battery was first explored, including MOFs interface layers for solid electrolytes as well as the use of MOFs materials directly as solid electrolytes after modification and compounding with ionic liquids. Meanwhile, the application of MOFs as fillers in composite polymer electrolytes was also reviewed. This review provides guidance in exploring the ionic conduction mechanism inside ILs@MOFs-based solid electrolyte materials, and brings significant suggestions for the future application orientation of MOFs materials in energy storage devices.

2. Ionic Liquids Hybrid MOFs as Electrolyte
2.1. Composition and Structure Introduction of MOFs Hybrid Electrolyte

Traditional ionic liquid electrolytes possess a low Li^+ transference number and weak dendrite resistance. To solve these problems, the innovative development of advanced electrolyte by applying MOFs as the main body for ionic conduction, as well as taking advantage of metal nodes in consideration of the distinctive features of MOFs, have been proposed by researchers. Generally, MOFs exhibit three types of features: tunable porous structure, multi metal node properties and modular nature. Firstly, since the pore size is determined by organic linkers, it is feasible to adjust the pore size to suit the desired application by inserting molecules into the MOFs cage or using MOFs as fillers for selective permeating, for example [12,26]. Additionally, high specific surface area allows a high density of charged species and therefore provides abundant Li$^+$ hopping sites in a small volume. The ordered porosity could suppress dendrite formation by promoting uniform Li$^+$ plating [27]. Secondly, the metal nodes of MOFs not only play a role in connecting organic links, but also serve as Lewis acidic active sites which prefer to bind with electronic cloud, and thus anions can be selectively absorbed by MOFs with alternative absorption strength through the proper regulation of metal nodes [28,29]. Finally, in contrast with inorganic compounds, organic segments are also easy to make post-synthesis modifications to, and are endowed with additional features to improve electrolyte behavior. For example, fluoric groups and amination groups can functionalize MOFs through simple aqueous reaction which would simultaneously promote the formation of stable anion-derived SEI [30].

Inspired by porous zeolites, which were investigated as fillers in solid polymer lithium electrolyte systems [31], Long et al. used electrolyte solution-contained MOFs as possible lithium superionic conductors [32,33]. They synthesized MOFs-74 material upon the graft of lithium alkoxide. After soaking the as-prepared MOFs in 1 M LiBF4 in 1:1 ethylene carbonate (EC) and diethyl carbonate (DEC) solution, a maximum conductivity of 4.4×10^{-4} S cm^{-1} at room temperature and a low Ea of 0.15 eV can be achieved for the obtained electrolyte pellets. It had been speculated that the lithium alkoxide anion binding with exposed metal sites can promote Li$^+$ transportation through one-dimensional hexagonal channels of MOFs and the robust structure prevented dendrite growing in organic solutions. After soaking in carbonate solution, the high density of charge carriers in channels can also facilitate Li$^+$ hopping. Different from the ionic conduction mechanism of inorganic solid-state electrolytes, in which metal ions hop through vacancies to enable the charge transfer process, the main conduction mechanism of MOFs lies in the adsorption of solution anions by metal nodes, which allows the dissociated Li$^+$ to complex and transport along [34]. Nonetheless, the flammable inner solution cannot be avoided, which leaves safety hazards unresolved, and the relatively low conductivity also limits the further application of MOFs.

2.2. Proposal and Development of ILs@MOFs

On the basis of extensive research on MOFs containing carbonate solutions [35–37], as well as the application of ILs liquid electrolytes [38], some researchers suggests integrating ILs that possess high stability and nonflammability with MOFs. The combination of ILs with MOF materials is often achieved by encapsulating ILs in porous MOFs using host-guest interactions. In this system, ILs serve as ion conduction electrolytes while MOFs act as a solid supporting framework. As suggested by group interactions in FT-IR and microporous adsorption properties in BET tests, the ILs in the pores of MOFs exists in the form of both physisorption and chemisorption [39]. The first example regarding the incorporation of ILs into the micropores of MOFs was presented by Kitagawa et al. [40]. Figure 2a shows that researchers used a simple mixing and heating method to incorporate ILs of EMI-TFSA (1-ethyl-3-methylimidazolium bis(trifluoromethylsulfonyl)amide) into the pores of MOFs material ZIF-8 ((Zn (MeIM)2; H(MeIM) = 2-methylimidazole)). Kitagawa et al. systematically investigated EMI-TFSA@ZIF-8 and verified that the addition of ionic liquids into the framework of MOFs can lower the melting point of ionic liquids and stabilize the liquid phase of ionic liquids at low temperatures (Figure 2b,c). They also found that ion dynamics can be controlled through this subject-object interaction, which exhibited great potential in realizing the actual ion conduction process [41]. To endow actual ionic conduction capability, researchers mixed EMI-TFSA with LiTFSA to obtain (EMI0.8Li0.2) TFSA, which was then heated and mixed with ZIF-8 to obtain the target product. Although the ionic conductivity of obtained (EMI0.8Li0.2) TFSA@ZIF-8 ions is two orders of magnitude lower than that of the pure (EMI0.8Li0.2) TFSA due to the lower fluidity, the activation energy in (EMI0.8Li0.2) TFSA is nearly as high as that in bulk (EMI0.8Li0.2) TFSA, which suggests that the diffusing mechanisms of Li$^+$ are the same

(Figure 2d). Experiments demonstrate that Li⁺ can dissociate from anions in the framework of MOFs and achieve ion conduction by diffusion through the internal micropores.

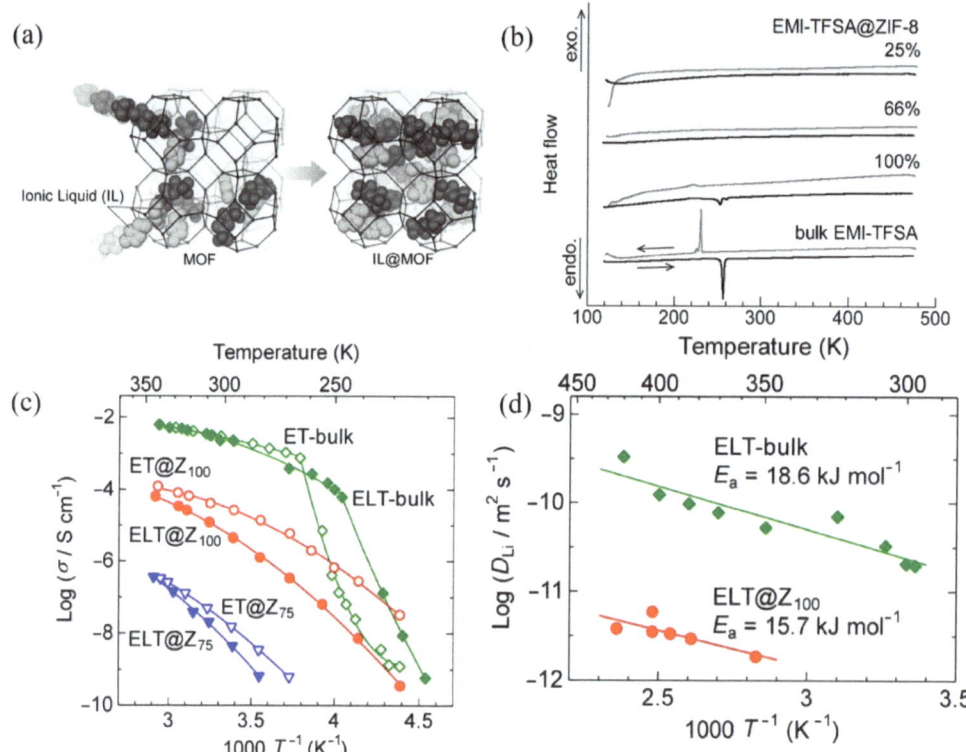

Figure 2. (**a**) The incorporation of ionic liquids into micro pores of MOFs. Reproduced with permission [40]. Copyright 2014, Wiley. (**b**) The DSC curves of bulk EMI-TFSA and EMI-TFSA@ZIF-8 at different volumetric occupancies. The sharp peak in 257 K and 231K represent melting and freezing of liquid state. Reproduced with permission [40]. Copyright 2014, Wiley. (**c**) Arrhenius plot of the ionic conductivity in heating process, the slope indicates the pseudoactivation energy changes around phase transition point, the green line stands for pure ILs and the red and blue lines stand for 100% and 75% ILs occupied MOFs respectively. Reproduced with permission [41]. Copyright 2015, American Chemical Society. (**d**) Arrhenius plot of the self-diffusion coefficient of lithium nucleus in ILs and ILs@MOFs. Reproduced with permission [41]. Copyright 2015, American Chemical Society.

2.3. Advantages and Function Mechanism of ILs@MOFs Electrolyte

Inspired by the research of Kitagawa's group, Pan's group further proposed the practical application of MOFs-based composite ionic liquids in battery systems, and summarized the three advantages brought by ILs@MOFs comprehensively, including high conductivity, mechanical support and dendrite suppression [42]. (EMI0.8Li0.2) TFSI ionic liquid and MOF-525 were selected for compounding based on the electrochemical stability and the appropriate pore size of the MOFs. Firstly, in terms of the high conductivity, it was found that the conductivity of the electrolyte increased substantially with the increase of ionic liquids loading, which proves that the ionic conductivity of the electrolyte is dependent on the bulk phase transport within the nanocrystal. In addition to conductor phase transport, the lithium-ion transport mechanism of ILs@MOF also includes intercrystal transport mechanism. Studies revealed that the mid-frequency spectrogram of EIS reflects liquid-like electrolyte properties, from which the researchers concluded that ionic liquids infiltrate the interface between nanocrystals at the atomic level. The nano-wetting interface enables the direct interfacial connection of the internal ILs, and X-ray photoelectron spectroscopy (XPS) characterization confirmed that the nature of interface lied in nanocrystals. Apart from that, MOFs play an important role in hindering the movement of large ions while the migration of lithium ions of small size remains unaffected, thus increasing the Li$^+$ transference number. Secondly, in terms of mechanical strength, the MOFs skeleton in ILs@MOF also provides the framework for alternative physical properties and presents a dry powder appearance rather than gel even with the encapsulation of ILs, which improves the overall mechanical strength. Finally, the ability to suppress dendrites' growth has been strengthened through the regulation of chemical environment and construction of physical barrier. Researchers discovered that ILs@MOFs possessing both high mechanical strength and nano-wetting properties are capable of not only impeding the penetration of dendrites, but also filling the gap between the electrolyte layer and metal electrode by utilizing dendrites for self-healing, ultimately resulting in reduced resistance after cycling.

The ability to suppress lithium dendrite is essential for ILs@MOFs, as dendrite may induce a short circuit and even thermal runaway. The factors influencing dendrite formation in solid-state batteries can be summarized as follows: (i) the high electric driving force for dendrite tips to extend into defects or grain boundaries of SSEs; (ii) the low dendrite consumption rate induced by the sluggish ionic transport kinetics of SSEs; and (iii) the high interface impedance of the solid-solid interphase in solid-state batteries that cause retardance in ionic transportation, which induces surface Li$^+$ deficiency and aggravates tip ion deposition and dendrite growth [43,44]. To overcome dendrite formation, three improvement strategies of ILs@MOFs have been proposed. First, a conformal coating layer of ILs@MOFs can be formed between the electrolyte and Li metal anode. Pan et al. prepared hybrid-sized ILs@MOF nanoparticles as novel electrolyte. The larger particles effectively suppressed Li dendrites' propagation while smaller particles filled the gap of larger particles to achieve better contact and further barricade dendrites [45]. Second, ILs@MOFs can inhibit dendrite through Li ion flux regulation. Zhang et al. constrained glymes ILs in Uio-66 to accelerate Li ions' transportation and confine ILs anion which relieved the difference between Li$^+$ ion diffusion and deposition rates and prevented inhomogeneous Li dendrite formation [46]. Finally, ILs can form a nanowetted interface between MOFs and Li metal to improve their compatibility, which ameliorates the detrimental point contact. Pan et al. found that in combination with ILs@MOFs electrolyte, the Li electrode surface was flat and composed by plate-like nanostructures after cycling [47]. The appearance of S and F elements belonging to ILs on Li surface after nanowetting ensured good contact between electrolyte and anode at first, and the formed stable interphase can also protect the battery from dendrite-induced short circuit.

Despite the many benefits of ILs@MOFs, the combination of solid-state frameworks and liquids also increases the research complexity in ion transport mechanisms. In pure ILs systems, the coordination environment and solvation pattern of ILs have been investigated by researchers [48,49]. Through molecular dynamics simulations, researchers calculated the free energy and transition barrier of alkali metal ions, as well as mapping the free energy landscape of alkali metal ions in ILs, which provided further instructions on restricting anion mobility. However, unlike pure ionic liquids, the ionic conduction of ILs@MOF contains both the contribution of highly mobile ILs at the nano-wetting interface and the conduction contribution of ILs in the pores of MOFs. Conventionally, EIS has been widely used to determine the transport mechanism of ILs, in which results showed that the ion transport was mainly determined by the migration and transport of ILs wrapped in the outer layer of MOFs, while the ILs in the pores would not participate [50]. In Figure 3a, the broad NMR lines above −20 °C indicated that there existed two lithium spin reservoirs, and the fast Li$^+$ fraction appears small. The result has shown that with ILs, most Li$^+$ did not change dynamics properties, which means that only Li$^+$ near surface can interact with surface ILs to facilitate transportation. Researchers also found that after flooding the MOFs crystals with excess Li-ILs to form a gel electrolyte, the overall Li-ion transport became faster and the measured conductivity also enhanced as expected (Figure 3b) [51]. However, if the isolated ILs on MOFs surface dominate the transport of lithium ions, the advantages of MOFs in avoiding ionic liquid leakage would be counterbalanced. Hence, the utilization of IL in the pores of MOFs and research on its intrinsic ion transport are of greater significance. Micaela et al. prepared layered ILs@MOF films and washed off the excess ILs, and then systematically investigated the mechanism of Li-based ionic liquid conduction in MOF [52]. They found that the conductivity of HKUST-1 MOF was many orders of magnitude higher when containing pure [BMIM][TFSI] ionic liquid inside, indicating that the measured conductivity was mainly attributed to the internal ions. The ion mobility in subsequent tests decreased by two orders of magnitude with the increase loading of ILs. Therefore, researchers believed that the internal anions and cations with a large radius would hinder ion conduction by ion bunching and pore blocking, as reproduced by the molecular dynamics simulations (Figure 3c). Meanwhile, if a more mobile Li^+ is added to the ionic liquid to form [Li0.2BMIM0.8] [TFSI], the full-load ionic conductivity is as high as 70 times that of the original ionic liquid. In addition, researchers demonstrated that the high ionic conductivity originated from the role of ionic liquids by loading samples with only LiTFSI. Li^+ in ILs attenuates ionic bunching as well as pore blocking through the effect of Li-TFSI neutrals. Simulations also revealed that Li^+ in ILs conducts via the typical Grotthuss mechanism, binding with $TFSI^-$ and then releasing to bind with $TFSI^-$ at the next site repeatedly.

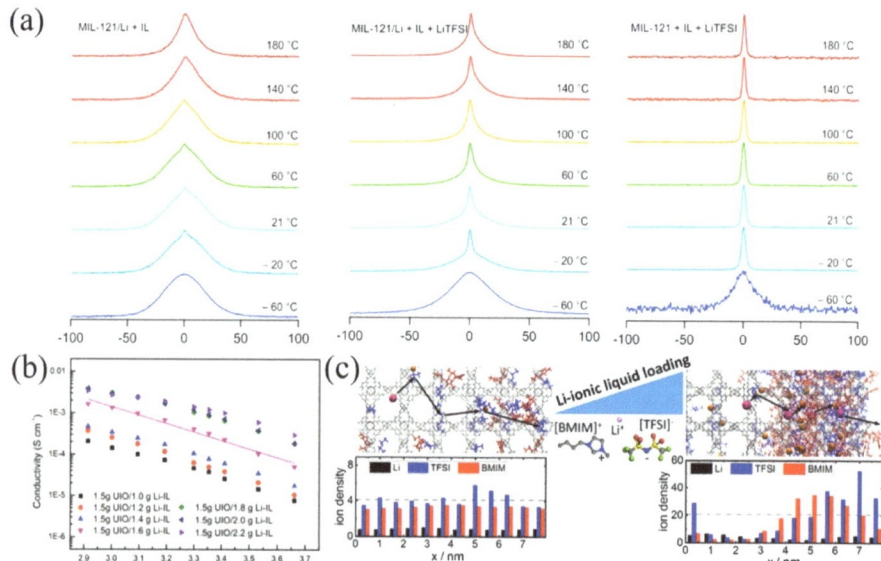

Figure 3. (**a**) The 7Li NMR lines of MOFs hybrid Li^+ with ILs, MOFs hybrid Li^+ with ILs and Li salt and pure MOFs with ILs and Li salt. Reproduced with permission [50]. Copyright 2021, Roman Zettl and Ilie Hanzu. (**b**) Arrhenius plots of the nanostructured MOFs/Li-ILs with different composition. Reproduced with permission [51]. Copyright 2019, Wiley. (**c**) The snapshot of the ions of [Li0.2BMIM0.8][TFSI] IL in HKUST-1 MOF at different loading and the ion distribution of Li-free reference system under identical conditions. Reproduced with permission [52]. Copyright 2021, American Chemical Society.

2.4. Improvement Direction of ILs@MOFs Electrolyte

Although research on the ionic conduction mechanism is not extensive, the general direction for improving ILs@MOFs is clear: enhancing the IL accommodation capacity of MOFs with high stability and striving to control the addition of ILs while ensuring overall conductivity. Among those efforts on the performance improvement of ILs@MOFs electrolyte, the synthesis of MOFs with core-shell structure has been extensively studied. Mai et al. prepared the first MOF-in-MOF core-shell structure of UiO-66@UiO-67 and filled the core-shell structure with the ionic liquid Li-[EMMI][TFSI] [53]. This unique structural design combined the advantages of different types of MOFs, using UiO-67 material with large pore size and high specific surface area as the external bulk of the ionic liquid load, and small pore size UiO-66 material as the inner core to restrict the movement of large ions inside ionic liquid (Figure 4a). Upon the same Li-IL addition, the compact granular electrolyte (CSIL) showed a high room-temperature ionic conductivity of 2.1×10^{-3} S cm^{-1}, which exceeded that of UiO-66 by nearly five times. The t_{Li+} is also twice as high as that of UiO-67, reaching 0.63 (Figure 4b). The researchers found that the fabricated materials did not form ILs@MOFs, which are prone to leakage, but formed nano-wetting interfaces between adjacent nanoparticles in a dry powder state. The prepared liquid-containing nanoparticles displayed excellent thermal, structural, and electrochemical stability and can withstand a compression pressure of 30 MPa, exhibited a thermal degradation temperature of over 360 °C, and showed an oxidative potential of up to 5.2 V. During a cycle stability test under rate of 2C, the Li/CSIL/LiFePO4 (LFP) battery exhibited specific capacity of 158 mAh g^{-1} and capacity retention of 99% after 100 cycles (Figure 4c). Although in-depth investigations regarding the maximum internal liquid capacity of MOFs before and after the formation of the core-shell structure are still required, their pioneering work demonstrating the potential of core-shell structured MOFs as high-performance solid-state electrolyte bodies remains relevant. Based on a similar thought, Tian et al. also proposed MOFs as cores to form the MOF@PIN (polymerized ionic net-work) structure [54]. PIN as a network-like polyconic liquid was used as a shell to adsorb IL to prevent leakage and provide conductive pathways, and HKUST MOFs as a core to support the shell and improve the structural ability to withstand pressure. Compared to solid PIN solids, HKUST@PIN provides internal frame space which obtained an ILs loading of 250% (mass ratio). In contrast to hollow PIN, HKUST@PIN revealed a stronger interaction tendency towards TFSI anions and thus exhibited a higher t_{Li+}. Other researchers have also used MOFs as shells and ceramic particles as cores inside the MOFs to increase the mechanical strength and provide internal ion pathways [55]. On the one hand, ILs were stored in the external HKUST-1 shell as well as in the voids of the reinforcement layer to facilitate the interfacial ion diffusion of the nanomaterials. On the other hand, Li6.75La3Zr1.75Nb0.25O12 (LLZN) ceramic cores not only provide additional Li$^+$ pathways that facilitate ion transport such as vacancy diffusion, interstitial atom diffusion and substitution diffusion [56,57], but also provide a thermal stability of 300 °C and a high modulus of 71.49 MPa for the electrolyte due to its excellent mechanical strength and heat endurance [58]. To illustrate the enhancing effect of ILs@MOFs, the performance of the above-mentioned ILs@MOFs in batteries has been summarized in Table 1.

However, as far as the current studies are concerned, several unsolved problems still exist, including the enhancement of mechanical strength, which remains speculative with indirect evidence, with the actual performance enhancement not as effective as claimed. In view of this, future efforts on improving comprehensive performances of ILs@MOFs need to focus more on the mechanism analysis and the use of more complete argumentation methods. From another point of view, using MOFs-in-Polymer (such as HKUST@PIN), which can combine the advantages of polymer, ILs and MOFs, and ILs@MOFs, as polymer electrolyte filler can also be promising.

Table 1. The conductivity and battery performance of different ILs@MOFs electrolytes.

Type of MOFs	Type of ILs	Room Temperature (R.T.) Conductivity (S/cm)	Electrochemical Oxidation Potential (V)	Li Symmetric Cell Performance	Cycling Capacity and Cathode in Battery	Cycling Performance in Battery	Ref.
MOF-525(Cu)	[EMIM][TFSI]	3.0×10^{-4}	4.1	800 h at 0.02 mA/cm^2 (R.T.)	142 mAh/g (LFP)	93%-100 cycles-0.1 C	[42]
Uio-66	[BMP][TFSI]	3.3×10^{-4}	5.2	2000 h at 0.1 mA/cm^2 (R.T.)	129 mAh/g (LFP)	94.8%-100 cycles-0.1 C	[45]
Uio-67	[EMIM][TFSI]	1.0×10^{-4}	5.2	40 days at 0.1 mA/cm^2 (R.T.)	170 mAh/g (LFP)	97%-150 cycles-0.1 C	[47]
Uio-66	[EMIM][TFSI]	5.0×10^{-4}	/	100 h at 0.2 mA/cm^2 (60 °C)	119 mAh/g (LFP)	94%-380 cycles-1 C	[51]
Uio-66@67	[EMIM][TFSI]	2.1×10^{-3}	5.2	1050 h at 1 mA/cm^2 (R.T.)	158 mAh/g (LFP)	99%-100 cycles-0.2 C	[53]
HKUST-1	[DEME][TFSI]	4.0×10^{-4}	5.5	300 h at 0.1 mA/cm^2 (R.T.)	120 mAh/g (LFP)	108%-100 cycles-0.5 C	[54]
HKUST-1	[EMIM][TFSI]	3.9×10^{-4}	5.2	1800 h at 0.1 mA/cm^2 (R.T.)	129.1 mAh/g (LCO)	98%-500 cycles-0.2 C	[55]

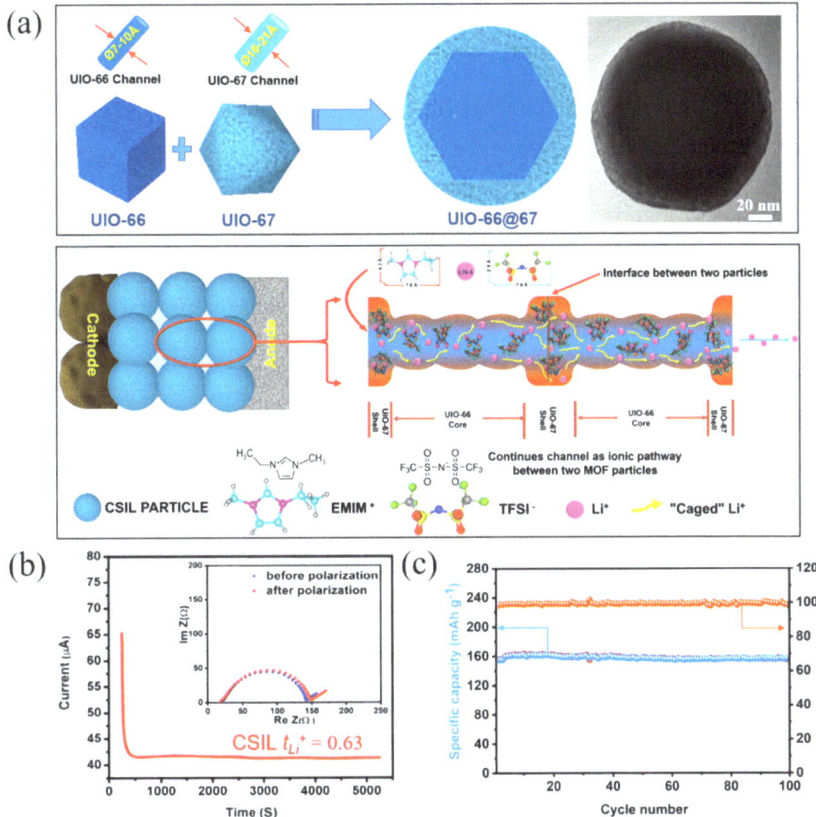

Figure 4. (**a**) Schematic diagram of the architecture UIO-66@67 with corresponding TEM image and continuous ionic pathways within nanowetted interface and particles in UIO-66@67 with Ils (CSIL). Reproduced with permission [53]. Copyright 2021, Wiley. (**b**) DC polarization curve and t_{Li+} of UIO-66@67 with Ils. Reproduced with permission [53]. Copyright 2021, Wiley. (**c**) Room temperature cycling stability (blue line) and Coulombic efficiencies (yellow line) of Li/CSIL/LFP under 0.2C. Reproduced with permission [53]. Copyright 2021, Wiley.

3. ILs@MOFs as Filler of Composite Electrolyte

3.1. Advantages and Bottlenecks of ILs/MOFs in Composite Electrolyte

As an important functional filler of CPEs, MOFs have been widely used [59–62]. On the one hand, the surface metal nodes as well as functional groups can interact with lithium salt and polymer, therefore facilitating the solvation of lithium salt and the Li ion transport through segment movement [63]. The defect of metal nodes may bring ion conduction to the crystal, making MOFs competitive fillers in enhancing ionic conduction efficiency for composite polymer electrolytes. On the other hand, as with traditional filler, rapid interphase conduction between polymer electrolyte and fillers can improve electrochemical performance [64].

Since MOFs have been extensively studied as CPE fillers, the interaction of polymers with ionic liquids has also attracted the attention of researchers. In general, ionic liquids are able to form ILs-Gel gel electrolytes with suitable polymers, acting as plasticizers [65]. To reduce the crystallinity of polyethylene oxide (PEO), ionic liquids [EMIM][TFSI] were added to PEO solid electrolytes to obtain a room-temperature ionic conductivity of up to 1.85×10^{-4} S cm^{-1} by Polu et al. [66]. Considering the narrow electrochemical window of

PEO, Hofmann added 1-ethyl-1-methylpyrrolidinium bis(trifluoromethylsulfonyl)azanide (EMPyrr-TFSA) ILs into PEO, which achieved an electrochemical stabilization window of 0~5.2 V while obtaining a room temperature ionic conductivity at the mS cm^{-1} level [67]. Further substituting PEO matrix with PVDF-HFP, a wider electrochemical stable window of 0~6.2 V can be obtained. Apart from that, researchers also used ionic liquids to replace flammable organic solutions, thus improving the thermal stability and safety of electrolytes [68].

Previous studies revealed that as framework structure compounds rich in OMS, MOFs materials can encapsulate organic solutions to form quasi-solid electrolytes [41,57]. Therefore, the application of MOFs can be extended into polymer electrolytes to act as fillers which would bind ILs (Figure 5). Firstly, the MOFs framework provided suppor to ILs, which prevents the leakage of ILs during tight battery assembling and also enhances the mechanical strength of electrolyte. Secondly, the ILs provide extra ionic transport pathways inside MOFs to enhance conductivity (Figure 6a). Finally, the interaction of MOFs particles and ILs reduces the addition of Ils to achieve approximate performance, which lowers the production cost. Guo et al. synthesized a series of PEO-n-UiO CPE containing Li-[EMIM][TFSI] for the first time in regards to reducing the mass fraction of Ils@MOFs fillers based on the study of Ils@MOFs electrolytes [42,57,69]. Combining SEM, XRD and DSC characterization results, the researchers found that the addition of ILs@MOFs still retained the plasticizer effect of ILs in reducing the crystallinity of PEO. The resulting CPE reached an optimized room temperature ionic conductivity of 1.7×10^{-5} S cm^{-1}, and the Li$^+$ transference number was increased to 0.35 due to the pore absorption effect of MOFs on anions.

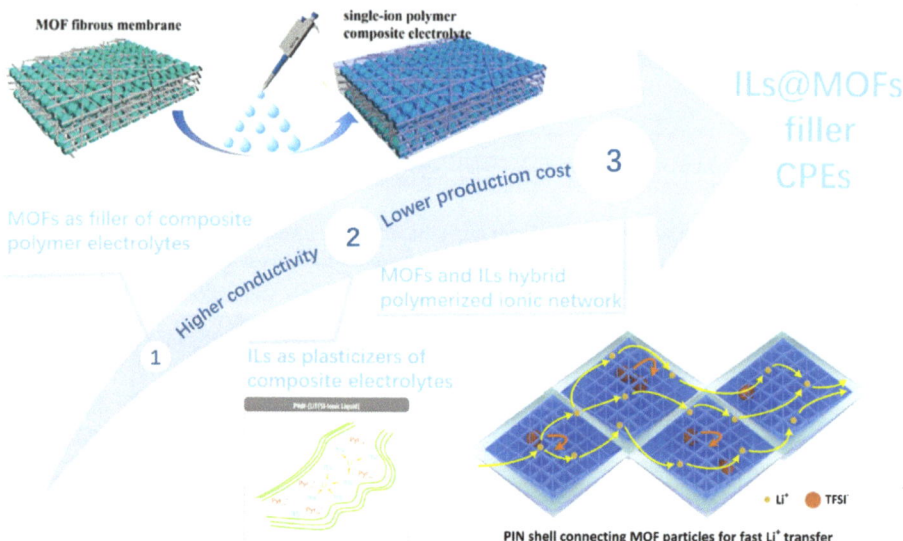

Figure 5. The development from MOFs filler CPEs, ILs plasticized CPEs and MOFs and ILs PIN to ILs @MOFs filler CPEs. Reproduced with permission [54]. Copyright 2023, Elsevier. Reproduced with permission [70]. Copyright 2016, Wiley. Reproduced with permission [71]. Copyright 2022, American Chemical Society.

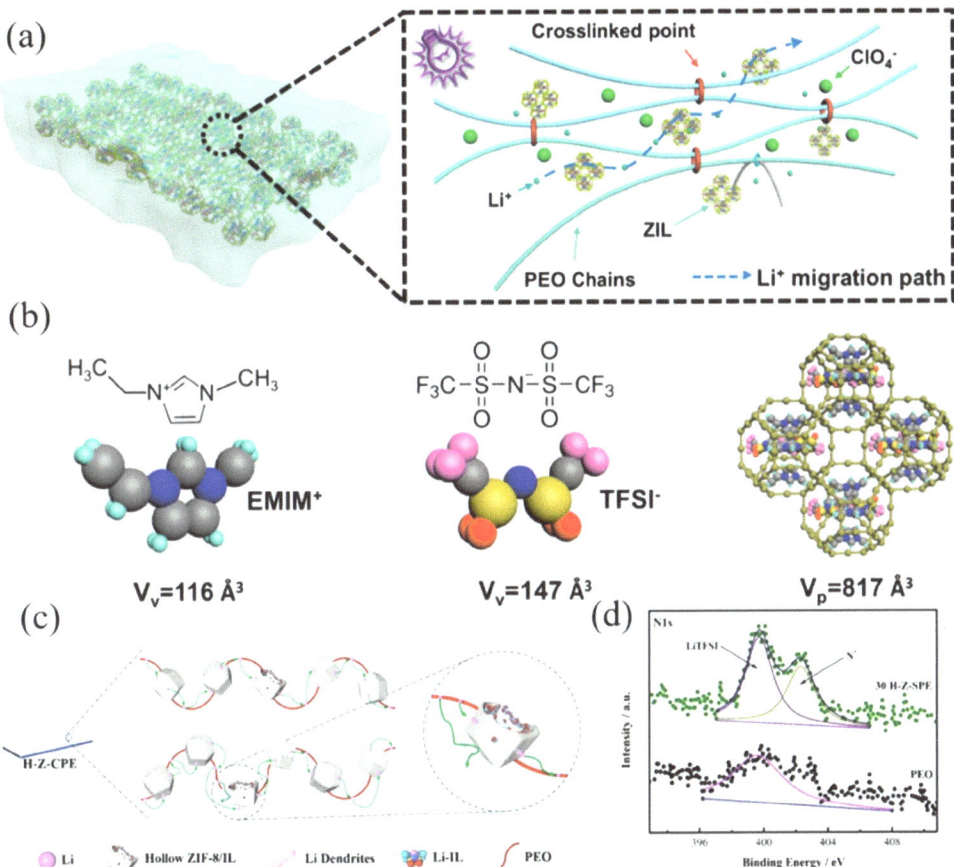

Figure 6. (**a**) The crosslinked composite solid electrolyte and Li$^+$ migration pathway. Reproduced with permission [72]. Copyright 2020, American Chemical Society. (**b**) The van der Waals volumes of EMIM$^+$ and TFSI$^-$ and structure of ZIF-based ionic conductor after incorporating (EMIM0.83Li0.17)TFSI. Reproduced with permission [72]. Copyright 2020, American Chemical Society. (**c**) Schematic illustration of structure of hollow ZIF-8/IL filler and the its function in storage LiTFSI. Reproduced with permission [73]. Copyright 2022, Wiley. (**d**) XPS N 1s spectra of PEO CPE and ILs@MOFs CPE, reflected the extra N$^+$ from [EMIM][TFSI] ILs. Reproduced with permission [73]. Copyright 2022, Wiley.

Mai et al. selected ZIF-8 materials with micropore sizes that matched the size of EMIM$^+$ ions as MOFs framework to accommodate ionic liquids, which obtained a t_{Li+} up to 0.67 [72]. Explanation for this performance enhancement lies in the confining effect of ZIF framework on EMIM$^+$ and TFSI$^-$ in lithium-containing ionic liquids with a large ionic radius, making them difficult to diffuse under the drive of electric field (Figure 6b). Meanwhile, the zinc ion in ZIF-8 which acts as a Lewis acid center, can adsorb TFSI$^-$ based on Lewis acid-base interaction, which immobilizes anion movement and promotes Li$^+$ dissociation at the same time. The Lewis acid site also reduced the crystallinity of polymer by the reduction of ion coupling; therefore, the ion migration ability is enhanced. Consequently, a room-temperature ionic conductivity of 4.26×10^{-4} and a t_{Li+} of 0.67 were achieved simultaneously. In addition to the advantages of ionic liquids themselves being retained, the characteristics of MOFs can be better utilized when combined with polymer. Luo et al. used tannic acid for etching to obtain ZIF-8 with larger size after the introduction of ionic liquids (Figure 6c) [73]. They suggested that the effect of ZIF-8 on ionic liquids not only lay in the adsorption of anions by Lewis acid, but also depended on the interaction between imidazole N of ZIF-8 with N^+ of $EMIM^+$ [74]. This can be confirmed from N 1s XPS spectra of $EMIM^+$ in Figure 6d. Through density functional theory (DFT), researchers found that the interaction made Li^+ inclined to combine with $TFSI^-$ site and thus promoted lithium-ion movement. The Li deposition and stripping stability was also found to be improved in the cycling of symmetric lithium batteries. Conclusions were drawn by the authors that, firstly, the insulate hollow ZIF-8 shell layer inhibits the generation of lithium dendrites; secondly, the hollow inner cavity can accommodate the ionic liquid as well as provide low-potential electron deposition of dead lithium, which alleviates the degradation of polymer matrix; and finally, the rigid shell structure can mechanically block lithium dendrites and prevent penetration.

However, as an inert filler, MOFs material not only provides insufficient conductivity, but is also likely to agglomerate after exceeding the percolation threshold, which would in reverse destroy the continuous transport pathways inside electrolyte [75]. Therefore, preparation methodologies regarding the mixing of ILs@MOFs and polymers have also been investigated by researchers. The traditional mixed coating technique inevitably suffers from particle aggregation, which is caused by the decrease tendency of surface energy. Moreover, a single ILs@MOFs filler may not achieve the complex performance requirements. Exploring novel composite approaches using ionic liquids, MOFs, and polymers can facilitate further enhancement of the performance of composite electrolytes. Tu et al. proposed an in-situ growth approach to prepare CPE containing MOFs material, followed by ionic liquid impregnation (Figure 7a). This ensured the uniform distribution of MOFs in CPE which can be considered a promising way to enhance the comprehensive performance of solid-state lithium batteries [76]. The Li^+ flux can be modified through uniform growth MOFs to obtain homogenized lithium deposition. Yang et al., on the other hand, used an electrostatic spinning technique to form hybrid composite fillers including ILs@MOFs with flame retardant materials [77]. They poured PEO hybrid ILs@MOFs composite polymer in the electrospinning framework structure constructed by polymer fiber and flame-retardant materials, achieving improved flame retardancy and relieved lithium ion concentration (Figure 7b). As shown in Figure 7c, Sun et al. used an electrospray technique combined with electrostatic spinning to prepare highly stable polymer backbones loaded with ILs@MOFs [78]. The electrospray technique can not only retain the integrity of the fiber skeleton, but also ensure the uniform distribution of nanoparticles. Consequently, more preparation methodologies such as in-situ growth, hybrid fillers and electrospray should be developed to solve problems of particle aggregation as well as performance deficiency. Although great promises have been delivered by these applications in the combination of ionic liquids with polymer electrolytes, shortcomings still exist. For example, the compatibility of ionic liquids with polymers generally limits the loading amount of ionic liquids. Excessive addition not only reduces the mechanical strength of the electrolyte, but also results in the leakage of ILs when the cell is assembled compactly with high pressure loading [59,79]. Moreover, even though the ionic conductivity of the electrolyte can rise to excellent levels, previous studies have demonstrated that the overall conductivity in ionic liquids containing lithium salts is mostly contributed by the ionic liquid anion and cation, while the Li^+ transference number is usually lower than 0.3 [80]. This would result in the concentration polarization of effectively conducted lithium ions in the electrolyte, limiting the charge/discharge capability of the battery, especially at a high C-rate [81].

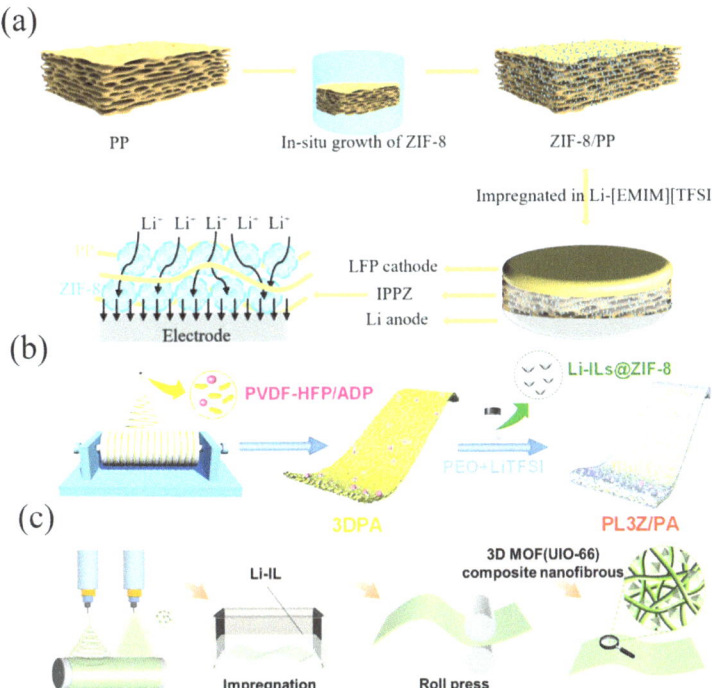

Figure 7. (**a**) The preparation method of in-situ growth ILs@MOFs CPEs, Reproduced with permission [76]. Copyright 2022, American Chemical Society. (**b**) Schematic diagram of electrostatic spinning ILs@MOFs CPEs. Reproduced with permission [82]. Copyright 2022, Elsevier. (**c**) The preparation process of electrospray and electrostatic spinning ILs@MOFs CPEs. Reproduced with permission [83]. Copyright 2022, Elsevier.

3.2. Ionic Transport Mechanism and Development Strategy of ILs@MOFs CPEs

Although many ILs@MOFs polymer electrolytes have been developed, the ion-dipole interaction and ionic transport mechanism behind it still remain to be discovered. In terms of factors influencing the Li$^+$ transport efficiency, the pore size of MOFs, as well as the loading amount of ionic liquid, are concerned, since only the proper pore size can restrict the movement of large ions and ILs exert great influence on polymer chain mobility and carrier concentration. However, although horrow-ILs-ZIF-8 adds more lithium salts to the ionic liquid than ILs-ZIF-8 in order to increase the lithium-ion migration number, the t_{Li+} was still lower than ILs-ZIF-8 (0.41 vs. 0.59) even though the etched hollow ZIF-8 exhibited a pore size nine times larger than the normal ZIF-8 [74,84]. Based on this irregular variation trend, future studies on the relationship between pore size of MOFs and Li$^+$ transfer number of electrolytes in the same framework system are particularly important. Moreover, the loading conditions of ionic liquids in the pores of MOFs mainly depend on the variation of particle pore volume and specific surface area, and if only the single specific surface area variation is considered, the possible impact of ionic liquids wrapping around the outside of MOFs will be ignored. Meanwhile, excess ionic liquids may exist outside MOFs and form a triple layer structure of ILs@MOFs@ILs, which can be identified by thermogravimetry analysis (TGA) in Figure 8a [77]. The exposed ionic liquid layer is not only prone to leaking when added to the polymer electrolyte, but certain decomposition may also occur for ionic liquids with narrow electrochemical windows (Figure 8b) [82]. In the study proposed by Yuan et al., it was found that although ILs@MOFs were able to increase the ionic conductivity of the PEO electrolyte, the lithium-ion transfer number decreased from 0.23 to 0.13 compared with pure PEO (Figure 8c). The reason for this phenomenon mainly lies in exceeding ILs addition in consideration of accommodation limit, which resulted in the release of internal uncoordinated anions accompanied by the increase of polarization degree [52,84]. Therefore, favorable ionic transport capabilities can only be achieved with the right amount of ionic liquid. Unlike traditional CPEs, the addition of ionic liquids increases the analysis complexity of the system from the perspective of multi-components. The role of pure MOFs materials without the addition of ILs acts as both solid plasticizer and Lewis acid site to disrupt the segmental regularity of polymer chain and adsorb anions, and no additional Li$^+$ conduction pathways would be formed inside MOFs due to its intrinsic inert nature. The crucial difference between MOFs and other inert fillers is the ion restriction effect of the MOFs framework structure and the enhancement of the Lewis acid-base interaction provided by the abundant metal sites. Upon the addition of ionic liquids, additional ion transport paths can be taken into account in the composite polymer electrolytes, which is similar to the percolation theory in the active filler contained-CPE category [75], and the highest ion transport capacity can be achieved at the optimum content of ILs@MOFs active fillers. The regulation is depicted as follows: at a low addition of active filler, ion transport mainly occurs in the polymer phase, in which an ion migration enhancement of the polymer chain with decreased crystallinity is achieved. When the filler reaches the optimum content, Li$^+$ conduction routes can interconnect along the continuous percolation interface on the MOFs surface and therefore enhance ionic conductivity through the fast pathway (Figure 8d) [82].

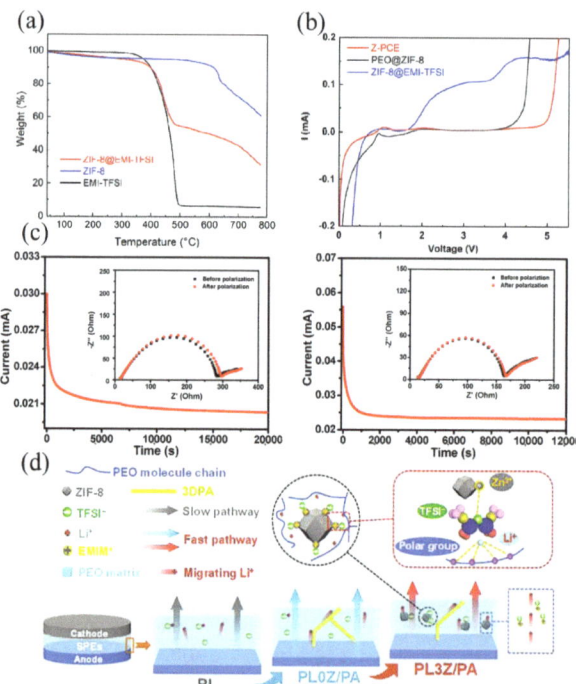

Figure 8. (**a**) TGA curves of ILs@MOFs and IL and MOF while the IL is EMI-TFSI and the MOFs is ZIF-8. Reproduced with permission [82]. Copyright 2022, Elsevier. (**b**) The LSV curve of MOFs ILs CPE, PEO@ZIF-8 and ZIF-8@EMI-TFSI. Reproduced with permission [82]. Copyright 2022, Elsevier. (**c**) DC curves of PEO-only electrolyte and ILS@MOFs CPEs (insert: AC impedance test of symmetric cells before and after polarization). Reproduced with permission [84]. Copyright 2020, American Chemical Society. (**d**) The Li$^+$ conductive mechanism of CPEs. Reproduced with permission [82]. Copyright 2022, Elsevier.

However, the highly conductive ionic transport paths would be blocked when the content of particles continuous to increase and unavoidable aggregation occurs, and the ion transport path through polymer matrix is also diluted by the aggregated fillers. Although ions can be transported through the matrix and filler, the polymer phase and the inorganic phase undergo mutual obstruction, which eventually leads to a decrease in ionic conductivity [85]. In the current study on the addition of ILs@MOFs, Sun et al. found a slight increase in polymer ionic conductivity with the addition of small amounts of ILs@MOFs, and a significant increase in conductivity can be obtained after the loading amount achieved a certain extent, which triggered the generation of continuous ion channels [39]. Mai et al., on the other hand, proposed a solid-liquid-like transport interface mechanism [72]. They concluded that the ILs@MOFs material added in the experiment acts as a high-speed migration path for lithium-ion transport between polymer chains through experimental and computational analysis of diffusion energy barriers. Lithium ions bounded by polymer chain segments are solvated on the surface of ILs@MOFs by $TFSI^-$ to enter the framework. Simultaneous desolvation of an equal amount of lithium ions is transferred to other PEO chains. The lithium ions in ILs@MOFs exhibit significantly stronger mobility, and the electron density in the micropores changes after the addition of ionic liquids. It is difficult to determine both the lithium ion and external lithium salt concentrations and energy levels inside the filler, which poses a new challenge to the application of space charge layer theory. More scientific evaluation methods for compatibility of ionic liquids, MOFs materials, and polymeric substrates, and Li^+ migration paths in the solid-liquid systems of ILs@MOFs are still required for future research applications.

4. Conclusions and Perspectives

In this review, the main advances in hybrid solid-state electrolytes associated with MOFs materials are highlighted. As an important candidate technology for next-generation electrochemical energy storage devices and two application aspect for MOFs-based solid electrolytes are discussed in this paper, including the employment of ILs@MOFs materials as conductive bodies, and ILs@MOFs as composite polymer electrolyte fillers. As an emerging fast ion transport material, ionic liquids in combination with MOFs in framework structures can combine the advantages of ILs and MOFs to enhance ion transport capacity and ion selectivity and improve interfaces while achieving liquid encapsulation to improve the overall structural stability. ILs@MOFs hybrid electrolytes utilize a robust framework to enhance dendrite resistance and restrict internal fluid flow to prevent leakage. The organic framework with metal ions offers an enhanced lithium-ion transference number due to the confining effect on both anions and cations, and the absorption of anions on metal nodes. The internal ionic liquid provides rapid ion transfers within the crystal and intergranular wetting interfaces, which also reduces system flammability. Therefore, the matching of pore size with ions and the adjustment of Lewis acidity can better perform the role of host and guest. Consequently, the possible ion transport mechanism is mainly one-dimensional ion transport within the framework of MOFs, where Li^+ ions dissociate and transfer at the solid-phase interface and the internal solvent. Although the complex ion transport mechanism and precise role of ILs in this system still need deeper investigation, the application of ILs@MOFs has shown promising prospects, which provides a structural reference for new solid electrolytes.

The use of ILs@MOFs materials as fillers for polymers originates from the combination of multifunctional MOFs framework structure and ILs plasticizer with high ionic conductivity. While using ILs@MOFs materials as polymer electrolyte fillers, novel low-energy paths for ion transportation can be formed, in which the solid-liquid-like interfacial conduction mode in the polymer greatly improves the Li^+ transport speed of the electrolyte. Therefore, the overall ionic conduction network is composed of Li^+ transport pathways formed at the interface between MOFs fillers as well as the solvate-desolvate ILs conduction routes inside the polymer matrix, originating from both chain segment movement and interfacial percolation interfacial theory. Moreover, compared to direct IL addition, ILs@MOF also plays a better role in limiting adsorption and stabilization, improving ion selectivity and structural stability effectively. In view of the above-mentioned advantages and prospects, the higher application feasibility of MOFs as solid electrolyte also sheds light on its possible future commercialization.

All in all, ILs@MOFs materials are a promising solid-state electrolyte candidate, either used as bulk material or fillers. Comprehensive performances of ILs@MOFs can be achieved through the complementary individual components including ILs and MOFs, in which ILs provide ion conductivity and surface wettability, and MOFs provide structural support and functional sites. However, the interaction mechanism between ILs and MOFs still requires further research, such as on solvation and desolation behavior at the ILs@MOFs surface. More importantly, the inner ion transportation mechanism of the ILs@MOFs system remains unclear, which needs more illustration in order to guide frontier technologies. Future research can address:

1. The development of new MOFs materials. MOFs materials have powerful modular properties due to the rich variety of inorganic metal centers and bridging organic ligands in combination with various grafting methods. For lithium conduction in Li-ion batteries, the composition of the MOFs material will determine the strength of the encapsulant-frame interaction. The reasonable regulation on Lewis acidity and charge density of MOFs material will effectively improve the ionic conductivity and selectivity of the hybrid electrolyte.

2. Developing evaluation methods for the performance of ILs@MOFs. While the encapsulation of different kinds of ILs in MOFs has become relatively common, the types of ILs and the generated transference properties are not yet well summarized and are still in the mapping stage. Therefore, theoretical calculations of ILs and systematic encapsulation schemes based on theoretical and practical phenomena are particularly important for the systematic development of ILs@MOFs hybrid electrolytes.

3. Deepening the study of key structural factors for MOFs. Due to the complex topology of MOFs materials, the specific surface area, particle size and pore size of the formed structures can have different effects on interaction with ILs and electrochemical properties. The study of these structure-related relationships can better serve the development of new ILs@MOFs materials.

4. Carrying out in-depth theoretical studies on ion transport mechanisms. The mechanism of lithium-ion transport can serve as an important guide for the design of both electrolyte systems with ILs@MOFs as the main body and composite polymer systems with ILs@MOFs as fillers. As a new type of solid electrolyte system, the transport mechanism of ILs@MOFs as the main electrolyte system has not been clarified in relation to the type and amount of inner or external ILs. Thus, the basis of the conduction theory still needs to be clarified. Nonetheless, the transport mechanism regarding MOFs in composite polymer systems can be explained using the common theory of composite electrolyte systems to a certain extent, while the influence of their metal sites and encapsulants on conduction still lacks empirical evidence and remains to be explored. Along with more extensive experiments, basic conduction theory research needs to be developed as soon as possible.

5. Deepening the analysis of electrolyte/electrode interfacial evolution processes. In electrolyte with ILs@MOFs as the main body, the stability of ILs@MOFs in contact with the lithium anode interface directly affects the SEI generation and the growth of dendrites. Therefore, ensuring good contact between the interface of rigid ILs@MOFs materials and electrodes as well as providing high mechanical strength and thermodynamics stability for electrolyte against electrode is urgent. A thorough theoretical study of the multiple interfaces related to ILs@MOFs can better integrate it with the existing theoretical system and facilitate further theoretical design and practical applications.

6. The integrated development of ILs@MOFs in the mass production of electrolytes. In industrialized battery design, factors such as raw material production process, matching degree with present manufacturing procedures, and overall battery performance

need to be considered. ILs@MOFs materials can be synthesized at room temperature or using hydrothermal methods, which can meet commercial large-scale production requirements. Due to the precedent of commercial polymers, electrolyte membranes can be easily prepared through solution-casting technique by the simple mixing and flowing of precursor polymer solution containing ILs@MOFs fillers. However, the current problem is the high price of raw materials such as ionic liquids and the difficulty of ensuring uniformity in large-scale production. The development of sustainable, environmentally friendly, low-cost mass production solutions is of great significance for practical production.

Author Contributions: Conceptualization, R.L.; formal analysis, Y.J.; investigation, R.L.; resources, R.L. and X.Z.; data summary, X.Z.; writing original draft preparation, R.L. and Y.L.; writing—review and editing, Y.J. and Y.X.; supervision, Y.X. All authors have read and agreed to the published version of the manuscript.

Funding: This research received no external funding.

Conflicts of Interest: The authors declare no conflict of interest.

References

1. Amirante, R.; Cassone, E.; Distaso, E.; Tamburrano, P. Overview on recent developments in energy storage: Mechanical, electrochemical and hydrogen technologies. *Energy Convers. Manag.* **2017**, *132*, 372–387. [CrossRef]
2. Lukatskaya, M.R.; Dunn, B.; Gogotsi, Y. Multidimensional materials and device architectures for future hybrid energy storage. *Nat. Commun.* **2016**, *7*, 12647. [CrossRef] [PubMed]
3. Jin, Y.; Zong, X.; Zhang, X.; Jia, Z.; Tan, S.; Xiong, Y. Cathode structural design enabling interconnected ionic/electronic transport channels for high-performance solid-state lithium batteries. *J. Power Sources* **2022**, *530*, 231297. [CrossRef]
4. Li, S.Q.; Wang, K.; Zhang, G.F.; Li, S.N.; Xu, Y.A.; Zhang, X.D.; Zhang, X.; Zheng, S.H.; Sun, X.Z.; Ma, Y.W. Fast Charging Anode Materials for Lithium-Ion Batteries: Current Status and Perspectives. *Adv. Funct. Mater.* **2022**, *32*, 2200796. [CrossRef]
5. Tan, S.J.; Zeng, X.X.; Ma, Q.; Wu, X.W.; Guo, Y.G. Recent Advancements in Polymer-Based Composite Electrolytes for Rechargeable Lithium Batteries. *Electrochem. Energy Rev.* **2018**, *1*, 113–138. [CrossRef]
6. Koerver, R.; Zhang, W.B.; de Biasi, L.; Schweidler, S.; Kondrakov, A.O.; Kolling, S.; Brezesinski, T.; Hartmann, P.; Zeier, W.G.; Janek, J. Chemo-mechanical expansion of lithium electrode materials—on the route to mechanically optimized all-solid-state batteries. *Energy Environ. Sci.* **2018**, *11*, 2142–2158. [CrossRef]
7. Xu, L.; Tang, S.; Cheng, Y.; Wang, K.Y.; Liang, J.Y.; Liu, C.; Cao, Y.C.; Wei, F.; Mai, L.Q. Interfaces in Solid-State Lithium Batteries. *Joule* **2018**, *2*, 1991–2015. [CrossRef]
8. Uemura, T.; Yanai, N.; Kitagawa, S. Polymerization reactions in porous coordination polymers. *Chem. Soc. Rev.* **2009**, *38*, 1228–1236. [CrossRef]
9. Zhang, Q.; Xiao, Y.; Li, Q.; Wang, J.; Guo, S.; Li, X.; Ouyang, Y.; Zeng, Q.; He, W.; Huang, S. Design of thiol-lithium ion interaction in metal-organic framework for high-performance quasi-solid lithium metal batteries. *Dalton Trans.* **2021**, *50*, 2928–2935. [CrossRef]
10. Tao, F.; Tian, L.; Liu, Z.; Cui, R.; Liu, M.; Kang, X.; Liu, Z. A novel lithium-impregnated hollow MOF-based electrolyte realizing an optimum balance between ionic conductivity and the transference number in solid-like batteries. *J. Mater. Chem. A* **2022**, *10*, 14020–14027. [CrossRef]
11. Yang, G.; Chanthad, C.; Oh, H.; Ayhan, I.A.; Wang, Q. Organic-inorganic hybrid electrolytes from ionic liquid-functionalized octasilsesquioxane for lithium metal batteries. *J. Mater. Chem. A* **2017**, *5*, 18012–18019. [CrossRef]
12. Xia, W.; Mahmood, A.; Zou, R.; Xu, Q. Metal-organic frameworks and their derived nanostructures for electrochemical energy storage and conversion. *Energy Environ. Sci.* **2015**, *8*, 1837–1866. [CrossRef]
13. Majid, M.F.; Mohd Zaid, H.F.; Kait, C.F.; Ahmad, A.; Jumbri, K. Ionic Liquid@Metal-Organic Framework as a Solid Electrolyte in a Lithium-Ion Battery: Current Performance and Perspective at Molecular Level. *Nanomaterials* **2022**, *12*, 1076. [CrossRef]
14. Zhou, T.H.; Zhao, Y.; Choi, J.W.; Coskun, A. Ionic Liquid Functionalized Gel Polymer Electrolytes for Stable Lithium Metal Batteries. *Angew. Chem. Int. Ed.* **2021**, *60*, 22791–22796. [CrossRef] [PubMed]
15. Yoo, D.J.; Kim, K.J.; Choi, J.W. The Synergistic Effect of Cation and Anion of an Ionic Liquid Additive for Lithium Metal Anodes. *Adv. Energy Mater.* **2018**, *8*. [CrossRef]
16. Lei, W.Y.; Zhang, C.F.; Qiao, R.; Ravivarma, M.; Chen, H.X.; Ajdari, F.B.; Salavati-Niasari, M.; Song, J.X. Stable Li|LAGP Interface Enabled by Confining Solvate Ionic Liquid in a Hyperbranched Polyanionic Copolymer for NASICON-Based Solid-State Batteries. *ACS Appl. Energy Mater.* **2023**, *6*, 4363–4371. [CrossRef]
17. Wu, S.H.; Guo, Z.W.; Sih, C.J. Enhancing the enantioselectivity of Candida lipase-catalyzed ester hydrolysis via noncovalent enzyme modification. *J. Am. Chem. Soc.* **1990**, *112*, 1990–1995. [CrossRef]

18. Banerjee, R.; Furukawa, H.; Britt, D.; Knobler, C.; O'Keeffe, M.; Yaghi, O.M. Control of Pore Size and Functionality in Isoreticular Zeolitic Imidazolate Frameworks and their Carbon Dioxide Selective Capture Properties. *J. Am. Chem. Soc.* **2009**, *131*, 3875–3877. [CrossRef]
19. Cho, S.H.; Ma, B.Q.; Nguyen, S.T.; Hupp, J.T.; Albrecht-Schmitt, T.E. A metal-organic framework material that functions as an enantioselective catalyst for olefin epoxidation. *Chem. Commun.* **2006**, *24*, 2563–2565. [CrossRef]
20. Torad, N.L.; Hu, M.; Kamachi, Y.; Takai, K.; Imura, M.; Naito, M.; Yamauchi, Y. Facile synthesis of nanoporous carbons with controlled particle sizes by direct carbonization of monodispersed ZIF-8 crystals. *Chem. Commun.* **2013**, *49*, 2521–2523. [CrossRef]
21. Wang, Z.; Zhou, H.; Meng, C.; Zhang, L.; Cai, Y.; Yuan, A. Anion-Immobilized and Fiber-Reinforced Hybrid Polymer Electrolyte for Advanced Lithium-Metal Batteries. *Chemelectrochem* **2020**, *7*, 2660–2664. [CrossRef]
22. Moggach, S.A.; Oswald, I.D.H. Crystallography Under High Pressures. In *21st Century Challenges in Chemical Crystallography I: History and Technical Developments*; Mingos, D.M.P., Raithby, P.R., Eds.; Springer: Cham, Switzerland, 2020; Volume 185, pp. 141–198.
23. AbdelSalam, H.; El-Maghrbi, H.H.; Zahran, F.; Zaki, T. Microwave-assisted production of biodiesel using metal-organic framework $Mg_3(bdc)_3(H_2O)_2$. *Korean J. Chem. Eng.* **2020**, *37*, 670–676. [CrossRef]
24. Zhang, X.; Qiao, J.; Zhao, J.; Xu, D.; Wang, F.; Liu, C.; Jiang, Y.; Wu, L.; Cui, P.; Lv, L.; et al. High-Efficiency Electromagnetic Wave Absorption of Cobalt-Decorated NH_2-UIO-66-Derived Porous ZrO_2/C. *ACS Appl. Mater. Interfaces* **2019**, *11*, 35959–35968. [CrossRef] [PubMed]
25. Krause, S.; Hosono, N.; Kitagawa, S. Chemistry of Soft Porous Crystals: Structural Dynamics and Gas Adsorption Properties. *Angew. Chem.-Int. Ed.* **2020**, *59*, 15325–15341. [CrossRef] [PubMed]
26. Furukawa, H.; Cordova, K.E.; O'Keeffe, M.; Yaghi, O.M. The Chemistry and Applications of Metal-Organic Frameworks. *Science* **2013**, *341*, 1230444. [CrossRef]
27. Miner, E.M.; Park, S.S.; Dinca, M. High Li^+ and Mg^{2+} Conductivity in a Cu-Azolate Metal-Organic Framework. *J. Am. Chem. Soc.* **2019**, *141*, 4422–4427. [CrossRef]
28. Li, H.F.; Wu, P.; Xiao, Y.W.; Shao, M.; Shen, Y.; Fan, Y.; Chen, H.H.; Xie, R.J.; Zhang, W.L.; Li, S.; et al. Metal-Organic Frameworks as Metal Ion Precursors for the Synthesis of Nanocomposites for Lithium-Ion Batteries. *Angew. Chem.-Int. Ed.* **2020**, *59*, 4763–4769. [CrossRef]
29. Li, F.L.; Shao, Q.; Huang, X.Q.; Lang, J.P. Nanoscale Trimetallic Metal-Organic Frameworks Enable Efficient Oxygen Evolution Electrocatalysis. *Angew. Chem.-Int. Ed.* **2018**, *57*, 1888–1892. [CrossRef]
30. Wang, T.; Zhang, X.; Yuan, N.; Sun, C. Molecular design of a metal-organic framework material rich in fluorine as an interface layer for high-performance solid-state Li metal batteries. *Chem. Eng. J.* **2023**, *451*, 138819. [CrossRef]
31. Ben Saad, K.; Hamzaoui, H.; Mohamed, M.M. Ionic conductivity of metallic cations encapsulated in zeolite Y and mordenite. *Mater. Sci. Eng. B* **2007**, *139*, 226–231. [CrossRef]
32. Wiers, B.M.; Foo, M.L.; Balsara, N.P.; Long, J.R. A solid lithium electrolyte via addition of lithium isopropoxide to a metal-organic framework with open metal sites. *J. Am. Chem. Soc.* **2011**, *133*, 14522–14525. [CrossRef] [PubMed]
33. Colomban, P.; Novak, A. Proton transfer and superionic conductivity in solids and gels. *J. Mol. Struct.* **1988**, *177*, 277–308. [CrossRef]
34. Zhao, T.; Kou, W.J.; Zhang, Y.F.; Wu, W.J.; Li, W.P.; Wang, J.T. Laminar composite solid electrolyte with succinonitrile-penetrating metal-organic framework (MOF) for stable anode interface in solid-state lithium metal. *J. Power Sources* **2023**, *554*, 232349. [CrossRef]
35. Colombo, V.; Galli, S.; Choi, H.J.; Han, G.D.; Maspero, A.; Palmisano, G.; Masciocchi, N.; Long, J.R. High thermal and chemical stability in pyrazolate-bridged metal-organic frameworks with exposed metal sites. *Chem. Sci.* **2011**, *2*, 1311–1319. [CrossRef]
36. Sumida, K.; Her, J.-H.; Dinca, M.; Murray, L.J.; Schloss, J.M.; Pierce, C.J.; Thompson, B.A.; FitzGerald, S.A.; Brown, C.M.; Long, J.R. Neutron Scattering and Spectroscopic Studies of Hydrogen Adsorption in $Cr_3(BTC)_2$-A Metal-Organic Framework with Exposed Cr^{2+} Sites. *J. Phys. Chem. C* **2011**, *115*, 8414–8421. [CrossRef]
37. Bloch, E.D.; Queen, W.L.; Krishna, R.; Zadrozny, J.M.; Brown, C.M.; Long, J.R. Hydrocarbon Separations in a Metal-Organic Framework with Open Iron(II) Coordination Sites. *Science* **2012**, *335*, 1606–1610. [CrossRef]
38. Zhou, Q.; Boyle, P.D.; Malpezzi, L.; Mele, A.; Shin, J.-H.; Passerini, S.; Henderson, W.A. Phase Behavior of Ionic Liquid–LiX Mixtures: Pyrrolidinium Cations and TFSI- Anions—Linking Structure to Transport Properties. *Chem. Mater.* **2011**, *23*, 4331–4337. [CrossRef]
39. Liu, L.; Sun, C. Flexible Quasi-Solid-State Composite Electrolyte Membrane Derived from a Metal-Organic Framework for Lithium-Metal Batteries. *Chemelectrochem* **2020**, *7*, 707–715. [CrossRef]
40. Fujie, K.; Yamada, T.; Ikeda, R.; Kitagawa, H. Introduction of an Ionic Liquid into the Micropores of a Metal-Organic Framework and Its Anomalous Phase Behavior. *Angew. Chem. Int. Ed.* **2014**, *53*, 11302–11305. [CrossRef]
41. Fujie, K.; Ikeda, R.; Otsubo, K.; Yamada, T.; Kitagawa, H. Lithium Ion Diffusion in a Metal–Organic Framework Mediated by an Ionic Liquid. *Chem. Mater.* **2015**, *27*, 7355–7361. [CrossRef]
42. Wang, Z.; Tan, R.; Wang, H.; Yang, L.; Hu, J.; Chen, H.; Pan, F. A Metal-Organic-Framework-Based Electrolyte with Nanowetted Interfaces for High-Energy-Density Solid-State Lithium Battery. *Adv. Mater.* **2018**, *30*, 1704436. [CrossRef]
43. Liu, H.; Cheng, X.-B.; Huang, J.-Q.; Yuan, H.; Lu, Y.; Yan, C.; Zhu, G.-L.; Xu, R.; Zhao, C.-Z.; Hou, L.-P.; et al. Controlling Dendrite Growth in Solid-State Electrolytes. *ACS Energy Lett.* **2020**, *5*, 833–843. [CrossRef]

44. Liu, Y.D.; Liu, Q.; Xin, L.; Liu, Y.Z.; Yang, F.; Stach, E.A.; Xie, J. Making Li-metal electrodes rechargeable by controlling the dendrite growth direction. *Nat. Energy* **2017**, *2*, 17083. [CrossRef]
45. Wang, K.; Yang, L.Y.; Wang, Z.Q.; Zhao, Y.; Wang, Z.J.; Han, L.; Song, Y.L.; Pan, F. Enhanced lithium dendrite suppressing capability enabled by a solid-like electrolyte with different-sized nanoparticles. *Chem. Commun.* **2018**, *54*, 13060–13063. [CrossRef]
46. Liu, Z.; Hu, Z.; Jiang, X.; Wang, X.; Li, Z.; Chen, Z.; Zhang, Y.; Zhang, S. Metal-Organic Framework Confined Solvent Ionic Liquid Enables Long Cycling Life Quasi-Solid-State Lithium Battery in Wide Temperature Range. *Small* **2022**, *18*, 2203011. [CrossRef] [PubMed]
47. Wang, Z.; Wang, Z.; Yang, L.; Wang, H.; Song, Y.; Han, L.; Yang, K.; Hu, J.; Chen, H.; Pan, F. Boosting interfacial Li+ transport with a MOF-based ionic conductor for solid-state batteries. *Nano Energy* **2018**, *49*, 580–587. [CrossRef]
48. Kachmar, A.; Carignano, M.; Laino, T.; Iannuzzi, M.; Hutter, J. Mapping the Free Energy of Lithium Solvation in the Protic Ionic Liquid Ethylammonuim Nitrate: A Metadynamics Study. *Chemsuschem* **2017**, *10*, 3083–3090. [CrossRef]
49. Kachmar, A.; Goddard, W.A. Free Energy Landscape of Sodium Solvation into Graphite. *J. Phys. Chem. C* **2018**, *122*, 20064–20072. [CrossRef]
50. Zettl, R.; Hanzu, I. The Origins of Ion Conductivity in MOF-Ionic Liquids Hybrid Solid Electrolytes. *Front. Energy Res.* **2021**, *9*, 714698. [CrossRef]
51. Wu, J.-F.; Guo, X. Nanostructured Metal-Organic Framework (MOF)-Derived Solid Electrolytes Realizing Fast Lithium Ion Transportation Kinetics in Solid-State Batteries. *Small* **2019**, *15*, 1804913. [CrossRef]
52. Vazquez, M.; Liu, M.; Zhang, Z.; Chandresh, A.; Kanj, A.B.; Wenzel, W.; Heinke, L. Structural and Dynamic Insights into the Conduction of Lithium-Ionic-Liquid Mixtures in Nanoporous Metal-Organic Frameworks as Solid-State Electrolytes. *Acs Appl. Mater. Interfaces* **2021**, *13*, 21166–21174. [CrossRef] [PubMed]
53. Abdelmaoula, A.E.; Shu, J.; Cheng, Y.; Xu, L.; Zhang, G.; Xia, Y.; Tahir, M.; Wu, P.; Mai, L. Core-Shell MOF-in-MOF Nanopore Bifunctional Host of Electrolyte for High-Performance Solid-State Lithium Batteries. *Small Methods* **2021**, *5*, 2100508. [CrossRef] [PubMed]
54. Tian, X.L.; Yi, Y.K.; Wu, Z.D.; Cheng, G.Y.; Zheng, S.T.; Fang, B.R.; Wang, T.; Shchukin, D.G.; Hai, F.; Guo, J.Y.; et al. Ionic liquid confined in MOF/polymerized ionic network core-shell host as a solid electrolyte for lithium batteries. *Chem. Eng. Sci.* **2023**, *266*, 118271. [CrossRef]
55. He, C.; Sun, J.; Hou, C.; Zhang, Q.; Li, Y.; Li, K.; Wang, H. Sandwich-structural ionogel electrolyte with core-shell ionic-conducting nanocomposites for stable Li metal battery. *Chem. Eng. J.* **2023**, *451*, 138993. [CrossRef]
56. Gao, Z.H.; Sun, H.B.; Fu, L.; Ye, F.L.; Zhang, Y.; Luo, W.; Huang, Y.H. Promises, Challenges, and Recent Progress of Inorganic Solid-State Electrolytes for All-Solid-State Lithium Batteries. *Adv. Mater.* **2018**, *30*, 1705702. [CrossRef]
57. Jiang, T.L.; He, P.G.; Wang, G.X.; Shen, Y.; Nan, C.W.; Fan, L.Z. Solvent-Free Synthesis of Thin, Flexible, Nonflammable Garnet-Based Composite Solid Electrolyte for All-Solid-State Lithium Batteries. *Adv. Energy Mater.* **2020**, *10*, 1903376. [CrossRef]
58. Lu, W.; Xue, M.; Zhang, C. Modified $Li_7La_3Zr_2O_{12}$ (LLZO) and LLZO-polymer composites for solid-state lithium batteries. *Energy Storage Mater.* **2021**, *39*, 108–129. [CrossRef]
59. Liu, S.L.; Liu, W.Y.; Ba, D.L.; Zhao, Y.Z.; Ye, Y.H.; Li, Y.Y.; Liu, J.P. Filler-Integrated Composite Polymer Electrolyte for Solid-State Lithium Batteries. *Adv. Mater.* **2023**, *35*, 2110423. [CrossRef]
60. Liu, X.; Li, X.; Li, H.; Wu, H.B. Recent Progress of Hybrid Solid-State Electrolytes for Lithium Batteries. *Chem.-A Eur. J.* **2018**, *24*, 18293–18306. [CrossRef]
61. Zheng, Y.; Yao, Y.; Ou, J.; Li, M.; Luo, D.; Dou, H.; Li, Z.; Amine, K.; Yu, A.; Chen, Z. A review of composite solid-state electrolytes for lithium batteries: Fundamentals, key materials and advanced structures. *Chem. Soc. Rev.* **2020**, *49*, 8790–8839. [CrossRef]
62. Xu, Y.; Zhao, F.; Fang, J.; Liang, Z.; Gao, L.; Bian, J.; Zhu, J.; Zhao, Y. Metal-organic framework (MOF)-incorporated polymeric electrolyte realizing fast lithium-ion transportation with high Li+ transference number for solid-state batteries. *Front. Chem.* **2022**, *10*, 1013965. [CrossRef]
63. Huo, H.; Wu, B.; Zhang, T.; Zheng, X.; Ge, L.; Xu, T.; Guo, X.; Sun, X. Anion-immobilized polymer electrolyte achieved by cationic metal-organic framework filler for dendrite-free solid-state batteries. *Energy Storage Mater.* **2019**, *18*, 59–67. [CrossRef]
64. Liu, W.; Lee, S.W.; Lin, D.C.; Shi, F.F.; Wang, S.; Sendek, A.D.; Cui, Y. Enhancing ionic conductivity in composite polymer electrolytes with well-aligned ceramic nanowires. *Nat. Energy* **2017**, *2*, 17035. [CrossRef]
65. Ye, Y.S.; Rick, J.; Hwang, B.J. Ionic liquid polymer electrolytes. *J. Mater. Chem. A* **2013**, *1*, 2719–2743. [CrossRef]
66. Polu, A.R.; Rhee, H.-W. Ionic liquid doped PEO-based solid polymer electrolytes for lithium-ion polymer batteries. *Int. J. Hydrog. Energy* **2017**, *42*, 7212–7219. [CrossRef]
67. Hofmann, A.; Schulz, M.; Hanemann, T. Gel electrolytes based on ionic liquids for advanced lithium polymer batteries. *Electrochim. Acta* **2013**, *89*, 823–831. [CrossRef]
68. Grewal, M.S.; Tanaka, M.; Kawakami, H. Solvated Ionic-Liquid Incorporated Soft Flexible Cross-Linked Network Polymer Electrolytes for Safer Lithium Ion Secondary Batteries. *Macromol. Chem. Phys.* **2022**, *223*, 2100317. [CrossRef]
69. Chen, L.; Xue, P.; Liang, Q.; Liu, X.; Tang, J.; Li, J.; Liu, J.; Tang, M.; Wang, Z. A Single-Ion Polymer Composite Electrolyte Via In Situ Polymerization of Electrolyte Monomers into a Porous MOF-Based Fibrous Membrane for Lithium Metal Batteries. *ACS Appl. Energy Mater.* **2022**, *5*, 3800–3809. [CrossRef]
70. Wu, J.-F.; Guo, X. MOF-derived nanoporous multifunctional fillers enhancing the performances of polymer electrolytes for solid-state lithium batteries. *J. Mater. Chem. A* **2019**, *7*, 2653–2659. [CrossRef]

71. Osada, I.; de Vries, H.; Scrosati, B.; Passerini, S. Ionic-Liquid-Based Polymer Electrolytes for Battery Applications. *Angew. Chem.-Int. Ed.* **2016**, *55*, 500–513. [CrossRef]
72. Xia, Y.; Xu, N.; Du, L.; Cheng, Y.; Lei, S.; Li, S.; Liao, X.; Shi, W.; Xu, L.; Mai, L. Rational Design of Ion Transport Paths at the Interface of Metal-Organic Framework Modified Solid Electrolyte. *ACS Appl. Mater. Interfaces* **2020**, *12*, 22930–22938. [CrossRef] [PubMed]
73. Hu, Y.; Feng, T.; Xu, L.; Zhang, L.; Luo, L. A Hollow Porous Metal-Organic Framework Enabled Polyethylene Oxide Based Composite Polymer Electrolytes for All-Solid-State Lithium Batteries. *Batter. Supercaps* **2022**, *5*, e202100303. [CrossRef]
74. Kinik, F.P.; Altintas, C.; Balci, V.; Koyuturk, B.; Uzun, A.; Keskin, S. [BMIM][PF6] Incorporation Doubles CO_2 Selectivity of ZIF-8: Elucidation of Interactions and Their Consequences on Performance. *ACS Appl. Mater. Interfaces* **2016**, *8*, 30992–31005. [CrossRef] [PubMed]
75. Wang, W.; Yi, E.; Fici, A.J.; Laine, R.M.; Kieffer, J. Lithium Ion Conducting Poly(ethylene oxide)-Based Solid Electrolytes Containing Active or Passive Ceramic Nanoparticles. *J. Phys. Chem. C* **2017**, *121*, 2563–2573. [CrossRef]
76. Qi, X.; Cai, D.; Wang, X.; Xia, X.; Gu, C.; Tu, J. Ionic Liquid-Impregnated ZIF-8/Polypropylene Solid-like Electrolyte for Dendrite-free Lithium-Metal Batteries. *ACS Appl. Mater. Interfaces* **2022**, *14*, 6859–6868. [CrossRef] [PubMed]
77. Zheng, X.; Wu, J.; Chen, J.; Wang, X.; Yang, Z. 3D flame-retardant skeleton reinforced polymer electrolyte for solid-state dendrite-free lithium metal batteries. *J. Energy Chem.* **2022**, *71*, 174–181. [CrossRef]
78. Sun, M.; Li, J.; Yuan, H.; Zeng, X.; Lan, J.; Yu, Y.; Yang, X. Fast Li$^+$ transport pathways of quasi-solid-state electrolyte constructed by 3D MOF composite nanofibrous network for dendrite- free lithium metal battery. *Mater. Today Energy* **2022**, *29*, 101117. [CrossRef]
79. Long, L.Z.; Wang, S.J.; Xiao, M.; Meng, Y.Z. Polymer electrolytes for lithium polymer batteries. *J. Mater. Chem. A* **2016**, *4*, 10038–10069. [CrossRef]
80. Frömling, T.; Kunze, M.; Schönhoff, M.; Sundermeyer, J.; Roling, B. Enhanced Lithium Transference Numbers in Ionic Liquid Electrolytes. *J. Phys. Chem. B* **2008**, *112*, 12985–12990. [CrossRef]
81. Diederichsen, K.M.; McShane, E.J.; McCloskey, B.D. Promising Routes to a High Li+ Transference Number Electrolyte for Lithium Ion Batteries. *ACS Energy Lett.* **2017**, *2*, 2563–2575. [CrossRef]
82. Shen, J.; Lei, Z.; Wang, C. An ion conducting ZIF-8 coating protected PEO based polymer electrolyte for high voltage lithium metal batteries. *Chem. Eng. J.* **2022**, *447*, 137503. [CrossRef]
83. Wang, Z.; Zhou, H.; Meng, C.; Xiong, W.; Cai, Y.; Hu, P.; Pang, H.; Yuan, A. Enhancing Ion Transport: Function of Ionic Liquid Decorated MOFs in Polymer Electrolytes for All-Solid-State Lithium Batteries. *ACS Appl. Energy Mater.* **2020**, *3*, 4265–4274. [CrossRef]
84. Zhang, X.-L.; Shen, F.-Y.; Long, X.; Zheng, S.; Ruan, Z.; Cai, Y.-P.; Hong, X.-J.; Zheng, Q. Fast Li$^+$ transport and superior interfacial chemistry within composite polymer electrolyte enables ultra-long cycling solid-state Li-metal batteries. *Energy Storage Mater.* **2022**, *52*, 201–209. [CrossRef]
85. Fu, J.L.; Li, Z.; Zhou, X.Y.; Guo, X. Ion transport in composite polymer electrolytes. *Mater. Adv.* **2022**, *3*, 3809–3819. [CrossRef]

Disclaimer/Publisher's Note: The statements, opinions and data contained in all publications are solely those of the individual author(s) and contributor(s) and not of MDPI and/or the editor(s). MDPI and/or the editor(s) disclaim responsibility for any injury to people or property resulting from any ideas, methods, instructions or products referred to in the content.

Article

Epoxy Resin-Reinforced F-Assisted Na$_3$Zr$_2$Si$_2$PO$_{12}$ Solid Electrolyte for Solid-State Sodium Metal Batteries

Yao Fu, Dangling Liu, Yongjiang Sun, Genfu Zhao and Hong Guo *

International Joint Research Center for Advanced Energy Materials of Yunnan Province, Yunnan Key Laboratory of Carbon Neutrality and Green Low-Carbon Technologies, School of Materials and Energy, Yunnan University, Kunming 650091, China
* Correspondence: guohong@ynu.edu.cn

Abstract: Solid sodium ion batteries (SIBs) show a significant amount of potential for development as energy storage systems; therefore, there is an urgent need to explore an efficient solid electrolyte for SIBs. Na$_3$Zr$_2$Si$_2$PO$_{12}$ (NZSP) is regarded as one of the most potential solid-state electrolytes (SSE) for SIBs, with good thermal stability and mechanical properties. However, NZSP has low room temperature ionic conductivity and large interfacial impedance. F$^-$ doped NZSP has a larger grain size and density, which is beneficial for acquiring higher ionic conductivity, and the composite system prepared with epoxy can further improve density and inhibit Na dendrite growth. The composite system exhibits an outstanding Na$^+$ conductivity of 0.67 mS cm^{-1} at room temperature and an ionic mobility number of 0.79. It also has a wider electrochemical stability window and cycling stability.

Keywords: sodium ion battery; epoxy-NZSPF$_{0.7}$; composite solid electrolyte; solid-state electrolyte

1. Introduction

Energy storage systems (ESSs) are gradually becoming essential and important in people's daily lives, as these can provide us with convenience in many aspects [1–4]. As a hopeful substitute for lithium-ion batteries (LIBs), SIBs have caught the attention of a number of researchers recently because of their rich sodium resources, low prices and excellent sustainability. However, because sodium is more reactive than lithium, it is more likely to form dendrites in the conventional liquid electrolyte battery system, posing serious efficiency problems and safety hazards. Hence, research on solid or quasi-solid sodium ion batteries is of great importance for improving battery efficiency and safety [5–16].

To date, different types of sodium materials, such as Na-β''-Al$_2$O$_3$, sulfides, polymers, and Na superionic conductor (NASICON) have been reported for use as sodium ion solid-state electrolytes. As a solid electrolyte, Na-β''-Al$_2$O$_3$ is now successfully used in Na-S batteries; however, it is sensitive to moisture [17] and has a high preparation temperature, which poses some limitations to its production applications. Most sulfide based solid electrolytes are limited in their application due to their instability in air [18] and narrow electrochemical stability window [19], despite their high ionic conductivity and good ductility. The polymer solid electrolyte is flexible, and the contact between it and the electrode is flexible, which makes it malleable and easy to process and shape. However, the ionic conductivity and ion transfer number dose not meet the imposed requirements when using at room temperature. To achieve ionic conductivity for battery applications, a temperature of 60 °C or higher is required. This temperature approaches the melting point of anode, Na (97 °C), which can cause safety problems. By comparing with the sodium-ion solid electrolyte above, a significant amount of attention has been devoted to NASICON (Na$_{1+x}$Zr$_2$Si$_x$P$_{3-x}$O$_{12}$) because of its high ionic conductivity, wide window of electrochemical stability and stability in air. Hong [20] and Goodenough [21] were the first to study Na$_{1+x}$Zr$_2$Si$_x$P$_{3-x}$O$_{12}$. The highest ionic conductivity was 10^{-4} S cm^{-1} (x = 2) at room temperature [20,22]. It has also been further improved by doping modifications

Citation: Fu, Y.; Liu, D.; Sun, Y.; Zhao, G.; Guo, H. Epoxy Resin-Reinforced F-Assisted Na$_3$Zr$_2$Si$_2$PO$_{12}$ Solid Electrolyte for Solid-State Sodium Metal Batteries. *Batteries* 2023, 9, 331. https://doi.org/10.3390/batteries9060331

Academic Editors: Chunwen Sun, Siqi Shi, Yongjie Zhao, Yong-Joon Park and Johan E. ten Elshof

Received: 11 April 2023
Revised: 5 June 2023
Accepted: 12 June 2023
Published: 19 June 2023

Copyright: © 2023 by the authors. Licensee MDPI, Basel, Switzerland. This article is an open access article distributed under the terms and conditions of the Creative Commons Attribution (CC BY) license (https://creativecommons.org/licenses/by/4.0/).

of NASICON. The rare earth element La-doped $Na_3Zr_2Si_2PO_{12}$ has an ionic conductivity comparable with that of ionic liquid electrolytes (10^{-1} S cm^{-1}) [23].

In recent years, several studies have reported anion-assisted ways to increase the room temperature's ionic conductivity [23–25]. Lu et al. modified the $Li_{6.25}Ga_{0.25}La_3Zr2O_{12}$ crystals via F$^-$ assisted synthesis to generate smoother and quicker diffusion channels for Li$^+$, thus improving the ionic conductivity [23]. Li et al. reported that anion-substituted $Li_{3x}La_{2/3-x}TiO_3$ has a higher ionic conductivity [24]. Goodenough reports that the substitution of F for OH, which allows $Li_2(OH)_{0.9}F_{0.1}Cl$ to have an improved Li$^+$ diffusion path [26]. Although anion-assisted NZSPs have high ionic conductivity, they still do not provide effective inhibition of dendrite growth, resulting in uneven Na plating or streaking, leading to poor cell performance. This is because voids still exist on the surface of the SSE, leading to the uneven deposition of Na on the electrode surface and promoting the growth of dendrites. Here, we investigated F$^-$-assisted NZSP solid electrolyte materials. It was found that the grain size and densities of NZSP increased with the introduction of F. However, it is still not enough to inhibit the growth of Na dendrites. Then, we prepared a composite system by depositing epoxy into the pores of NZSPF via vacuum adsorption. The composite system greatly modified the cycling performance of the cell with almost no decrease in ionic conductivity and no increase in impedance. The ionic conductivity of epoxy-NZSPF$_{0.7}$ is 0.67 mS cm^{-1} (NZSPF$_{0.7}$ is 0.95 mS cm^{-1}). In addition, epoxy-NZSPF$_{0.7}$ has a wider electrochemical stability window. Finally, we also assembled a Na | epoxy-NZSPF$_{0.7}$ | $Na_3V_2(PO_4)_3$ quasi-solid SIB, with good cycling performance at 40 °C.

2. Materials and Methods

2.1. Synthesis of NZSPF$_x$ Solid Electrolyte

Traditional solid-phase reactions were adopted to synthesize F$^-$-assisted NZSPF$_x$ materials. Na_2CO_3, ZrO_2, SiO_2, and $NH_4H_2PO_4$ were weighed according to certain stoichiometric ratios and then x mol of NaF (x = 0, 0.1, 0.3, 0.5, 0.7, and 1.0) was added separately. The final results were labeled NZSPF$_x$. Balls, materials and ethanol were added to the ball mill tank in the ratio of 32:5:8. The milling was then carried out with a planetary ball mill at 400 rpm for 10 h. The 12 h drying of precursors was carried out under vacuum at 80 °C; then, the mixtures were transferred to a muffle furnace for 12 h preheating at 900 °C, followed by 3 h sintering at 1050 °C. The sintered powders were further ball-milled at 200 rpm for 10 h to obtain a uniform powder size. During the ball-milling process, a 5% mass fraction of PVA solution was added to act as a binder. The pressing of powers into pellets (diameter 16 mm, thickness 1 mm) was carried out at 120 MPa. Finally, the 4 h sintering of pellets was carried out at 800 °C to remove the PVA solution. After cooling to room temperature, the 24 h sintering of pellets was performed at 1100 °C to obtain NZSPF$_x$ solid electrolyte.

2.2. Synthesis of Epoxy-NZSPF$_{0.7}$ Solid Electrolyte

The DGEBA (diglycidyl ether of bisphenol-A) and PACM (4,4'-diaminodicyclohexylmethane) were dissolved in THF in stoichiometric ratios to obtain 1 mol L^{-1} solutions, respectively. Then, the two solutions were mixed in a volume ratio of 2:1 (DGEBA:PACM) and stirred for 1 h. The sintered NZSPF ceramic sheets in 2.1 were completely immersed in the mixed solution, kept in a vacuum environment for 20 min and cycled three times. The sintering of pellets was carried out at 150 °C for 24 h to obtain an epoxy-NZSPF$_x$ solid electrolyte.

2.3. Characterization and Measurements

A Bruker D8 Advance X-ray diffractometer configured with Cu Kα radiation in the 2θ scope of 10–60° was adopted to collect the X-ray diffraction (XRD) patterns, and the collection of data was carried out at 5° min^{-1}. The chemical component in an Escalab 250Xi instrument from Thermo Scientific configured with an Al Kα micro-focused X-ray source was demonstrated via X-ray photoelectron spectroscopy (XPS) tests. The collection of Fourier transform infrared (FTIR) spectra were carried out on a Nicoletis 10 spectrometer from Thermo Scientific. The spectrum was from 4000 cm^{-1} to 400 cm^{-1}. The microstructures of SSE were studied using scanning electron microscopy (SEM) (QUNATA-FEG). The measurement of thermostability was carried out via thermogravimetric analysis (TGA) on a Netzch STA449F3 analyzer under N_2 conditions at a heating rate of 10 °C/min from 25 to 700 °C.

A CHI660E (ChenHua) electrochemical workstation with an AC amplitude of 10 Mv and a frequency scope from 10 Hz to 10^6 Hz was adopted to make electrochemical impedance spectroscopy (EIS) tests at room temperature. A combination of direct-current (DC) polarization and alternating-current (AC) impedance was used for the ion transference number test. A CR2032 cell case was chosen to assemble a Na/NZSPF$_x$/Na symmetric battery. AN AC impedance test is performed on the battery before the DC polarization test; then, a small bias voltage (5 mV) is added to the battery for the DC polarization test, and after the planned current of the battery is stabilized, the battery is then tested for AC impedance again. The electrochemical stability windows of NZSPF$_{0.7}$ and epoxy-NZSPF$_{0.7}$ solid electrolytes were evaluated with cyclic voltammetry (CV) and linear sweep voltammetry (LSV). The tests were performed on the Au | NZSPF$_{0.7}$ | Na and Au | epoxy-NZSPF$_{0.7}$ | Na cells and the scanning rate was 0.2 mV s^{-1} at room temperature. The electrochemical stability test was carried out to test the interfacial stability and Na stripping–plating behavior. The testing of assembled Na | NZSPF$_x$ | Na and Na | epoxy-NZSPF$_{0.7}$ | Na symmetric cells was carried out on a Blue Power Test System CT2001A with a constant current density of 0.1 mA cm^{-1} at 40 °C.

The assembly of the full cell was carried out in an argon atmosphere glove box using Na metal as the anode, NZSPF$_{0.7}$ and epoxy-NZSPF$_{0.7}$ as the solid electrolyte, and $Na_3V_2(PO_4)_3$ as the cathode using a 2032 cell case. For the preparation of the cathode electrode sheet, $Na_3V_2(PO_4)_3$, the mixture of acetylene black carbon and PVDF was carried out in NMP with a mass ratio of 7:2:1. Using Al foil as a fluid collector, the coasting of the slurry was carried out on the surface of Al foil, followed by 12 h OF drying under vacuum at 80 °C to eliminate the NMP solvent. The Na metal, solid electrolyte and cathode pole piece were assembled in the order of Na | epoxy-NZSPF0.7 | $Na_3V_2(PO_4)_3$. The addition of 10 μL liquid electrolyte (ethylene carbonate (EC) as the solvent and $NaClO_4$ as the salt) was conducted between the solid electrolyte and the pole pieces to wet the contact surface. The constant current charge/discharge test of the batteries was performed on the LAND CT2001A test system at 40 °C to evaluate the long-cycle performance.

3. Results and Discussion

To investigate the influence of adding F$^-$, we used XRD to measure the lattice structure of samples with different fluorine contents. According to Figure S1, the main diffraction peaks of NZSPF$_x$ (x = 0, 0.1, 0.3, 0.5, 0.7, and 1.0) are consistent with the monoclinic NZSP structure [27]. This phenomenon proves that the introduction of F$^-$ did not destroy the crystal structure. However, with the increase in NaF content, a ZrO_2 secondary phase was detected. This may be due to the reaction of Si^{4+}/P^{5+} with F during the process of high-temperature sintering, as it leads to a reduction in Si^{4+}/P^{5+} content and, finally, the deposition of the ZrO_2 secondary phase. In addition, no fluorine-related phases were observed in the XRD patterns of all samples, which could be due to F occupying positions in the lattice dot matrix or high-temperature volatilization.

These featured peaks of P-O and Si-O groups (Figure S2) were analyzed with FTIR spectroscopy. All samples displayed the same stretching or bending vibration pattern. The peaks at 598 and 1128 cm^{-1} were ascribed to P-O stretching in tetrahedral PO_4^{3-} units [25,27], while the peaks at 501 and 863 cm^{-1} were ascribed to the Si-O stretching in tetrahedral SiO_4^{4-} units [25,28]. These outcomes definitively prove that SiO_4^{4-} and PO_4^{3-} units are in the NZSPF$_x$ crystal structure. These peaks are shifted to varying degrees, presumably due to F occupying the position of the O element in the NZSPF$_x$ lattice, causing the conformity of positive and negative centers in the lattice [26].

To further investigate the elemental composition of NZSPF$_x$, XPS was used to characterize NZSP, NZSPF$_{0.7}$. Figure S3a shows the full XPS spectra of NZSP and NZSPF$_{0.7}$. Different peak areas corresponding to Na, Zr, Si, P, O are obviously found, which proves that the major components of the ceramics is NZSPF$_{0.7}$. High-resolution XPS (Figure S3b) is adopted to detect the F 1s peak, which proves the successful doping of the F element. The binding energy at 684 eV corresponds to the F-Si/P bond [29,30], showing that the F$^-$ takes up the O^{2-} place in the NZSPF$_x$ lattice. The intensities of the P 2p and Si 2p peaks reduce slightly under the help of F$^-$ (Figure S3c,d), indicating a slight decrease in the concentration of P^{5+} and Si^{4+} because of the reaction of P^{5+}/Si^{4+} with F and the volatilization of the results at high temperatures. In addition, a slight shift in P 2p and Si 2p to a higher binding energy was observed in NZSPF$_{0.7}$ by comparing with the undoped sample, and the presence of P/Si-F bonds [31] can explain this shift, which also proves that F$^-$ takes up part of the O^{2-} sites in the NZSPF$_{0.7}$ lattice.

Although the doping of NaF can reduce the resistance of NZSP (Figure S5) and improve the ionic conductivity (Figure S6), the cycling stability is deemed unsatisfactory (Figure S7). This is because the NZSPF$_{0.7}$ grain boundaries are not dense, which also leads to the ability of growing and forming Na dendrites in the crevices (Figure S9). Therefore, it is necessary to modify NZSPF$_{0.7}$ to strengthen the interface between the SSE and the electrode, and to stop the development of Na dendrites.

Figure 1a shows the FTIR spectra of the epoxy, NZSPF$_{0.7}$, and epoxy-NZSPF$_{0.7}$ composite system. The FTIR spectra of epoxy-NZSPF$_{0.7}$ do not show new peaks, only a simple superposition of the two monomers. This indicates that no new substances are produced in the composite system and no reaction occurs between the two monomers. Figure 1b shows the TGA curves of epoxy, NZSPF$_{0.7}$, and epoxy-NZSPF$_{0.7}$. It can be seen that in the tested temperature range, NZSPF$_{0.7}$ does not undergo mass loss and is thermally stable, while the mass loss of epoxy starts to occur at 300 °C and almost completely decomposes after reaching 600 °C. This coincides with the weight loss range of the composite system, and it can be determined that the mass loss of the composite system is caused by epoxy. This indicates that epoxy was successfully filled into the voids of NZSPF$_{0.7}$, further improving the densities of the ceramics (Figure S10).

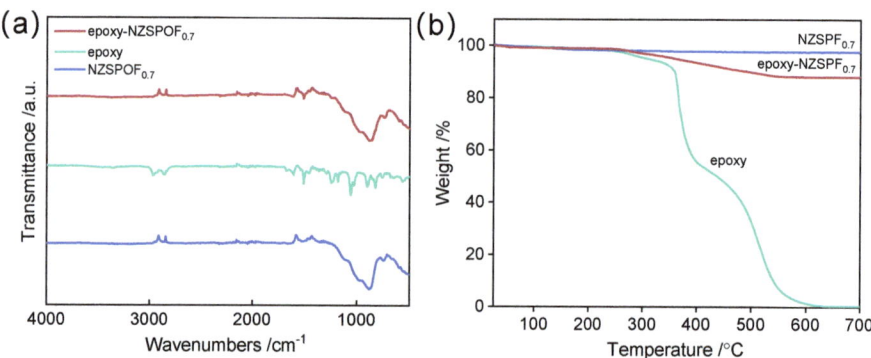

Figure 1. (**a**) FTIR spectra and (**b**) TGA analysis for NZSPF$_{0.7}$, epoxy and epoxy−NZSPF$_{0.7}$.

To observe the surface difference between NZSPF$_{0.7}$ and epoxy-NZSPF$_{0.7}$, and the distribution of epoxy resin in NZSPF$_{0.7}$, SEM and EDS tests were performed on both NZSPF$_{0.7}$ (Figure S4) and epoxy-NZSPF$_{0.7}$ (Figure 2). Figure S4 shows the SEM image of NZSPF$_{0.7}$. The SEM images show that with the growth in NaF content, the size of NZSPF$_x$ grains gradually grows larger. However, F promotes grain growth, reduces grain boundary concentration and increases ceramic density to a certain extent, which can efficiently decrease the grain boundary concentration, increase the ionic conductivity and reduce the grain boundary resistance [32,33]. There are still a large number of interstices in the middle of the ceramic after high-temperature sintering, and the existence of these interstices leads to uneven deposition of dendrites here and reduces the electrochemical performance. Figure 2 shows the surface of epoxy-NZSPF$_{0.7}$ composite solid electrolyte after polishing; compared with NZSPF$_{0.7}$, the gap in the composite system is significantly reduced, which is a good verification of the increased densities. The denseness of epoxy-NZSPF$_{0.7}$ is 97.6% and NZSPF$_{0.7}$ is 93.4%, while that of un-assisted NZSP is only 79.8%. This shows that the filling of epoxy can raise the densities of the composite system because the prior condition for a solid electrolyte to have a high ionic conductivity is a high density [34,35]. Additionally, Figure 2c displays the EDS discussion on the epoxy-NZSPF$_{0.7}$ ceramic sample. Because epoxy contains a large number of C elements, by observing the distribution of C elements, we can find that epoxy is evenly distributed in the composite system. The maps also confirm that O, P, Si, F, Na and Zr factors generally show an even homogeneous distribution in the composite system, and the content of the F element is essentially the same as that of NASPF$_{0.7}$.

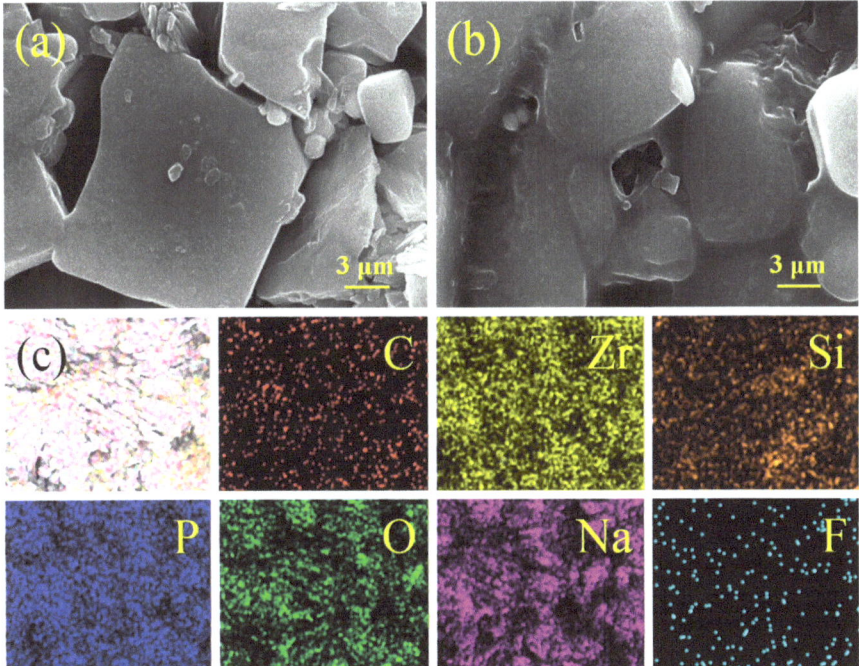

Figure 2. SEM image of (**a**) NZSPF$_{0.7}$ and (**b**) epoxy-NZSPF$_{0.7}$; (**c**) EDS mapping of epoxy-NZSPF$_{0.7}$.

The electrical conductivities of the solid electrolyte were investigated with EIS. Figure 3a displays the impedance spectra of NZSPF$_{0.7}$ and epoxy-NZSPF$_{0.7}$. According to the Nyquist curve, the resistance of NZSPF$_{0.7}$ and epoxy-NZSPF$_{0.7}$ are calculated as 370 Ω and 420 Ω. The overall conductivity measures from the EIS are 0.95 and 0.67 mS cm^{-1}. The impedance of epoxy-NZSPF$_{0.7}$ becomes larger and the ionic conductivity decreases are due to the fact that the epoxy impedance of the pure phase is extremely large; therefore, filling the gap to NZSPF$_{0.7}$ results in a growth in the impedance and a decline in the ionic conductivity of epoxy-NZSPF$_{0.7}$.

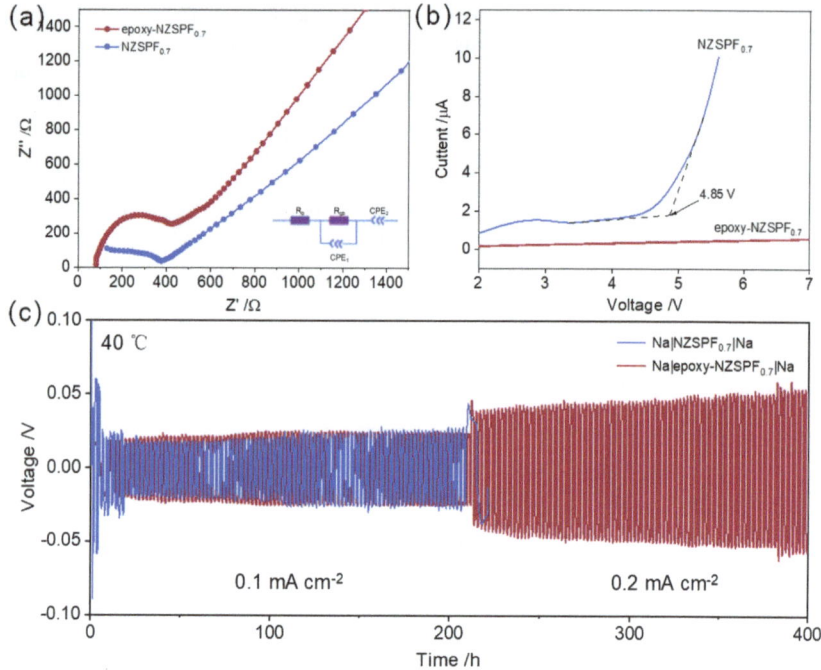

Figure 3. (a) EIS measurements at room temperature; (b) LSV of the Au l NZSPF$_{0.7}$ l Na and Au l epoxy−NZSPF$_{0.7}$ l Na cells at room temperature; (c) variable current cycling of Na l NZSPF$_{0.7}$ l Na and Na l epoxy−NZSPF$_{0.7}$ l Na symmetric cells (the current density is 0.1 mA cm^{-2} and becomes 0.2 mA cm^{-2} after 200 h).

To verify the conduction behavior of sodium ions, the sodium ion transfer number (t_{Na}^+) was determined. The t_{Na}^+ (40 °C) of NZSPF$_{0.7}$ was 0.84, while that of epoxy-NZSPF$_{0.7}$ was 0.79. This proves that the filling of epoxy does not significantly affect the t_{Na}^+ and Na$^+$ is still transported by NZSPF$_{0.7}$ as a transport channel rather than epoxy.

In addition to having a high ionic conductivity, a wide-range electrochemical stability window is also a basic necessity for SSE in actual use. The electrochemical stability windows of NZSPF$_{0.7}$ and epoxy-NZSPF$_{0.7}$ is assessed via LSV with Au l NZSPF$_{0.7}$ l Na and Au l epoxy-NZSPF$_{0.7}$ l Na batteries with a scan rate of 0.2 mV s^{-1} at 30 °C (Figure 3b). The electrochemical stability window of the NZSPF$_{0.7}$ is 4.85 V, while the current intensity of epoxy-NZSPF$_{0.7}$ remains constant in the range of 2–7 V. This indicates that the electrochemical steadiness window of theepoxy-NZSPF$_{0.7}$ up to 7 V. It is expected that epoxy-NZSPF$_{0.7}$ can be used as a high-voltage sodium battery. The electrochemical properties of NZSPF$_{0.7}$, NZSPF$_{0.7}$ and epoxy-NZSPF$_{0.7}$ are listed in Table 1.

Table 1. Electrochemical properties of NZSPF$_0$, NZSPF$_{0.7}$ and epoxy-NZSPF$_{0.7}$.

Sample	Impedance at Room Temperature	Ionic Conductivity	Ion Transfer Number	Electrochemical Stabilization Window
NZSPF$_0$	1527	0.45	0.68	-
NZSPF$_{0.7}$	370 Ω	0.95 mS cm^{-1}	0.84	4.85 V
epoxy-NZSPF$_{0.7}$	420 Ω	0.67 mS cm^{-1}	0.79	7 V

Figure 3c shows that epoxy-NZSPF$_{0.7}$ can be cycled steadily for above 200 h (100 cycles) at a current density of 0.1 mA cm^{-2} with no fluctuation in potential, and remains stable for nearly 200 h when the current density grows to 0.2 mA cm^{-2}. This indicates that epoxy-NZSPF$_{0.7}$ can promote the even deposition of Na$^+$ and stop the Na dendrites from developing. In contrast, NZSPF$_{0.7}$ could only keep steady at 0.1 mA cm^{-2} and immediately short-circuited at 0.2 mA cm^{-2}. It can be noted that NZSPF$_{0.7}$ needs 5–10 h to adapt to the electrode with significant potential fluctuations, while epoxy-NZSPF$_{0.7}$ hardly needs this process. Compared to this, epoxy-NZSPF$_{0.7}$ requires a shorter activation time and does not show significant short-circuiting after increasing the current, indicating that it has higher electrode stability.

Finally, Na | NZSPF$_{0.7}$ | Na$_3$V$_2$(PO$_4$)$_3$ and Na | epoxy-NZSPF$_{0.7}$ | Na$_3$V$_2$(PO$_4$)$_3$ solid SIBs were assembled. Figure 4 shows the performances of batteries. The polarization of NZSPF$_{0.7}$ is bigger with the gradual current rate growth (Figure 4a), while the charge/discharge curve of epoxy-NZSPF$_{0.7}$ is smooth and the polarization is basically unchanged (Figure 4b). The discharge specific capacities of epoxy-NZSPF$_{0.7}$ are 96.9, 80.1, 68.9 and 48.9 mAh g^{-1} at 0.1C, 0.5C, 1C and 2C. In the case of the reversion of the current density to 1C and 0.5C, the given capacities are 60.8 and 70.5 mAh g^{-1} (Figure 4c). This shows that when the current returns to the same rates, it has excellent reversibility and steadiness, although the reversible capacity is slightly reduced. Figure 4d shows the cycling performance of cells assembled with two different electrolytes at a current rate of 0.1C at 40 °C. The discharge-specific capacity remains 93.7 mAh g^{-1} after 150 cycles, which is 95.8% capacity retention of initial capacity (97.8 mAh g^{-1}). The mean coulombic efficiency is close to 99%, which is only 79.1% for Na$_3$V$_2$(PO$_4$)$_3$ | NZSPF$_{0.7}$ | Na battery.

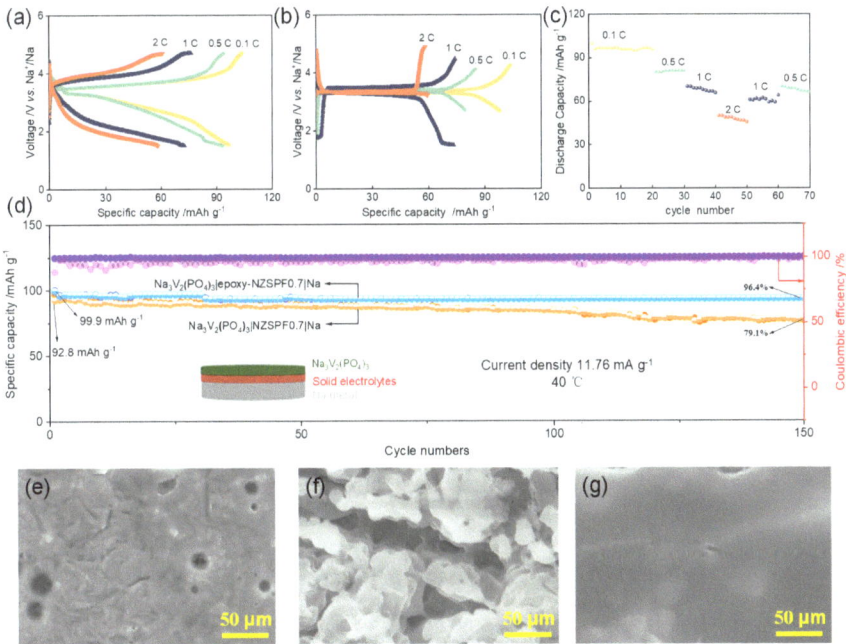

Figure 4. (**a**) Charging–discharging profiles of Na$_3$V$_2$(PO$_4$)$_3$ | NZSPF$_{0.7}$ | Na at different ates; (**b**,**c**) charging–discharging profiles of Na$_3$V$_2$(PO$_4$)$_3$ | epoxy-NZSPF$_{0.7}$ | Na at different rates; (**d**) cycling performance of the Na$_3$V$_2$(PO$_4$)$_3$ | epoxy−NZSPF$_{0.7}$ | Na battery at a current density of 11.76 mA g^{-1}; (**e**) SEM image of sodium metal surface before cycling; SEM image of sodium metal surface of (**f**) Na$_3$V$_2$(PO$_4$)$_3$ | NZSPF$_{0.7}$ | Na battery and (**g**) Na$_3$V$_2$(PO$_4$)$_3$ | epoxy−NZSPF$_{0.7}$ | Na battery after cycling.

Figure 4e–g shows the SEM images of the Na electrode surface. Obviously, after 150 cycles, Na dendrites were formed at the boundaries and voids of the NZSPF0.7 solid electrolyte (Figure 4f), resulting in inhomogeneous Na plating/striping with a corresponding aggravation of the cell behavior [30]. However, the Na electrode surface of $Na_3V_2(PO_4)_3$ | epoxy-NZSPF$_{0.7}$ | Na battery was clean and smooth (Figure 4g), and almost no Na deposition was found, indicating that the epoxy-filled dielectric material can effectively prevent the formation of dendrites and enhance the battery cycling behavior.

4. Conclusions and Outlook

In conclusion, the grain boundary concentration of NZSP was effectively reduced by introducing NaF into the NZSP solid electrolyte and a denser NZSPF$_{0.7}$ with improved ionic conductivity was obtained (0.98 mS cm^{-1}); however, NZSPF$_{0.7}$ did not stop the development of sodium dendrites well and had poor interfacial properties. For this reason, a composite solid electrolyte epoxy-NZSPF$_{0.7}$ was prepared via simply vacuum adsorption and by filling the interstitial space of NZSPF$_{0.7}$ with epoxy and curing it. Epoxy-NZSPF$_{0.7}$ combines the great ionic conductivity of inorganic electrolytes as well as the good interfacial contact from organic polymer electrolytes, with high ionic conductivity (0.67 mS cm^{-1}) as well as excellent interfacial performance and cycling behavior (the initial capacity of cells is 97.9 mAh g^{-1} with a 96.4% retention rate after 150 turns). According to the above-presented results, this study may offer a novel idea for research concerning solid electrolytes for sodium-ion batteries.

Supplementary Materials: The following supporting information can be downloaded at: https://www.mdpi.com/article/10.3390/batteries9060331/s1. Figure S1. (a) XRD patterns of NZSPFx (x = 0, 0.1, 0.3, 0.5, 0.7, and 1.0); Figure S2. FTIR spectras of NZSPFx (x = 0, 0.1, 0.3, 0.5, 0.7, and 1.0); Figure S3. XPS spectra of (a) survey spectra, (b) F 1s, (c) P 2p and (d) Si 2p of NZSPF0.7; Figure S4. SEM images of NZSPFx ceramic pellets, (a) x = 0, (b) x = 0.1, (c) x = 0.3, (d) x = 0.5, (e) x = 0.7, and (f) x = 1.0; (g,h) the corresponding elemental mapping in the square of (e) image; (h) densities and relative densities of NZSPFx; Figure S5. EIS measurements performed of NZSPFx (x = 0, 0.1, 0.3, 0.5, 0.7, and 1.0) solid electrolytes; Figure S6. Ion conductivity (red) and ion transfer number (blue) of NZSPFx (x = 0, 0.1, 0.3, 0.5, 0.7, and 1.0) solid electrolytes; Figure S7. cycling performance of Na3V2(PO4)3 | NZSPF0.7 | Na battery; Figure S8. Arrhenius plots of NZSPFx (x = 0, 0.1, 0.3, 0.5, 0.7, and 1.0) solid electrolytes; Figure S9. SEM images of the sodium metal surface of cell Na3V2(PO4)3 | NZSPF0.7 | Na (a) before and (b) after cycling; Figure S10. The densities of NZSPF0.7 and epoxy- NZSPF0.7; Figure S11. Variable current cycling of Na | NZSPF0.7 | Na symmetric cells (the current density is 0.1 mA cm^{-2} and becomes 0.2 mA cm^{-2} after 200 h); Figure S12. Variable current cycling of Na | epoxy-NZSPF0.7 | Na symmetric cells (the current density is 0.1 mA cm^{-2} and becomes 0.2 mA cm^{-2} after 200 h); Table S1. Chemical composition for NZSPFx (x = 0, 0.1, 0.3, 0.5, 0.7, 1.0).

Author Contributions: Y.F.: conceptualization, data analysis, and writing of the original draft. D.L.: experimental execution, data analysis and discussion. Y.S.: experimental execution and discussion. G.Z.: writing-review and editing. H.G.: research design, funding supporting and supervision. All authors have read and agreed to the published version of the manuscript.

Funding: The authors acknowledge financial support provided the Key National Natural Science Foundation of Yunnan Province (2019FY003023).

Institutional Review Board Statement: Not applicable, as studies on humans and animals are not involved.

Informed Consent Statement: Not applicable, as studies on humans are not involved.

Data Availability Statement: The data are not publicly available due to the data required to reproduce these findings forming part of an ongoing study.

Acknowledgments: The authors acknowledge financial support provided by the National Natural Science Foundation of China (No. 52064049), the National Natural Science Foundation of Yunnan Province (202301AS070040), Key Laboratory of Solid-State Ions for Green Energy of Yunnan University, the Electron Microscope Center of Yunnan University for the support of this work.

Conflicts of Interest: The authors declare no conflict of interest.

References

1. Chu, S.; Majumdar, A. Opportunities and challenges for a sustainable energy future. *Nature* **2012**, *7411*, 294–303. [CrossRef]
2. Xu, K. Electrolytes and interphases in Li-ion batteries and beyond. *Chem. Rev.* **2014**, *23*, 11503–11618. [CrossRef]
3. Yoo, H.D.; Liang, Y.L.; Li, Y.F.; Yao, Y. High areal capacity hybrid magnesium-lithium-ion battery with 99.9% coulombic efficiency for large-scale energy storage. *ACS Appl. Mater. Interfaces* **2015**, *12*, 7001–7007. [CrossRef]
4. Hao, F.; Liang, Y.L.; Zhang, Y.; Chen, Z.Y.; Zhang, J.B.; Ai, Q.; Guo, H.; Fan, Z.; Lou, J.; Yao, Y. High-energy all-solid-state organic-lithium batteries based on ceramic electrolytes. *ACS Energy Lett.* **2021**, *1*, 201–207. [CrossRef]
5. Yabuuchi, N.; Kubota, K.; Dahbi, M.; Komaba, S. Research Development on sodium-ion batteries. *Chem. Rev.* **2014**, *23*, 11636–11682. [CrossRef] [PubMed]
6. Che, H.Y.; Chen, S.L.; Xie, Y.Y.; Wang, H.; Amine, K.; Liao, X.Z.; Ma, Z.F. Electrolyte design strategies and research progress for room-temperature sodium-ion batteries. *Energy Environ. Sci.* **2017**, *5*, 1075–1101. [CrossRef]
7. Cheng, X.B.; Zhang, R.; Zhao, C.Z.; Zhang, Q. Toward safe lithium metal anode in rechargeable batteries: A Review. *Chem. Rev.* **2017**, *15*, 10403–10473. [CrossRef]
8. Cohn, A.P.; Muralidharan, N.; Carter, R.; Share, K.; Pint, C.L. Anode-free sodium battery through in situ Plating of sodium metal. *Nano Lett.* **2017**, *2*, 1296–1301. [CrossRef]
9. Fu, K.; Gong, Y.H.; Hitz, G.T.; McOwen, D.W.; Li, Y.J.; Xu, S.M.; Wen, Y.; Zhang, L.; Wang, C.W.; Pastel, G.; et al. Three-dimensional bilayer garnet solid electrolyte based high energy density lithium metal-sulfur batteries. *Energy Environ. Sci.* **2017**, *7*, 1568–1575. [CrossRef]
10. Li, H.S.; Ding, Y.; Ha, H.; Shi, Y.; Peng, L.L.; Zhang, X.G.; Ellison, C.J.; Yu, G.H. An all-stretchable-component sodium-ion full battery. *Adv. Mater.* **2017**, *29*, 1700898. [CrossRef]
11. Zhao, Y.; Goncharova, L.V.; Lushington, A.; Sun, Q.; Yadegari, H.; Wang, B.Q.; Xiao, W.; Li, R.Y.; Sun, X.L. Superior stable and long life sodium metal anodes achieved by atomic layer dposition. *Adv. Mater.* **2017**, *29*, 1606663. [CrossRef]
12. Liu, H.; Guo, H.; Liu, B.H.; Liang, M.F.; Lv, Z.L.; Adair, K.R.; Sun, X.L. Few-layer MoSe$_2$ nanosheets with expanded (002) planes confined in hollow carbon nanospheres for ultrahigh-performance Na-ion batteries. *Adv. Funct. Mater.* **2018**, *28*, 1707480. [CrossRef]
13. Zhao, C.L.; Liu, L.L.; Qi, X.G.; Lu, Y.X.; Wu, F.X.; Zhao, J.M.; Yu, Y.; Hu, Y.S.; Chen, L.Q. Solid-state sodium batteries. *Adv. Energy Mater.* **2018**, *17*, 1601196. [CrossRef]
14. Kwak, H.; Lyoo, J.; Park, J.; Han, Y.; Asakura, R.; Remhof, A.; Battaglia, C.; Kim, H.; Hong, S.T.; Jung, Y.S. Na$_2$ZrCl$_6$ enabling highly stable 3 V all-solid-state Na-ion batteries. *Energy Storage Mater.* **2021**, *37*, 47–54. [CrossRef]
15. He, X.Z.; Ji, X.; Zhang, B.; Rodrigo, N.D.; Hou, S.; Gaskell, H.; Deng, T.; Wan, H.L.; Liu, S.F.; Xu, J.J.; et al. Tuning interface lithiophobicity for lithium metal solid-state batteries. *ACS Energy Lett.* **2022**, *7*, 131–139. [CrossRef]
16. Bay, M.C.; Grissa, R.; Egorov, K.V.; Asakura, R.; Batrtaglia, C. Low Na-β″-alumina electrolyte/cathode interfacial resistance enabled by a hydroborate electrolyte opening up new cell architecture designs for all-solid-state sodium batteries. *Mater. Futures* **2022**, *1*, 031001. [CrossRef]
17. Will, F.G. Effect of water on beta alumina conductivity. *J. Electrochem. Soc.* **1976**, *6*, 834–836. [CrossRef]
18. Zhang, L.; Yang, K.; Mi, J.L.; Lu, L.; Zhao, L.R.; Wang, L.M.; Li, Y.M.; Zeng, H. Na$_3$PSe$_4$: A novel chalcogenide solid electrolyte with high Ionic conductivity. *Adv. Energy Mater.* **2015**, *5*, 1501294. [CrossRef]
19. Chi, X.W.; Liang, Y.L.; Hao, F.; Zhang, Y.; Whiteley, J.; Dong, H.; Hu, P.; Lee, S.; Yao, Y. Tailored organic electrode material compatible with sulfide electrolyte for stable all-solid-state sodium batteries. *Angew. Chem. Int. Edit.* **2018**, *10*, 2630–2634. [CrossRef]
20. Hong, H.Y.P. Crystal-structures and crystal-chemistry in system Na$_{1+X}$Zr$_2$Si$_X$P$_{3-X}$O$_{12}$. *Mater. Res. Bull.* **1976**, *2*, 173–182. [CrossRef]
21. Goodenough, J.B.; Hong, H.Y.P.; Kafalas, J.A. Fast Na$^+$-ion yransport in skeleton structures. *Mater. Res. Bull.* **1976**, *2*, 203–220. [CrossRef]
22. Wang, H.; Zhao, G.F.; Wang, S.M.; Liu, D.L.; Mei, Z.Y.; An, Q.; Jiang, J.W.; Guo, H. Enhanced ionic conductivity of a Na$_3$Zr$_2$Si$_2$PO$_{12}$ solid electrolyte with Na$_2$SiO$_3$ obtained by liquid phase sintering for solid-state Na$^+$ batteries. *Nanoscale* **2022**, *14*, 823–832. [CrossRef] [PubMed]
23. Lu, Y.; Meng, X.Y.; Alonso, J.A.; Fernandez-Diaz, M.T.; Sun, C.W. Effects of fluorine doping on structural and electrochemical properties of Li$_{6.25}$Ga$_{0.25}$La$_3$Zr$_2$O$_{12}$ as electrolytes for solid-state lithium batteries. *ACS Appl. Mater. Interfaces* **2019**, *2*, 2042–2049. [CrossRef]
24. Li, J.X.; Wen, Z.Y.; Xu, X.X.; Zhu, X.J. Lithium-ion conduction in the anion substituted La$_{2/3-x}$Li$_{3x-y}$TiO$_{3-y}$F$_y$ electrolyte with perovskite-type structure. *Solid State Ionics* **2005**, *29–30*, 2269–2273. [CrossRef]
25. Li, Y.T.; Zhou, W.D.; Xin, S.; Li, S.; Zhu, J.L.; Lu, X.J.; Cui, Z.M.; Jia, Q.X.; Zhou, J.S.; Zhao, Y.S.; et al. Fluorine-doped antiperovskite electrolyte for all-solid-state lithium-ion batteries. *Angew. Chem. Int. Edit.* **2016**, *34*, 9965–9968. [CrossRef]
26. He, S.N.; Xu, Y.L.; Chen, Y.J.; Ma, X.N. Enhanced ionic conductivity of an F$^-$-assisted Na$_3$Zr$_2$Si$_2$PO$_{12}$ solid electrolyte for solid-state sodium batteries. *J. Mater. Chem. A* **2020**, *25*, 12594–12602. [CrossRef]
27. Liu, C.; Wen, Z.Y.; Rui, K. High ion conductivity in garnet-type F-doped Li$_7$La$_3$Zr$_2$O$_{12}$. *Int. J. Inorg. Mater.* **2015**, *9*, 995–1000.
28. Song, S.F.; Duong, H.M.; Korsunsky, A.M.; Hu, N.; Lu, L. A Na$^+$ superionic conductor for room-temperature sodium batteries. *Sci. Rep.* **2016**, *6*, 32330. [CrossRef]

29. Lee, S.H. Surface properties of fluoroethylene carbonate-derived solid electrolyte interface on graphite negative electrode by narrow-range cycling in cell formation process. *Appl. Surf. Sci.* **2014**, *322*, 64–70. [CrossRef]
30. Kanezashi, M.; Matsutani, T.; Wakihara, T.; Nagasawa, H.; Okubo, T.; Tsuru, T. Preparation and has permeation properties of fluorine-silica membranes with controlled amorphous silica structures: Effect of fluorine source and calcination temperature on network size. *ACS Appl. Mater. Interfaces* **2017**, *29*, 24625–24633. [CrossRef]
31. Dalavi, S.; Guduru, P.; Lucht, B.L. Performance enhancing electrolyte additives for lithium ion batteries with silicon anodes. *J. Electrochem. Soc.* **2012**, *5*, A642–A646. [CrossRef]
32. Ihlefeld, J.F.; Gurniak, E.; Jones, B.H.; Wheeler, D.R.; Rodriguez, M.A.; McDaniel, A.H. Scaling effects in sodium zirconium silicate phosphate ($Na_{1+x}Zr_2Si_xP_{3-x}O_{12}$) ion-conducting thin films. *J. Am. Ceram. Soc.* **2016**, *8*, 2729–2736. [CrossRef]
33. Wang, H.; Sun, Y.J.; Liu, Q.; Mei, Z.Y.; Yang, L.; Duan, L.Y.; Guo, H. An asymmetric bilayer polymer-ceramic solid electrolyte for high-performance sodium metal batteries. *J. Enerdy Chem.* **2022**, *74*, 18–25. [CrossRef]
34. Xu, X.X.; Wen, Z.Y.; Yang, X.L.; Chen, L.D. Dense nanostructured solid electrolyte with high Li-ion conductivity by spark plasma sintering technique. *Mater. Res. Bull.* **2008**, *8–9*, 2334–2341. [CrossRef]
35. Fuentes, R.O.; Figueiredo, F.M.; Marques, F.M.B.; Franco, J.I. Influence of microstructure on the electrical properties of NASICON materials. *Solid State Ionics* **2001**, *1–2*, 173–179. [CrossRef]

Disclaimer/Publisher's Note: The statements, opinions and data contained in all publications are solely those of the individual author(s) and contributor(s) and not of MDPI and/or the editor(s). MDPI and/or the editor(s) disclaim responsibility for any injury to people or property resulting from any ideas, methods, instructions or products referred to in the content.

Review

Progress on Designing Artificial Solid Electrolyte Interphases for Dendrite-Free Sodium Metal Anodes

Pengcheng Shi [1,†], Xu Wang [2,†], Xiaolong Cheng [1] and Yu Jiang [1,*]

[1] School of Materials Science and Engineering, Anhui University, Hefei 230601, China; spchfut@sina.com (P.S.); chengxl@ahu.edu.cn (X.C.)
[2] School of Computer Science and Engineering, Anhui University of Science and Technology, Huainan 232001, China; wangxu0304@sina.com
[*] Correspondence: jiangyu21@ahu.edu.cn
[†] These authors contributed equally to this work.

Abstract: Nature-abundant sodium metal is regarded as ideal anode material for advanced batteries due to its high specific capacity of 1166 mAh g^{-1} and low redox potential of −2.71 V. However, the uncontrollable dendritic Na formation and low coulombic efficiency remain major obstacles to its application. Notably, the unstable and inhomogeneous solid electrolyte interphase (SEI) is recognized to be the root cause. As the SEI layer plays a critical role in regulating uniform Na deposition and improving cycling stability, SEI modification, especially artificial SEI modification, has been extensively investigated recently. In this regard, we discuss the advances in artificial interface engineering from the aspects of inorganic, organic and hybrid inorganic/organic protective layers. We also highlight key prospects for further investigations.

Keywords: sodium metal; artificial SEI; dendrite formation; batteries

1. Introduction

To date, sodium (Na) ion batteries have been commercialized as a supplemental technology for lithium (Li) ion batteries due to natural-abundant Na resources and low costs [1]. However, the energy density of Na ion batteries appears to be unsatisfactory as compared to updated Li ion batteries [2,3]. To meet the rapidly growing demands for the energy density of Na ion batteries, the development of advanced electrode materials with high capacity is highly desired [4].

Among various materials, the metal Na has been proposed as an ideal candidate due to its high specific capacity (1166 mAh g^{-1}) and low redox potential (−2.71 V) [5–7]. In this regard, investigations regarding Na-based batteries, including Na-S, Na-O$_2$ and Na-CO$_2$ batteries, have been widely reported [5]. However, the cycling performances and safety issues of Na anodes remain unsatisfactory. It has been reported that growth of dendrites may be the root reason. The spontaneous reaction between Na and electrolytes can form a chemically/mechanically unstable solid electrolyte interphase (SEI), which cannot maintain long-term cycling of the Na anode [8,9]. During plating/stripping, the SEI would be thickened, broken and collapsed [7,9], inducing dendrite formation. Additionally, the thickness change during Na plating/stripping can lead to great local stress, making the SEI much more unstable and more easily cracked [10,11]. In particular, the dendritic Na can penetrate through the separator and detach from the matrix easily to form "dead" Na, leading to battery short circuits and a short cycle life [11–14]. Therefore, effective efforts to modify Na metal anodes are highly necessary.

Under this background, several approaches have been proposed to stabilize Na anodes: for instance, constructing a 3D host to resolve infinite volume expansion [6,14,15], coating the separator to block Na dendrites [16,17] and employing an Na alloy to build

stable anodes [18–20]. Although these approaches have some positive effects in suppressing dendritic Na formation, the properties of SEI films remain unsatisfactory, and the irreversible side reactions cannot be totally suppressed. The electrolyte modification seems to be promising for increasing the stability of the SEI interphase. However, the additives, salts and solvents cannot hold for long-term cycling due to continuous consumption [11,21]. Accordingly, the ideal SEI for Na metal should possesses excellent chemical/electrochemical stability, good ionic conductivity, even Na^+ flux/electric field distribution, sufficient Young's module, good flexibility and robustness [22]. In this regard, artificial interphase engineering is of vital importance, since the protective layer can be precisely designed and easily adjusted. More importantly, the artificial SEI boasts most of the above-mentioned merits of an ideal SEI. So far, extensive research has been conducted on artificial interphase configuration to improve the stability of the SEI [23,24]. Therefore, it is necessary to summarize the research progress in artificial SEI design in recent years.

In this review, we discuss the advances in artificial interface engineering from the aspects of inorganic, organic and hybrid inorganic/organic protective layers, as shown in Figure 1. The specific modified materials, synthetic processes and properties of the artificial SEI layers are systematically reviewed. Meanwhile, the working mechanism of these artificial SEIs is also briefly analyzed. We also conclude by outlining future directions of artificial interphase chemistry for advanced Na metal anodes. We hope this review can deepen the understanding of artificial SEI layers by exploring stable and dendrite-free Na anodes.

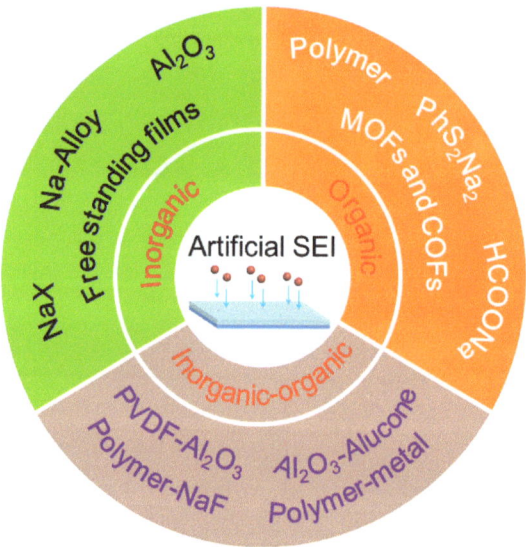

Figure 1. Schematic illustration of artificial interface engineering from the aspects of inorganic, organic and hybrid inorganic/organic protective layers. Different colors represent different artificial SEI layers. For each artificial SEI layer, the typical materials are showed correspondingly.

2. Challenges for Dendrite-Free Na Metal Anodes

Like other alkali metals, Na is thermodynamically unstable; this is the root cause of uncontrollable parasitic reactions and the formation of chemically/mechanically unstable SEIs [23,24]. Figure 2a shows the main challenges of Na metal anodes. As compared with Li metal, Na metal is more prone to deposits in dendritic morphology and suffers from severe volume expansion [25,26]. During plating/stripping, the SEI can be cracked and form "dead" and isolated Na. Meanwhile, the growth of dendrites can lead to battery short-circuiting. The overall challenges regarding dendrite-free Na metal anodes are discussed below.

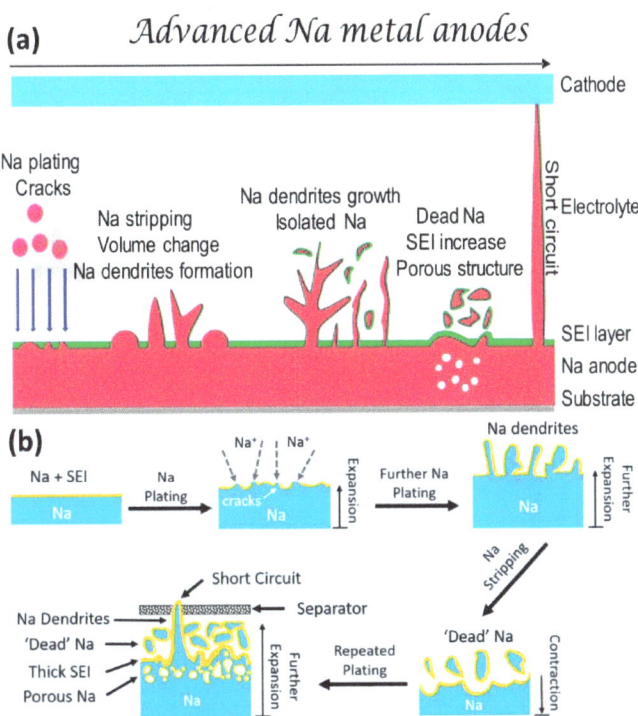

Figure 2. (**a**) Schematic illustration of challenges for Na metal anodes [25]. Copyright 2018, Elsevier. (**b**) The growth of dendrites and formation of "dead" Na [6]. Copyright 2020, Wiely-VCH.

2.1. High Reactivity

The Na atom can lose electrons to form Na$^+$ easily. In dry air, the Na metal can react with O_2 and CO_2. When contacting water or moist air, the Na metal can form flammable H_2 to cause fire or even explosions. Due to high reactivity, the Na metal will induce unavoidable side reactions with liquid electrolytes, resulting in SEI formation, Na corrosion and poor cycling performance, as shown in Figure 2b. Even worse, the leakage or breakage of batteries can cause safety issues.

2.2. Unstable SEIs

It is expected that the ideal SEI layer is dense and inert so as to effectively isolate electron transfer and prevent further parasitic reactions [24,27]. Nevertheless, the structure of the SEI layer formed in common electrolytes is demonstrated to be porous and fragile.

As it is recognized, the properties of the SEI layer formed in common electrolytes depend on the solvents, additives and Na salts. Typically, the SEI layer is mainly composed of inorganic species (e.g., NaF, Na$_2$O and Na$_2$CO$_3$) and organic species (e.g., RONa, ROCO$_2$Na and RCOONa; where R is the alkyl group) [25]. The possible formation mechanism is summarized in the following equations [28,29].

$$C_3H_4O_3(EC) + 2Na^+ + 2e^- \rightarrow Na_2CO_3\downarrow + C_2H_4\uparrow \quad (1)$$

$$C_3H_6O_3(PC) + 2Na^+ + 2e^- \rightarrow Na_2CO_3\downarrow + C_3H_6\uparrow \quad (2)$$

$$C_3H_3FO_3(FEC) + Na^+ + e^- \rightarrow NaF\downarrow + CO_2\uparrow + CH_2CHO\uparrow \quad (3)$$

$$PF_6^- + 3Na^+ + 2e^- \rightarrow 3NaF\downarrow + PF_3 \quad (4)$$

$$Na + 3Na^+ + 2e^- \rightarrow 3NaF\downarrow + PF_3 \quad (5)$$

Meanwhile, the reduction of solvents can supply a large amount of oxygen atoms, leading to the formation of Na_2O. Owing to the lack of advanced characterization techniques, the formation mechanism and detailed composition of the SEI layer remain controversial. Further investigations are needed for understanding the mechanism. Additionally, the SEI layer formed on the Na metal is dissolved in electrolytes more easily than that of Li [30,31]. Due to the non-uniform distribution of compositions, the ionic conductivity of the SEI layer is spatial varying, resulting in uneven distribution of the Na^+ flux. Meanwhile, due to the "host-less" nature of the Na matrix, the SEI layer cracks easily during repeated Na^+ plating/stripping, which in turn accelerates dendrite growth due to increased Na^+ flux and preferential Na^+ plating around the cracks. Furthermore, the repeated breakage of the SEI layer also leads to uncontrollable electrolyte consumption, followed by low coulombic efficiency and high SEI impedance [32,33]. As a result, Na metal with unsatisfied SEI properties inevitably suffers from poor performance.

Based on previous research [34,35], further progress on building ideal SEI layers for dendrite-free Na metal should be centered around the following characteristics: firstly, high Na^+ conductivity so as to facilitate uniform Na^+ deposition and regulate preferential Na plating; secondly, electrochemical stability and electronic insulation to prevent further side reactions; thirdly, sufficiently robust to maintain long-term large volume expansion and dendrite propagation; finally, homogeneous in composition to decentralize the Na^+ flux.

2.3. Uncontrollable Dendritic Na Formation

Dendrite growth is also a serious problem, as shown in Figure 2b. The dendrite growth can penetrate the separator and form "dead" Na, leading to battery short circuiting and poor cycling stability. The morphology of Na dendrites can be divided into needle-like, tree-like and mossy-like types; however, it is difficult to distinguish them clearly. In most case, these types of dendrites can co-exist in rechargeable batteries [36,37].

Based on previous research [38], it is widely accepted that the concentration of Na^+ will decrease to zero near the surface in Sand's time. Due to the spatial variation in ionic conductivity and the localized electric field, the rough surface will induce uneven Na^+ plating/stripping, resulting in dendrite formation. Subsequently, the tips of dendrites become hot sites for further dendritic Na nucleation and growth due to their larger electric field and ionic concentration gradients. Once the dendrite is nucleated, the growth rate of dendrites is a key parameter to determine the lifetime of Na anode. According to Sand's law, the speed of dendrite formation is inversely proportional to the square of the deposition current [32,37,39,40].

Dendrite growth can expose the fresh Na surface to depletion of electrolytes and active Na. Meanwhile, the unstable dendrites detach from the matrix to form "dead" Na. Through microscopy observation, it has been proven that the porous Na dendrites can break away from the bulk Na matrix easily, as compared with Li dendrites. The dendrites intrinsically exhibit much higher chemical reactivity and weaker mechanical stability [39].

2.4. Severe Volume Expansion

The severe volume expansion can be regarded as the root cause of the continuous side reactions. Theoretically, the thickness would increase by 8.86 μm with 1 mAh cm^{-2} Na. To satisfy industrial requirements, the deposited capacity would be above 3.5 mAh cm^{-2} [34,41]. Due to uneven deposition, the practical volume variation would be more evident than theoretically expected. In addition, due to the host-less nature, the volume expansion is considered to be relatively infinite [42]. Meanwhile, due to lack of flexibility, the SEI can be cracked easily during volume expansion, which accelerates the formation of "dead" Na and consumption of electrolytes, as shown in Figure 2b.

To alleviate the volume expansion and mitigate the inner strain, nanostructured hosts such as Cu foam [43–45], carbon matrix [42,46,47] and Mxene [48,49] are proposed to accommodate Na. Nevertheless, these hosts increase the total weight and volume of the Na anode at the expense of total energy density. The recent development of hosts for dendrite-free Na metal has been discussed in several reviews [26,50].

3. Advances in Artificial SEI Interphase

A stable SEI is the ultimate pursuit for achieving dendrite-free Na metal anodes. With a deep understanding of the plating/stripping mechanism, several strategies (e.g., chemical pretreatment, building protective film by advanced deposition technologies and free-standing protective layers) for building artificial interphase have been proposed [36,37,39]. Typically, the chemical composition, structure and thickness of artificial SEI layers can be precisely controlled by optimizing the reagent species, concentration, and reaction temperature, time, etc. [36]. As reported, the artificial SEI can be classified into inorganic rich or organic rich or their hybrids [24,51,52]. The characteristics of the inorganic rich and organic rich SEI are schematically presented in Figure 3a,b. In this section, we will discuss the recent advances in constructing artificial SEIs for stable Na metal anodes.

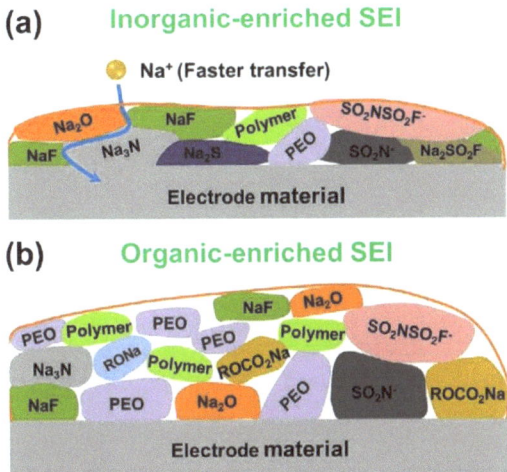

Figure 3. Schematic illustration of (**a**) inorganic-enriched SEI and (**b**) organic-enriched SEI on Na metal [51]. Different colors represent different SEI species. Copyright 2021, Wiely-VCH.

3.1. Inorganic Interphase

Adopting the experiences and knowledge of LiX (X = F, Cl, Br, I) for dendrite suppression in Li metal batteries, NaX are proposed for inorganic interphase configuration through chemical pretreatment methods [22,36,53]. In the early stage, Wang et al. proposed a simple chemical pretreatment of Na with a SbF_3/DMC solution. Through an exchange reaction, an inorganic SEI rich in NaF and Na_3Sb alloy is formed. By taking advantage of the synergistic effect of NaF and Na_3Sb, the hybrid NaF/Na_3Sb interphase greatly reduces the surface reactivity and interfacial impedance [54,55]. Recently, the NaF-rich interphase has also been reported by reaction with 1-butyl-2,3-dimethylimidazolium tetrafluoroborate ($BdmimBF_4$) [56], CoF_2 [57], AlF_3-coated solid-state electrolytes [58] and triethylamine trihydrofluoride [59]. The shear modulus of NaF is 31.4 GPa, which is much higher than that of metallic Na (3.3 GPa); thus, it plays an important role in suppressing Na dendrite growth [57,60]. Inspired by these works, NaCl-rich interphases have also been investigated. For instance, Huang et al. adopted $SnCl_2$ to treat Na with the formation of the NaCl/Na-Sn alloy interphase [61]. As expected, both rapid ion transportation and suppressed parasitic reactions were obtained, which jointly achieved a nondendritic morphology over 500 h in Na||Na batteries. Similar treatment methods have also been reported using $ZnCl_2$ and $SnCl_4$ [62–64]. Analogous to NaF and NaCl, the NaI- and NaBr-rich interphases were reported by reaction with 1-iodopropane and 1-bromopropane, respectively [65,66]. In Na||Na cells, the NaI-coated Na was stable for 500 h under 0.25 mA cm^{-2} and 0.75 mAh cm^{-2}, while the NaBr-coated Na was stable for 250 h under 1.0 mA cm^{-2} and 1.0 mAh cm^{-2}. According to the density functional theory calculations in Figure 4a, the energy barriers for interface ion diffusion decrease in the following order: NaF > NaCl ≈ NaI > NaBr [66]. The lower energy barrier is more favorable for nondendritic deposition.

The S-containing protective layer is also attractive for nondendritic Na plating/stripping due to its high ionic conductivity. Sun et al. synthesized Na_3PS_4 as an artificial protective layer by reacting Na with P_4S_{16} in diethylene glycol dimethyl ether. By controlling the concentrations of P_4S_{16} and the reacting time, the thickness and composition of the Na_3PS_4 can be optimized. The thin Na_3PS_4 layer can reduce unwanted side reactions and uniform Na^+ flux during plating/stripping [67]. The Mo_6S_8 and MoS_2 were also used for building $Na_xMo_6S_8$ and Na_2S protective layers, as shown in Figure 4b,c, respectively [68,69]. In addition to NaX and the S-containing protective layer, the Na_3N layer is also attractive due to its high ionic conductivity. In 2021, Sun et al. directly embedded $NaNO_3$ into the Na matrix to form Na_3N and NaN_xO_y. As shown in Figure 4d,e, the Na_3N and NaN_xO_y provide good SEI stability and Na^+ conductivity, while the remaining $NaNO_3$ works as an SEI stabilized for long-term cycling [70].

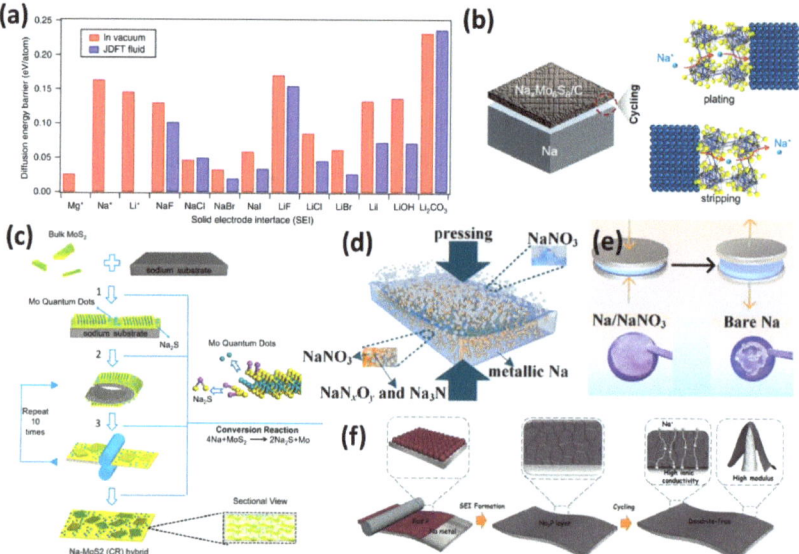

Figure 4. (**a**) Calculated energy barriers for Mg, Na and Li atom diffusion on the surface with noted chemistry [66]. In this paper, the * symbol marks the data points obtained from ref. [14]. Copyright 2017, Nature. (**b**) The function of Mo_6S_8-formed $Na_xMo_6S_8$ protective layer [68]. Copyright 2019, American Chemistry Society. (**c**) The fabrication of Mo_2S-based protective layer and the corresponding conversion reaction [69]. Copyright 2017, American Chemistry Society. (**d**) Mechanically fabricated $NaNO_3$-derived Na_3N/NaN_xO_y protective layer. (**e**) Image of the Na anodes with and without $NaNO_3$ [70]. Copyright 2021, American Chemistry Society. (**f**) Schematic illustration of red phosphorus formed Na_3P layer and the dendrite suppression mechanism [71]. Copyright 2021, Wiely-VCH.

Recently, Yu's group built a Na_3P protective layer to protect Na metal by treating it with red phosphorus. As shown in Figure 4f, the Na_3P layer can provide high ionic conductivity of ~0.12 mS cm^{-1} and high Young's modulus of 8.6 GPa, which regulates uniform Na^+ flux and prevents the dendrite growth. As proven by cryo-TEM, the Na_3P phase can remain after repeated plating/stripping, which is highly attractive for achieving stable Na anodes. Benefiting from these advantages, the Na||Na cells with a Na_3P artificial layer present a nondendritic morphology for 780 h at 1.0 mA cm^{-2} and 1.0 mAh cm^{-2}. In addition, the artificial phosphorus derived protection layer was also applied for dendrite-free potassium metal, with satisfactory performance [71]. More recently, Yu's group also proved that the Na_2Te artificial interfacial layer showed similar advantages [72].

At 1.0 mA cm^{-2} and 1.0 mAh cm^{-2}, the Na@Na$_2$Te provides excellent cyclic stability for 700 h. As interfaces with a single component cannot meet all the requirements of an ideal SEI, Wu et al. and Ji et al. further developed a hybrid Na$_3$P/NaBr interphase with faster ionic conductivity compared with Na$_3$P [73,74]. Rui et al. also adopted V$_2$S$_3$ [75], VN [13], VSe [76], VP$_2$ [77] and BiOCl [78] as precursors to build artificial heterogeneous interphase layers. With vanadium and Na$_3$Bi, a more uniform deposition of Na$^+$ is promoted and better cycling performance is achieved.

The Na alloy interphases are also attractive. Due to the low reduction potential of Na, the metal cations dissolved in solvents can spontaneously alloy with Na. For Li metal, Li et al. immersed Li in Mg(TFSI)$_2$ containing electrolytes, with the formation of Li-Mg alloy [79]. The pre-alloying with Mg avoids the nucleation of Li at the hot points for dendrite growth and prevents electrolyte corrosion. This approach also applies to Na metal anodes. By taking advantage of Sn(TFSI)$_2$, a Na-Sn alloy interphase rich in Na$_9$Sn$_4$ and Na$_{14.7}$Sn$_4$ can be obtained. Under 0.25 mAh cm^{-2}, the Na||Na cells can physically mitigate dendrite growth for 1700 h due to their fast ion transport properties. Despite the surface alloy, some cations can be reduced as metals, usually acting as nucleation seeds for dendrite suppression [80]. Chen's group used Bi(SO$_3$CF$_3$)$_3$ to treat Na with the formation of Bi. Under 0.5 mA cm^{-2}, the Na-Bi anode can cycle for 1000 h without overpotential increase [81]. Analogous AgTFSI and AgCF$_3$SO$_3$ were also achieved with the formation of Ag seeds [82,83]. Notably, these species are also powerful for using as additives for electrolyte modification [39]. More recently, Yu et al. also alloyed the Na surface with Ga liquid metal and Sn foil via in-suit rolling [84–86].

In addition to chemical pretreatment methods for inorganic SEI configuration, the physical deposition method is also proposed. Among the different technologies, the atomic layer deposition (ALD) technology is most attractive, since it has long been used for building advanced protective layers for batteries. Early in 2017, a thin Al$_2$O$_3$ protective layer was achieved on Na through low-temperature plasma-enhanced ALD, as shown in Figure 5a [87–89]. The low-temperature ALD can avoid the melting of Na (98 °C) due to its low working temperature of 75 °C. Based on the growth rate of Al$_2$O$_3$, the thickness of 10, 25 and 50 cycles of ALD Al$_2$O$_3$ is confirmed to be 1.4, 3.5 and 7 nm, respectively. Attractively, the Al$_2$O$_3$ can convert into highly conductive NaAlO$_x$ during cycling. The Na@Al$_2$O$_3$ displayed an island-like morphology up to 500 cycles, even at 3 mA cm^{-2}. Analogous to ALD, the molecular layer deposition (MLD) technology is also proposed for building hybrid inorganic/organic protective layers [90]. The MLD will be discussed in the following hybrid interphase section.

Another type of inorganic SEI is designed by using prefabricated free-standing films. Typically, these free-standing films can improve the surface sodiophilicity with functional groups. Meanwhile, with these free-standing films, the average plating/stripping coulombic efficiency and cycling stability of Na||Cu batteries is increased during long-term cycling, indicating reduced side reactions. Peng et al. presented an O-functionalized 3D carbon nanotube film (O$_f$-CNT), as shown in Figure 5b [91]. According to DFT calculation, the oxygen function group strongly interacts with Na$^+$, as shown in Figure 5c, which provides a robust sodiophilic interphase. Benefiting from the sodiophilic nature, the O$_f$-CNT offers preferential Na nucleation with a reduced overpotential and improves the reaction kinetics. Similar free-standing films are also proposed with O and N functioning 3D nanofibers (ONCNFs) [92]. Despite 3D carbon nanofibers, 2D materials such as MXenes, graphene, silicene, germanene, phosphorene, h-BN, SnS, SnSe and g-C$_3$N$_4$ film have also attracted tremendous attention [93–95]. In order to accelerate surface Na$^+$ transfer and improve the ionic conductivity of the protective layer, the introduction of defects, the increase in bond length and the proximity effect should be seriously considered, as confirmed by first-principles calculations. Meanwhile, their balance with surface stiffness for dendrite suppression is also a critical factor. In this regard, Chen et al. used MXene and carbon nanotubes (CNTs) to construct a 3D MXene/CNTs sodiophilic layer for rapid Na$^+$ diffusion and dendrite suppression [96]. Li et al. also prepared Sn^{2+} pillared Ti$_3$C$_2$ MXene [97]. The

Sn^{2+} can act as sodiophilic seeds and form highly conductive Na$_{15}$Sn$_4$ alloy to balance the electric field. Tian et al. also reported Mg^{2+}-decorated Ti$_3$C$_2$ MXene as a protective layer for Na metal [98]. In addition to these, Wang et al. prepared a 3D sodiophilic Ti$_3$C$_2$ MXene@g-C$_3$N$_4$ hetero-interphase in Figure 5d, in which MXene acts as the highly conductive substrate, and the g-C$_3$N$_4$ acts as an interfacial modulation layer to regulate Na$^+$ deposition. As shown in Figure 5e, the Ti$_3$C$_2$ MXene@g-C$_3$N$_4$ hetero-interphase shows the largest adsorption energy, contributing to the formation of a sodiophilic surface [99]. In conclusion, these free-standing protective layers possess tuned electronic properties, strong sodiophilicity and structural robustness.

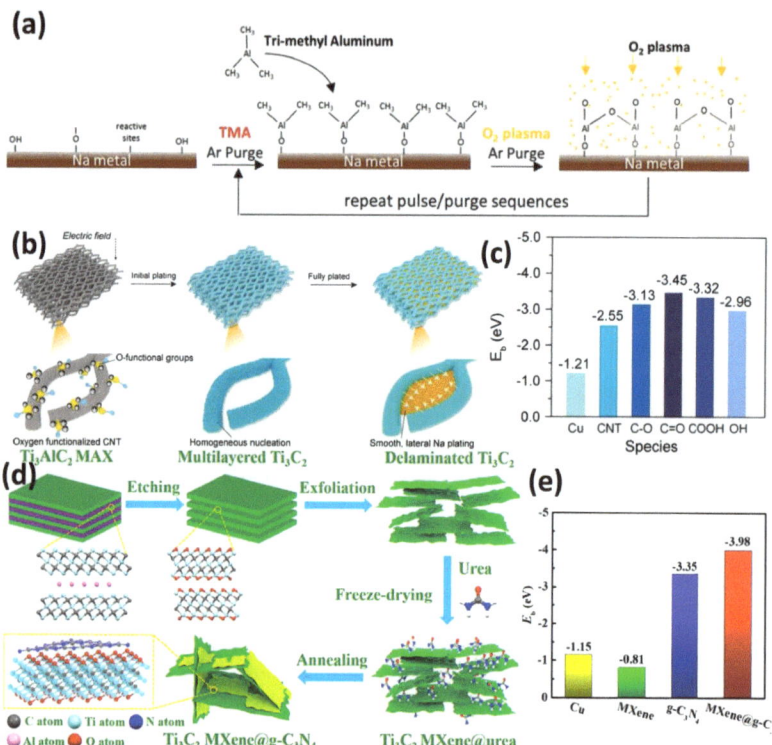

Figure 5. (**a**) Schematic of the ALD deposition of Al$_2$O$_3$ with TMA and O$_2$ plasma [87]. Copyright 2017, Wiely-VCH. (**b**) The function of the O$_f$−CNT network in homogeneous nucleation and smooth Na deposition. (**c**) Binding energies of Na with various functional groups [91]. Copyright 2019, Wiely-VCH. (**d**) The fabrication process of Ti$_3$C$_2$ MXene@g−C$_3$N$_4$ hetero−interphase for dendrite suppression. (**e**) Binding energies of Na with Cu, Ti$_3$C$_2$ MXene, g−C$_3$N$_4$ and Ti$_3$C$_2$ MXene@g−C$_3$N$_4$ [99]. Copyright 2022, American Chemistry Society.

3.2. Organic Interphase

Apart from the inorganic interphase, the organic interphase is also attractive, since the precursor can be precisely designed and optimized at the molecular level [24,100,101]. The organic SEI layer is capable of alleviating the volume expansion and preventing dendrite growth due to its excellent flexibility.

Previously, the polar polymers (poly(dimethylsiloxane)(PDMS), polyacrylic acid(PAA), etc.) were proven to be strongly interacting with Li$^+$, which would be effective for regulating uniform distribution of ion flux [102–104]. Inspired by these works, Ma's group prepared a fibrillar poly(1,1-difluoroethylene) (PVDF) fiber film (f-PVDF) with non-through pores by electro-spinning. By working as a blocking interlayer for dendrite suppression, the f-PVDF film is superior to the conventional compact PVDF film, PVDF film with through pores, polyethylene oxide (PEO) film, and polytetrafluoroethylene (PTFE) film. It is noticed that the polar C-F group affinity to Na$^+$ is stronger than C-O groups in PEO, which provides a better environment for uniform Na$^+$ deposition. Meanwhile, the f-PVDF shows better electrolyte uptake for faster ion conductivity. More recently, Lu et al. protected Na metal anodes by soaking them in 1,3-dioxolane (DOL), as shown in Figure 6a. The polar C-O of DOL can break with the formation of poly(DOL), which enables a faster interfacial transport and a lower interfacial resistance. In detail, the polymerization of DOL forms Na alkoxides (CH$_3$OCH$_2$CH$_2$ONa and CH$_3$CH$_2$OCH$_2$ONa) and HCOONa. Then, the Na alkoxides transform into RONa by further reacting with Na. Finally, the RONa and HCOONa in turn react with DOL continuously. With protected poly(DOL), a cycling life over 2800 h at 1mA cm^{-2} can be obtained in symmetric cells. Lu et al. also proposed spraying DOL for large-scale manufacturing [105]. Meanwhile, as shown in Figure 6b, Wei et al. also used imidazolium ionic liquid monomers to prepare ionic membranes through in-suit electro-polymerization. The obtained ionic membrane (about 50 nm thick) can regulate the electric field and stabilize the Na anode [106].

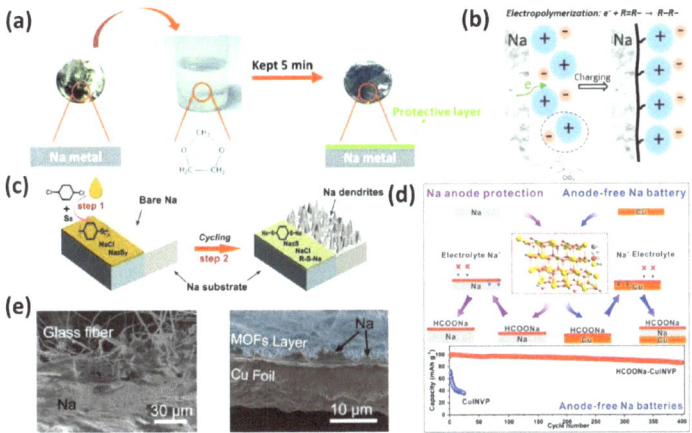

Figure 6. (a) Process of preparing poly(DOL)−protected Na metal [105]. Copyright 2021, Royal Society of Chemistry. (b) In−suit polymerization of imidazolium ionic liquid monomers on Na [106]. Copyright 2017, Wiely-VCH. (c) The fabrication of a PhS$_2$Na$_2$−rich protection layer on Na [107]. Copyright 2019, Wiely-VCH. (d) The HCOONa protective layer on Na metal and Cu foil for Na | | Na$_3$V$_2$(PO$_4$)$_3$ and HCOONa-Cu | | Na$_3$V$_2$(PO$_4$)$_3$ batteries [108]. Copyright 2023, Wiely-VCH. (e) Deposition of Na morphology on Cu foil without and with the MOFs layer [109]. Copyright 2019, Elsevier.

In addition to polymer-based SEI, organic Na benzenedithiolate (PhS$_2$Na$_2$) and HCOONa have also been reported, as shown in Figure 6c,d. In 2020, Wu et al. reported a PhS$_2$Na$_2$ rich protection layer for Na metal. They first chemically treated Na metal with S$_8$ and para-dichlorobenzene (p-DB) in tetrahydrofuran solution, along with the formation of poly(phenylene sulfides) (PPS), NaCl, and Na$_2$S$_y$. Then, it was converted into PhS$_2$Na$_2$ upon cycling. Using DFT calculations, they established the function of PhS$_2$Na$_2$ species. Since the binding energy of Na$^+$ in PhS$_2$Na$_2$ (−2.3 eV) and Ph-S-Na (−2.13 eV) is much lower than that of CH$_3$ONa (−2.49 eV), CH$_3$OCO$_2$Na (−2.497 eV) and Na$_2$CO$_3$ (−3.5 eV), a higher ionic conductivity is proven for the PhS$_2$Na$_2$-based SEI [107]. More recently, Zheng et al. treated Na with formic acid vapor via a solid–gas reaction strategy. After 10 s, the silvery-white Na surface changed into dark-red HCOONa, as confirmed by X-ray diffraction and Raman spectroscopy. Then, the organic HCOONa layer could work as a robust interfacial layer with a low Na$^+$ diffusion barrier. Additionally, the HCOONa interface could also extend to anode-free batteries with format-modified collectors [108].

Recently, metal–organic frameworks (MOFs) and covalent organic frameworks (COFs) have been reported to serve as ionic sieves to control uniform Na$^+$ plating. In 2019,

Chen et al. prepared MOF-199 and ZIF-8 as a coating layer on a Cu substrate [109]. By acting as a compact and robust shield, the MOF-199 layer can physically prevent dendrite growth, thus regulating dense Na deposition and producing less excess SEI formation, as shown in Figure 6e. Similar Mg-based MOF-74 has also been proposed by Yang et al. They first prove that the main group II metals (Be, Mg, and Ba) can act as nucleation seeds for homogeneous Na deposition. Benefiting from these merits, the Mg-based MOF-74 is used to control Na deposition. With eliminated nucleation barriers, a uniform morphology can be obtained [110]. The liquid MOF of ZIF-62 has also been proposed for building protective layers for solid batteries [111]. The ZIF-62 interlayer is synthesized from the high-temperature monophase of liquid MOF. The uniform ZIF-62 layer can increase interfacial sodiophilicity and improve e^-/Na^+ transport kinetics. More recently, the sp^2 carbon COF (sp2c-COF) functional separator has also been built to induce a robust SEI [16]. The high-polarity architecture shows a good affinity toward Na^+, which helps to achieve a uniform ion flux and a nondendritic morphology during plating/stripping [112,113]. To date, reports on applying MOFs and COFs to prevent dendrite growth of Na metal remain limited.

3.3. Hybrid Interphase

To combine the advantages of artificial inorganic SEI and organic SEI, researchers have proposed a hybrid organic–inorganic SEI, in which the inorganic components offer sufficient mechanical strength to suppress dendrites and the organic components provide a certain flexibility to alleviate the volume expansion. In 2017, Kim et al. presented a free-standing inorganic/organic protective layer composed of mechanically robust Al_2O_3 and flexible PVDF polymer (FCPL). The FPCL has a high shear modulus, which is critical for dendrite suppression. Nevertheless, the FCPL could not enhance cycling stability due to its low ionic conductivity [114]. In order to further improve the ionic conductivity, Jiao et al., using NaF and PVDF, prepared a similar free-standing and implantable artificial film (FIAPL) to protect Na [115]. In FIAPL, the organic PVDF film could accommodate the volume expansion and thereby maintain the integrity of the interface, while the inorganic NaF particles could improve ionic conductivity and mechanical strength, resulting in uniform Na nucleation and deposition. The same PVDF/NaF layer was also coated on Cu substrate for Na deposition [116]. Inspired by the PVDF/NaF layer, Yu et al. further treated Na with PTFE via in-suit rolling with the formation of NaF/organic carbon species, which function with C=C and C-F groups [117]. They experimentally verified the high mechanical strength, fast ionic kinetics and good sodiophilicity of this protective layer [117–120]. As reported by Tao et al., the PTFE-derived NaF/carbon layer can be rapidly induced by pressure and a diglyme-induced defluorination reaction, as shown in in Figure 7a. It is explained that the diglyme can bond with Na easily to form chains of O-Na-O, which react with PTFE film rapidly. Benefiting from these merits, the NaF/organic carbon protective layer shows a long life of 1800 h under 3mAh cm^{-2}. The authors also confirmed a similar H_nC-O-H_nC chain could be obtained using other solvents [121].

The polymer/metal interphases have also been proposed. The polymer film is flexible to accommodate surface expansion, whereas the sodiophilic metal can offer sufficient Na^+ ions and high mechanical modulus for dendrite-free plating/stripping. In 2020, Huang et al. reported a well-designed artificial protective layer consisting of PVDF and Sn by coating a Cu collector. [122]. With the PVDF–Sn protective layer, a high average CE of 99.73% can be obtained for 2800 h at 2 mA cm^{-2}. Li et al. also proposed a polyacrylonitrile (PAN) film with a thin Sn layer coated on the bottom. As shown in Figure 7b, benefiting from the low nucleation barrier of Sn seeds, the PAN–Sn protective layer can regulate Na deposition with a controlled location and orientation [123]. More recently, in 2022, Li et al. constructed a similar polymer PVDF and metal Bi layer on Cu substrate (PB@Cu). The cyclic voltammetry and galvanostatic discharge curves in Figure 7c,d confirm the alloying/dealloying of Bi. With Bi metal, the deposition kinetics of Na are increased. At 1 mA cm^{-2} and 1 mAh cm^{-2}, the PVDF-Bi layer provides a high utilization of Na and a long lifetime of 2500 h, as shown in Figure 7e. The superior electrochemical performance of the PVDF-Bi layer is revealed to originate from flexible PVDF, which could accommodate severe volume change induced by Na⁺ plating/stripping. Meanwhile, the Bi and/or sodiated Na$_3$Bi can offer high ionic conductivity and sufficient mechanical strength [124].

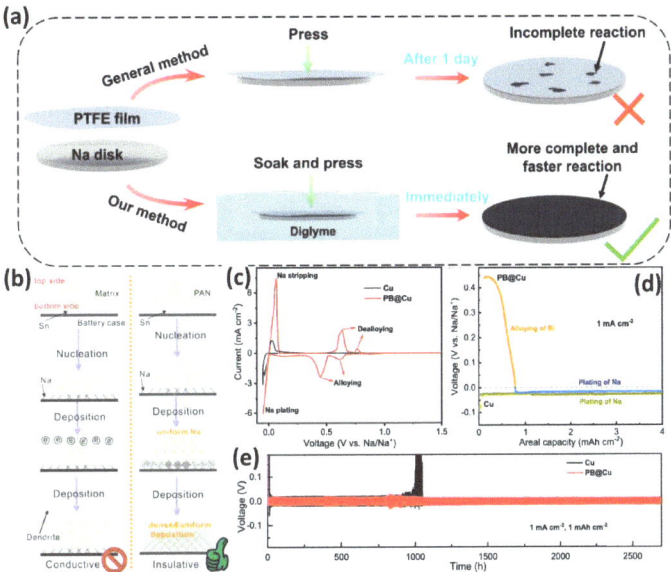

Figure 7. (**a**) Schematic of pressure and diglyme−induced defluorination reaction for preparing PTFE-derived NaF/carbon layer [121]. Copyright 2023, Wiely-VCH. (**b**) Schematic illustration of PAN−Sn guiding Na deposition with a controlled location and orientation [123]. Copyright 2020, Wiely-VCH. (**c**) CV curves and (**d**) the first discharge curves of Na||Cu batteries with bare Cu and PB@Cu. (**e**) The voltage−time curves of Cu and PB@Cu with Na anode at 1 mA cm^{-2} and 1 mAh cm^{-2} [124]. The plating of Na on Cu and Pb@Cu are highlighted with different colors, respectively. Copyright 2022, Elsevier.

In contrast to stiff and dense inorganic ALD coatings, MLD coatings are confirmed to release volume expansion due to the reduced density and increased flexibility of hybrid organic–inorganic layers [90]. Meanwhile, the hybrid layers provide higher tune ability, since the integration of organic bonds in MLD coatings provides attractive chemical/electrochemical, mechanical and electrical performances. As expected, the MLD technologies show significant improvements in stabilizing Na metal without dendrite growth. As shown in Figure 8a, in 2017, Zhao et al. used trimethylaluminum and ethylene glycol (Alucone) to introduce an organic–inorganic composite layer on the Na anode via MLD at 85 °C. During experimental testing, thicknesses of 10, 25 and 40 MLD cycles were performed. It was proven that 25 MLD cycles of AlEG (Na@25Alucone) were optimal. As reported, the SEI on Na@25Alucone showed higher contents of beneficial NaF and Na$_2$O [125].

The MLD technology is also beneficial for solid Na batteries. In 2020, Sun et al. also coated the same Alucone via MLD between Na and solid Na$_3$SbS$_3$ and Na$_3$PS$_4$ electrolytes, in which the Alucone layer worked as an interfacial stabilizer [126]. As confirmed, the type of artificial SEI layer is dependent on the ALD and MLD depositions cycles. If the deposition cycles of ALD and MLD are small, it will form a nano-alloy interface; if the deposition cycles of ALD and MLD are large, it will form full monolayer. More recently, in 2023, Sun et al. formed nano-hybrid interfaces with nano-alloy and nano-laminated structures (from Al$_2$O$_3$ to Alucone) through ALD-deposited inorganic Al$_2$O$_3$ and MLD-deposited organic Alucone for alkali metal anodes, as shown in Figure 8b [127]. Time-of-flight secondary ion mass spectrometry (TOF-SIMS) results are shown in Figure 8c–e; the Na⁻, Al⁻, CAL⁻ and AlO$_2$⁻ are probed on the Na surface, which is realized to be robust and chemically/electrochemically stable upon plating/stripping. In this study, three types of nano-hybrid interfaces are investigated: 1 layer of Al_2O_3 with 1 layer Alucone (1ALD-1MLD); 2 layers of Al_2O_3 with 2 layers of Alucone (2ALD-2MLD) and 5 layers of Al_2O_3 with 5 layers of Alucone (5ALD-5MLD). At the same time, the total thickness of the nano-hybrid interfaces can be controlled by deposited ALD/MLD cycles (mainly including 5, 10 and 25 cycles). The corresponding samples are donated as (1ALD-1MLD)5, (1ALD-1MLD)10 and (1ALD-1MLD)25. Among all samples, the (1ALD-1MLD)10 alloy interface shows the best performance at 3 mA cm^{-2} and 1 mAh cm^{-2} for Na metal. The mossy/dendritic Na growth and "dead" Na formation are effectively suppressed, which would account for the improved performance. Finally, the optimal thickness of the Al_2O_3–Alucone alloy interface for Na metal is 4 nm.

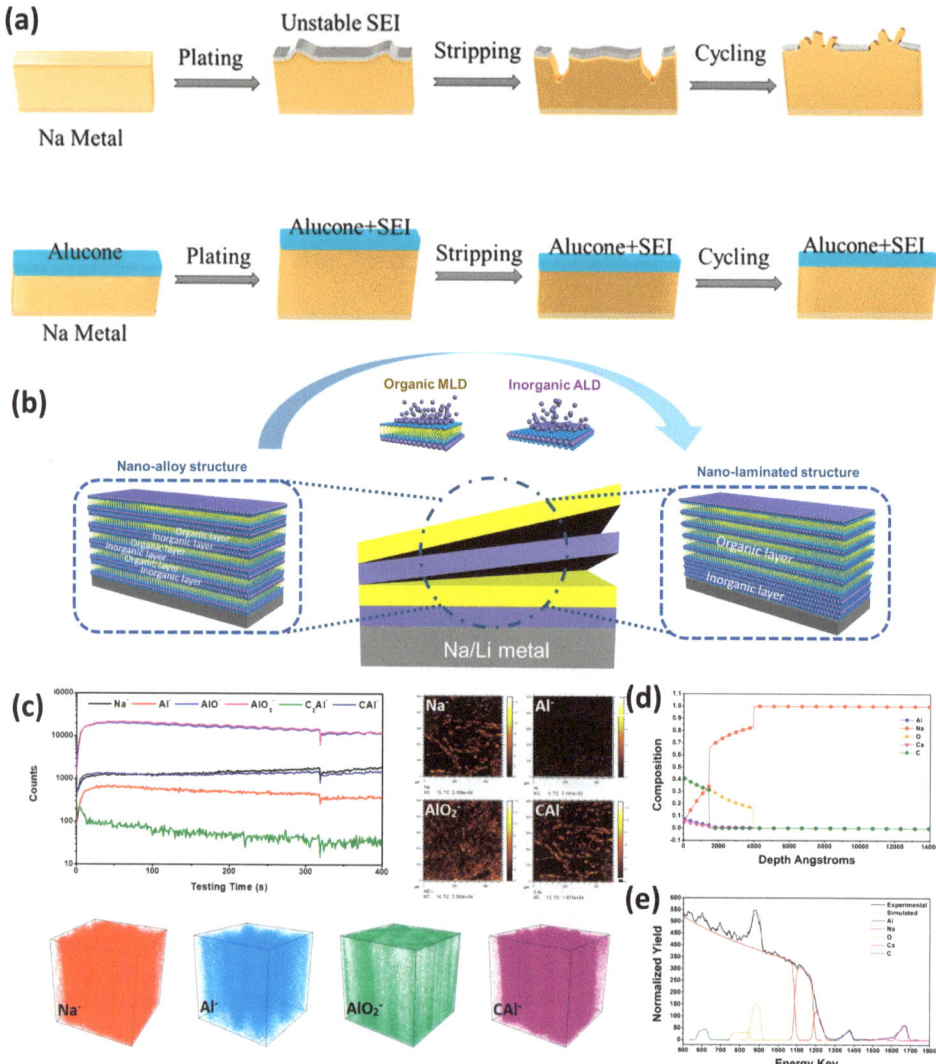

Figure 8. (**a**) Na plating/stripping on bare Na and MLD−coated alucone Na [125]. Copyright 2017, American Chemistry Society. (**b**) Schematic illustration of the fabrication of the nano−alloy and nano−laminated interfacial structures by ALD and MLD deposition. (**c**) The TOF−SIMS images and depth profiles of $Na^−$, $Al^−$, $CAL^−$ and $AlO_2^{2−}$ ions. (**d**) The Rutherford backscattering spectrometry and the (**e**) calculated depth profiles of the (1ALD−1MLD)10 alloy interface [127]. Copyright 2023, Wiely-VCH.

4. Conclusions and Perspectives

In this review, we summarize recent progress on artificial SEI design for Na metal anodes. Some related studies are summarized in Table 1. The configuration of advanced artificial SEI layers has been proposed by several researchers; this includes chemical coating, physical deposition, ex-suit conversion reactions and free-standing films. Based on experimental understanding, the artificial SEI layer can be precisely designed by optimizing the composition, thickness and morphology. Different types of artificial SEI films have their own advantages and disadvantages in suppressing dendrite growth. The inorganic artificial SEI layers usually show high ionic conductivity, good mechanical strength, high Young's modulus and excellent stability. However, the inorganic artificial SEI is brittle, which makes them rupture easily during huge volume expansion. On the contrary, the organic artificial SEI layers are usually highly elastic, which encourages intimate contact with the Na matrix and the effective maintenance of volume expansion. Meanwhile, the organic artificial SEI can be processed easily. However, the mechanical strength, Young's modulus and ionic conductivity of organic artificial SEI layers are much lower than those of inorganic artificial SEI layers. As a result, the long-term cycling performance of the Na anode with an organic artificial SEI layer is not very good. For the hybrid organic–inorganic SEI, the inorganic components offer sufficient mechanical strength to suppress dendrites, and the organic components provide a certain flexibility to alleviate the volume expansion. At present, the hybrid artificial SEI layers are mainly prepared via ex-suit coatings, which are limited in controlling the distribution and connection of inorganic–organic components. At the same time, the transport of Na^+ in the hybrid interphase is still limited. These protective layers are either highly conductive or demonstrate mechanical stiffness/flexibility. Benefiting from these merits, the protective layers are highly effective in regulating uniform Na^+ deposition and suppressing dendritic Na formation.

Despite the advantages mentioned above, some challenges still need to be explored for Na metal: (1) the effect of the physical structure, chemical composition, optimization method and Na^+ diffusion mechanism on dendrite suppression should be further investigated; (2) the evolution and failure of artificial SEIs during plating/stripping need to be studied; (3) advanced characterization technologies should be used to reveal the inner relationship between the SEI and Na metal; (4) the design of SEIs should meet practical conditions, especially with limited Na and lean electrolytes; (5) the formation/growth of dendritic Na and the dynamic evolution of the interphase layer need to be fully understood for better SEI configuration.

Table 1. A comparison of the electrochemical performances of various artificial SEI layers for Na metal. In the table, EC, PC, FEC, DMC, DEC, EMC, DME, TEGDME and NaTFSI represent ethylene carbonate, propylene carbonate, fluoroethylene carbonate, dimethyl carbonate, diethyl carbonate, ethyl methyl carbonate, 1,2-dimethoxyethane, tetraethylene glycol dimethyl ether, and Na bis(trifluoromethanesulfonyl)imide, respectively.

Interphase	Technique	Electrolyte	Current (mA cm^{-2}) Capacity (mAh cm^{-2})	Lifetime (h)	Ref.
NaF/Na$_3$Sb	In-suit reaction	1 M NaClO$_4$ in EC/PC/2%FEC	0.5, 0.25	650	[54]
NaF	In-suit reaction	1 M NaTFSI in DME	2, 2	1000	[59]
NaI	In-suit reaction	1 M NaClO$_4$ in EC/DEC/5%FEC	0.25, 0.75	500	[65]
NaBr	In-suit reaction	1 M NaPF$_6$ in EC/PC	1, 1	250	[66]
Na$_x$Mo$_6$S$_8$	In-suit sodiation	1 M NaPF$_6$ in EC/DMC	0.5, /	1200	[68]
Na$_3$P	Rolling	1 M NaTFSI in FEC/EMC	1, 1	780	[71]
Na$_2$Te	Rolling	1 M NaClO$_4$ in EC/DEC/5%FEC	1, 1	700	[72]
Na$_3$P/NaBr	In-suit reaction	1 M NaPF$_6$ in EC/DEC/5%FEC	1, 1	700	[73]
Bi	In-suit reaction	1 M NaSO$_3$CF$_3$ in Diglyme	0.5, 1	1000	[81]
Al$_2$O$_3$	ALD	1 M NaClO$_4$ in EC/DEC	0.25, 0.125	450	[87]

Table 1. Cont.

Interphase	Technique	Electrolyte	Current (mA cm^{-2}) Capacity (mAh cm^{-2})	Lifetime (h)	Ref.
O$_f$-CNT	Free-standing films	1 M NaSO$_3$CF$_3$ in Diglyme	1, 1	4500	[91]
Poly(DOL)	In-suit reaction	1 M NaPF$_6$ in TEGDME	1, 1	2800	[105]
PhS$_2$Na$_2$	Self-activation	1 M NaPF6 in EC/PC	1, 1	800	[107]
HCOONa	In-suit reaction	1 M NaPF$_6$ in Diglyme	2, 1	2200	[108]
NaF/PVDF	Rolling	1 M NaClO$_4$ in EC/DEC/2%FEC	1, 1	770	[117]
PBDF/Bi	Coating on Cu	1 M NaPF$_6$ in Diglyme	1, 1	2700	[124]
Alucone	MLD	1 M NaPF$_6$ in EC/PC	1, 1	270	[125]
Al$_2$O$_3$-Alucone	ALD-MLD	1 M NaPF$_6$ in EC/DEC/FEC	3, 1	1500	[127]

Artificial SEI layers with high ionic conductivity, high Young's modulus and mechanical flexibility are effective for suppressing dendritic Na formation. However, artificial SEIs alone are insufficient to address all the existing issues of Na metal anodes. For these reasons, multiple approaches with specific objectives are necessary for promoting the realization of metal Na. We expect this review will promote a deeper understanding of the SEI of Na.

Author Contributions: P.S. prepared and revised the raw manuscript. X.W. edited the figures. Y.J. and X.C. revised this manuscript. All authors have read and agreed to the published version of the manuscript.

Funding: This work was supported by the National Natural Science Foundation of China (No. 52002083) and the start-up grants from Anhui University (No. S020318031/001).

Data Availability Statement: There is no data support for this review.

Conflicts of Interest: The authors declare no competing financial interest.

References

1. Hwang, J.-Y.; Myung, S.-T.; Sun, Y.-K. Sodium-ion batteries: Present and future. *Chem. Soc. Rev.* **2017**, *46*, 3529–3614. [CrossRef] [PubMed]
2. Abraham, K.M. How Comparable are sodium-ion batteries to lithium-ion counterparts? *ACS Energy Lett.* **2020**, *5*, 3544–3547. [CrossRef]
3. Hirsh, H.S.; Li, Y.X.; Tan, D.H.S.; Zhang, M.H.; Zhao, E.Y.; Meng, Y.S. Sodium-ion batteries paving the way for grid energy storage. *Adv. Energy Mater.* **2020**, *10*, 2001274. [CrossRef]
4. Liu, T.; Zhang, Y.; Jiang, Z.; Zeng, X.; Ji, J.; Li, Z.; Gao, X.; Sun, M.; Lin, Z.; Ling, M.; et al. Exploring competitive features of stationary sodium ion batteries for electrochemical energy storage. *Energy Environ. Sci.* **2019**, *12*, 1512–1533. [CrossRef]
5. Zhao, Y.; Yang, X.F.; Kuo, L.Y.; Kaghazchi, P.; Sun, Q.; Liang, J.N.; Wang, B.Q.; Lushington, A.; Li, R.Y.; Zhang, H.M.; et al. High capacity, dendrite-free growth, and minimum volume change Na metal anode. *Small* **2018**, *14*, 1703717. [CrossRef]
6. Sun, B.; Xiong, P.; Maitra, U.; Langsdorf, D.; Yan, K.; Wang, C.Y.; Janek, J.; Schröder, D.; Wang, G.X. Design strategies to enable the efficient use of sodium metal anodes in high-energy batteries. *Adv. Mater.* **2019**, *32*, 1903891. [CrossRef]
7. Cao, R.G.; Mishra, K.; Li, X.L.; Qian, J.F.; Engelhard, M.H.; Bowden, M.E.; Han, S.H.; Mueller, K.T.; Henderson, W.A.; Zhang, J.-G. Enabling room temperature sodium metal batteries. *Nano Energy* **2016**, *30*, 825–830. [CrossRef]
8. Zheng, J.M.; Chen, S.R.; Zhao, W.G.; Song, J.H.; Mueller, K.T.; Zhang, J.-G. Extremely stable sodium metal batteries enabled by localized high-concentration electrolytes. *ACS Energy Lett.* **2018**, *3*, 315–321. [CrossRef]
9. Lee, Y.; Lee, J.; Lee, J.M.; Kim, K.; Cha, A.; Kang, S.J.; Wi, T.; Kang, S.J.; Lee, H.-W.; Choi, N.-S. Fluoroethylene carbonate-based electrolyte with 1 M sodium bis (fluorosulfonyl) imide enables high-performance sodium metal electrodes. *ACS Appl. Mater. Interfaces* **2018**, *10*, 15270–15280. [CrossRef]
10. Ma, B.; Bai, P. Fast charging limits of ideally stable metal anodes in liquid electrolytes. *Adv. Energy Mater.* **2022**, *12*, 2102967. [CrossRef]
11. Ji, Y.Y.; Li, J.B.; Li, J.L. Recent development of electrolyte engineering for sodium metal batteries. *Batteries* **2022**, *8*, 157. [CrossRef]
12. Zheng, X.Y.; Cao, Z.; Gu, Z.Y.; Huang, L.Q.; Sun, Z.H.; Zhao, T.; Yu, S.J.; Wu, X.L.; Huang, Y.H. Toward high temperature sodium metal batteries via regulating the electrolyte/electrode interfacial chemistries. *ACS Energy Lett.* **2022**, *7*, 2032–2042. [CrossRef]
13. Xia, X.M.; Lv, X.; Yao, Y.; Chen, D.; Tang, F.; Liu, N.; Feng, Y.Z.; Rui, X.H.; Yu, Y. A sodiophilic VN interlayer stabilizing a Na metal anode. *Nanoscale Horiz.* **2022**, *7*, 899–907. [CrossRef] [PubMed]

14. Jäckle, M.; Groß, A. Microscopic properties of lithium, sodium, and magnesium battery anode materials related to possible dendrite growth. *J. Chem. Phys.* **2014**, *141*, 174710. [CrossRef]
15. Wang, H.; Bai, W.L.; Wang, H.; Kong, D.Z.; Xu, T.Q.; Zhang, Z.F.; Zang, J.H.; Wang, X.C.; Zhang, S.; Tian, Y.T.; et al. 3D printed Au/rGO microlattice host for dendrite-free sodium metal anode. *Energy Storage Mater.* **2023**, *55*, 631–641. [CrossRef]
16. Kang, T.; Sun, C.; Li, Y.; Song, T.; Guan, Z.; Tong, Z.; Nan, J.; Lee, C.S. Dendrite-free sodium metal anodes via solid electrolyte interphase engineering with a covalent organic framework separator. *Adv. Energy Mater.* **2023**, *13*, 2204083. [CrossRef]
17. Li, M.H.; Lu, G.J.; Zheng, W.K.; Zhao, Q.N.; Li, Z.P.; Jiang, X.P.; Yang, Z.G.; Li, Z.Y.; Qu, B.H.; Xu, C.H. Multifunctionalized safe separator toward practical sodium-metal batteries with high-performance under high mass loading. *Adv. Funct. Mater.* **2023**, *33*, 2214759. [CrossRef]
18. Wang, Y.X.; Dong, H.; Katyal, N.; Hao, H.C.; Liu, P.C.; Celio, H.; Henkelman, G.; Watt, J.; Mitlin, D. Sodium-antimony-telluride intermetallic allows sodium metal cycling at 100% depth of discharge and as anode-free metal battery. *Adv. Mater.* **2022**, *34*, 2106005. [CrossRef]
19. Liu, H.; Cheng, X.B.; Huang, J.-Q.; Kaskel, S.; Chou, S.L.; Park, H.S.; Zhang, Q. Alloy anodes for rechargeable alkali-metal batteries: Progress and challenge. *ACS Mater. Lett.* **2019**, *1*, 217–229. [CrossRef]
20. Tang, S.; Zhang, Y.Y.; Zhang, X.G.; Li, J.T.; Wang, X.Y.; Yan, J.W.; Wu, D.Y.; Zheng, M.S.; Dong, Q.F.; Mao, B.W. Stable Na plating and stripping electrochemistry promoted by in situ construction of an alloy-based sodiophilic interphase. *Adv. Mater.* **2019**, *31*, 1807495. [CrossRef]
21. Wang, H.; Wang, C.L.; Matios, E.; Luo, J.M.; Lu, X.; Zhang, Y.W.; Hu, X.F.; Li, W.Y. Enabling ultrahigh rate and capacity sodium metal anodes with lightweight solid additives. *Energy Storage Mater.* **2020**, *32*, 244–252. [CrossRef]
22. Bao, C.Y.; Wang, B.; Liu, P.; Wu, H.; Zhou, Y.; Wang, D.L.; Liu, H.K.; Dou, S.X. Solid electrolyte interphases on sodium metal anodes. *Adv. Funct. Mater.* **2020**, *30*, 2004891. [CrossRef]
23. Gao, L.; Chen, J.; Chen, Q.L.; Kong, X.Q. The chemical evolution of solid electrolyte interface in sodium metal batteries. *Sci. Adv.* **2022**, *8*, eabm4606. [CrossRef]
24. Wang, T.; Hua, Y.B.; Xu, Z.W.; Yu, J.S. Recent advanced development of artificial interphase engineering for stable sodium metal anodes. *Small* **2022**, *18*, 2102250. [CrossRef] [PubMed]
25. Zhao, C.L.; Lu, Y.X.; Yue, J.M.; Pan, D.; Qi, Y.R.; Hu, Y.S.; Chen, L.Q. Advanced Na metal anodes. *J. Energy Chem.* **2018**, *27*, 1584–1596. [CrossRef]
26. Li, Z.P.; Zhu, K.J.; Liu, P.; Jiao, L.F. 3D confinement strategy for dendrite-free sodium metal batteries. *Adv. Energy Mater.* **2022**, *12*, 2100359. [CrossRef]
27. Chen, X.; Bai, Y.K.; Shen, X.; Peng, H.-J.; Zhang, Q. Sodiophilicity/potassiophilicity chemistry in sodium/potassium metal anodes. *J. Energy Chem.* **2020**, *51*, 1–6. [CrossRef]
28. Lu, Z.Y.; Yang, H.J.; Guo, Y.; He, P.; Wu, S.C.; Yang, Q.H.; Zhou, H.S. Electrolyte sieving chemistry in suppressing gas evolution of sodium-metal batteries. *Angew Chem. Int. Ed.* **2022**, *61*, e202206340. [CrossRef] [PubMed]
29. Wang, E.H.; Wan, J.; Guo, Y.-J.; Zhang, Q.Y.; He, W.H.; Zhang, C.H.; Chen, W.-P.; Yan, H.-J.; Xue, D.-J.; Fang, T.T.; et al. Mitigating electron leakage of solid electrolyte interface for stable sodium-ion batteries. *Angew Chem. Int. Ed.* **2023**, *62*, e202216354.
30. Mandl, M.; Becherer, J.; Kramer, D.; Mönig, R.; Diemant, T.; Diemant, T.; Behm, J.; Hahn, M.; Böse, O.; Danzer, M.A. Sodium metal anodes: Deposition and dissolution behaviour and SEI formation. *Electrochim. Acta* **2020**, *354*, 136698. [CrossRef]
31. Lee, B.; Paek, E.; Mitlin, D.; Lee, S.W. Sodium metal anodes: Emerging solutions to dendrite growth. *Chem. Rev.* **2019**, *119*, 5416–5460. [CrossRef] [PubMed]
32. Zhao, Y.; Adair, K.R.; Sun, X.L. Recent developments and insights into the understanding of Na metal anodes for Na-metal batteries. *Energy Environ. Sci.* **2018**, *11*, 2673–2695. [CrossRef]
33. Lei, D.N.; He, Y.B.; Huang, H.J.; Yuan, Y.F.; Zhong, G.M.; Zhao, Q.; Hao, X.G.; Zhang, D.F.; Lai, C.; Zhang, S.W.; et al. Cross-linked beta alumina nanowires with compact gel polymer electrolyte coating for ultra-stable sodium metal battery. *Nat. Commun.* **2019**, *10*, 4244. [CrossRef] [PubMed]
34. Liu, T.F.; Yang, X.K.; Nai, J.W.; Wang, Y.; Liu, Y.J.; Liu, C.T.; Tao, X.Y. Recent development of Na metal anodes: Interphase engineering chemistries determine the electrochemical performance. *Chem. Eng. J.* **2021**, *409*, 127943. [CrossRef]
35. Ji, Y.Y.; Sun, H.C.; Li, Z.B.; Ma, L.; Zhang, W.G.; Liu, Y.M.; Pan, L.K. Salt engineering toward stable cation migration of Na metal anodes. *J. Mater. Chem. A* **2022**, *10*, 25539–25545. [CrossRef]
36. Liu, W.; Liu, P.C.; Mitlin, D. Review of emerging concepts in SEI analysis and artificial SEI membranes for lithium, sodium, and potassium metal battery anodes. *Adv. Energy Mater.* **2020**, *10*, 2002297. [CrossRef]
37. Matios, E.; Wang, H.; Wang, C.L.; Li, W.Y. Enabling safe sodium metal batteries by solid electrolyte interphase engineering: A review. *Ind. Eng. Chem. Res.* **2019**, *58*, 9758–9780. [CrossRef]
38. Xia, X.M.; Du, C.F.; Zhong, S.E.; Jiang, Y.; Yu, H.; Sun, W.P.; Pan, H.G.; Rui, X.H.; Yu, Y. Homogeneous Na deposition enabling high-energy Na-metal batteries. *Adv. Funct. Mater.* **2022**, *32*, 2110280. [CrossRef]
39. Lee, J.; Kim, J.; Kim, S.; Jo, C.S.; Lee, J. A review on recent approaches for designing the SEI layer on sodium metal anodes. *Mater. Adv.* **2020**, *1*, 3143–3166. [CrossRef]
40. Liu, W.; Liu, P.C.; Mitlin, D. Tutorial review on structure-dendrite growth relations in metal battery anode supports. *Chem. Soc. Rev.* **2020**, *49*, 7284–7300. [CrossRef]

41. Lin, Z.; Liu, T.F.; Ai, X.P.; Liang, C.D. Aligning academia and industry for unified battery performance metrics. *Nat. Commun.* **2018**, *9*, 5262. [CrossRef] [PubMed]
42. Lee, K.; Lee, Y.J.; Lee, M.J.; Han, J.H.; Lim, J.; Ryu, K.; Yoon, H.; Kim, B.H.; Kim, B.J.; Lee, S.W. A 3D hierarchical host with enhanced sodiophilicity enabling anode-free sodium-metal batteries. *Adv. Mater.* **2022**, *34*, 2109767. [CrossRef] [PubMed]
43. Yang, W.; Yang, W.; Dong, L.B.; Shao, G.J.; Wang, G.X.; Peng, X.W. Hierarchical ZnO nanorod arrays grown on copper foam as an advanced three-dimensional skeleton for dendrite-free sodium metal anodes. *Nano Energy* **2021**, *80*, 105563. [CrossRef]
44. Wang, C.L.; Wang, H.; Matios, E.; Hu, X.F.; Li, W.Y. A chemically engineered porous copper matrix with cylindrical core-shell skeleton as a stable host for metallic sodium anodes. *Adv. Funct. Mater.* **2018**, *28*, 1802282. [CrossRef]
45. Ma, Y.; Gu, Y.T.; Yao, Y.Z.; Jin, H.D.; Zhao, X.H.; Yuan, X.T.; Lian, Y.L.; Qi, P.W.; Shah, R.; Peng, Y.; et al. Alkaliphilic Cu_2O nanowires on copper foam for hosting Li/Na as ultrastable alkali-metal anodes. *J. Mater. Chem. A* **2019**, *7*, 20926–20935. [CrossRef]
46. Mubarak, N.; Ihsan-Ul-Haq, M.; Huang, H.; Cui, J.; Yao, S.S.; Susca, A.; Wu, J.X.; Wang, M.Y.; Zhang, X.H.; Huang, B.L.; et al. Metal organic framework-induced mesoporous carbon nanofibers as ultrastable Na metal anode host. *J. Mater. Chem. A* **2020**, *8*, 10269–10282. [CrossRef]
47. Yue, L.; Qi, Y.R.; Niu, Y.B.; Bao, S.J.; Xu, M.W. Low-barrier, dendrite-free, and stable Na plating/stripping enabled by gradient sodiophilic carbon skeleton. *Adv. Energy Mater.* **2021**, *11*, 2102497. [CrossRef]
48. Wang, Z.X.; Huang, Z.X.; Wang, H.; Li, W.D.; Wang, B.Y.; Xu, J.M.; Xu, T.T.; Zang, J.H.; Kong, D.Z.; Li, X.J.; et al. 3D-printed sodiophilic V_2CT_x/rGO-CNT MXene microgrid aerogel for stable Na metal anode with high areal capacity. *ACS Nano* **2022**, *16*, 9105–9116. [CrossRef]
49. Shi, H.D.; Yue, M.; Zhang, C.F.; Dong, Y.F.; Liu, P.F.; Zheng, S.H.; Huang, H.J.; Chen, J.; Wen, P.C.; Xu, Z.C.; et al. 3D flexible, conductive, and recyclable $Ti_3C_2T_x$ MXene-melamine foam for high-areal-capacity and long-lifetime alkali-metal anode. *ACS Nano* **2020**, *14*, 8678–8688. [CrossRef]
50. Chu, C.X.; Li, R.; Cai, F.P.; Bai, Z.C.; Wang, Y.X.; Xu, X.; Wang, N.; Yang, J.; Dou, S.X. Recent advanced skeletons in sodium metal anodes. *Energy Environ. Sci.* **2021**, *14*, 4318–4340. [CrossRef]
51. Yang, C.; Xin, S.; Mai, L.Q.; You, Y. Materials design for high-safety sodium-ion battery. *Adv. Energy Mater.* **2021**, *11*, 2000974. [CrossRef]
52. Wang, H.; Yu, D.; Kuang, C.; Cheng, L.; Li, W.; Feng, X.; Zhang, Z.; Zhang, X.; Zhang, Y. Alkali metal anodes for rechargeable batteries. *Chem* **2019**, *5*, 313–338. [CrossRef]
53. Zhang, X.Q.; Cheng, X.B.; Zhang, Q. Advances in interfaces between Li metal anode and electrolyte. *Adv. Mater. Interfaces* **2018**, *5*, 1701097. [CrossRef]
54. Xu, Z.X.; Yang, J.; Zhang, T.; Sun, L.M.; Nuli, Y.N.; Wang, J.L.; Hirano, S. Stable Na metal anode enabled by a reinforced multistructural SEI layer. *Adv. Funct. Mater.* **2019**, *29*, 1901924. [CrossRef]
55. Fang, W.; Jiang, H.; Zheng, Y.; Zheng, H.; Liang, X.; Chen, C.H.; Xiang, H.F. A bilayer interface formed in high concentration electrolyte with SbF_3 additive for long-cycle and high-rate sodium metal battery. *J. Power Sources* **2020**, *455*, 227956. [CrossRef]
56. Wang, G.; Xiong, X.H.; Xie, D.; Fu, X.X.; Lin, Z.H.; Yang, C.H.; Zhang, K.L.; Liu, M.L. A scalable approach for dendrite-free alkali metal anodes via room-temperature facile surface fluorination. *ACS Appl. Mater. Interfaces* **2019**, *11*, 4962–4968. [CrossRef]
57. Zhou, X.F.; Liu, F.F.; Wang, Y.J.; Yao, Y.; Shao, Y.; Rui, X.H.; Wu, F.X.; Yu, Y. Heterogeneous interfacial layers derived from the in situ reaction of CoF_2 nanoparticles with sodium metal for dendrite-free Na metal anodes. *Adv. Energy Mater.* **2022**, *12*, 2202323. [CrossRef]
58. Miao, X.G.; Di, H.X.; Ge, X.L.; Zhao, D.Y.; Wang, P.; Wang, R.T.; Wang, C.X.; Yin, L.W. AlF_3-modified anode-electrolyte interface for effective Na dendrites restriction in NASICON-based solid-state electrolyte. *Energy Storage Mater.* **2020**, *30*, 170–178. [CrossRef]
59. Cheng, Y.F.; Yang, X.M.; Li, M.H.; Li, X.Y.; Lu, X.Z.; Wu, D.J.; Han, B.; Zhang, Q.; Zhu, Y.M.; Gu, M. Enabling ultrastable alkali metal anodes by artificial solid electrolyte interphase fluorination. *Nano Lett.* **2022**, *22*, 4347–4353. [CrossRef]
60. Xie, Y.Y.; Hu, J.X.; Zhang, L.Y.; Wang, A.N.; Zheng, J.Q.; Li, H.X.; Lai, Y.Q.; Zhang, Z.A. Stabilizing Na metal anode with NaF interface on spent cathode carbon from aluminum electrolysis. *Chem. Commun.* **2021**, *57*, 7561–7564. [CrossRef]
61. Zheng, X.Y.; Fu, H.Y.; Hu, C.C.; Xu, H.; Huang, Y.; Wen, J.Y.; Sun, H.B.; Luo, W.; Huang, Y.H. Toward a stable sodium metal anode in carbonate electrolyte: A compact, inorganic alloy interface. *J. Phys. Chem. Lett.* **2019**, *10*, 707–714. [CrossRef] [PubMed]
62. Kumar, V.; Eng, A.Y.S.; Wang, Y.; Nguyen, D.T.; Ng, M.F.; Seh, Z.W. An artificial metal-alloy interphase for high-rate and long-life sodium-sulfur batteries. *Energy Storage Mater.* **2020**, *19*, 1–8. [CrossRef]
63. Lu, Q.Q.; Omar, A.; Hantusch, M.; Oswald, S.; Ding, L.; Nielsch, K.; Mikhailova, D. Dendrite-free and corrosion-resistant sodium metal anode for enhanced sodium batteries. *Appl. Surf. Sci.* **2022**, *600*, 154168. [CrossRef]
64. Chen, Q.W.; He, H.; Hou, Z.; Zhuang, W.M.; Zhang, T.X.; Sun, Z.Z.; Huang, L.M. Building an artificial solid electrolyte interphase with high-uniformity and fast ion diffusion for ultralong-life sodium metal anodes. *J. Mate. Chem. A* **2020**, *8*, 16232–16237. [CrossRef]
65. Tian, H.J.; Shao, H.Z.; Chen, Y.; Fang, X.Q.; Xiong, P.; Sun, B.; Nottenc, P.H.L.; Wang, G.X. Ultra-stable sodium metal-iodine batteries enabled by an in-situ solid electrolyte interphase. *Nano Energy* **2019**, *57*, 692–702. [CrossRef]
66. Choudhury, S.; Wei, S.Y.; Ozhabes, Y.; Gunceler, D.; Zachman, M.J.; Tu, Z.Y.; Shin, J.H.; Nath, P.; Agrawall, A.; Kourkoutis, L.F.; et al. Designing solid-liquid interphases for sodium batteries. *Nat. Commun.* **2017**, *8*, 898. [CrossRef]

67. Zhao, Y.; Liang, J.W.; Sun, Q.; Goncharova, L.V.; Wang, J.W.; Wang, C.H.; Adair, K.R.; Li, X.N.; Zhao, F.P.; Sun, Y.P.; et al. In-situ formation of highly controllable and stable Na_3PS_4 as protective layer for Na metal anode. *J. Mate. Chem. A* **2019**, *47*, 4119–4125. [CrossRef]
68. Lu, K.; Gao, S.Y.; Li, G.S.; Kaelin, J.; Zhang, Z.C.; Cheng, Y.W. Regulating interfacial Na-ion flux via artificial layers with fast ionic conductivity for stable and high-rate Na metal batteries. *ACS Mater. Lett.* **2019**, *1*, 303–309. [CrossRef]
69. Zhang, D.; Li, B.; Wang, S.; Yang, S.B. Simultaneous formation of artificial SEI film and 3D host for stable metallic sodium anodes. *ACS Appl. Mater. Interfaces* **2017**, *9*, 40265–40272. [CrossRef]
70. Wang, X.C.; Fu, L.; Zhan, R.M.; Wang, L.Y.; Li, G.C.; Wan, M.T.; Wu, X.L.; Seh, Z.W.; Wang, L.; Sun, Y.M. Addressing the low solubility of a solid electrolyte interphase stabilizer in an electrolyte by composite battery anode design. *ACS Appl. Mater. Interfaces* **2021**, *13*, 13354–13361. [CrossRef]
71. Shi, P.C.; Zhang, S.P.; Lu, G.X.; Wang, L.F.; Jiang, Y.; Liu, F.F.; Yao, Y.; Yang, H.; Ma, M.Z.; Ye, S.F.; et al. Red phosphorous-derived protective layers with high ionic conductivity and mechanical strength on dendritefree sodium and potassium metal anodes. *Adv. Energy Mater.* **2021**, *11*, 2003381. [CrossRef]
72. Yang, H.; He, F.X.; Li, M.H.; Huang, F.Y.; Chen, Z.H.; Shi, P.C.; Liu, F.F.; Jiang, Y.; He, L.X.; Gu, M.; et al. Design principles of sodium/potassium protection layer for high-power high-energy sodium/potassium-metal batteries in carbonate electrolytes: A case study of Na_2Te/K_2Te. *Adv. Mater.* **2021**, *33*, 2106353. [CrossRef] [PubMed]
73. Luo, Z.; Tao, S.S.; Tian, Y.; Xu, L.Q.; Wang, Y.; Cao, X.Y.; Wang, Y.P.; Deng, W.T.; Zou, G.Q.; Liu, H.; et al. Robust artificial interlayer for columnar sodium metal anode. *Nano Energy* **2022**, *97*, 107203. [CrossRef]
74. Zhang, Y.J.; Huang, Z.Y.; Liu, H.X.; Chen, H.F.; Wang, Y.Y.; Wu, K.; Wang, G.Y.; Wu, C. Amorphous phosphatized hybrid interfacial layer for dendrite-free sodium deposition. *J. Power Sources* **2023**, *569*, 233023. [CrossRef]
75. Jiang, Y.; Yang, Y.; Ling, F.X.; Lu, G.X.; Huang, F.Y.; Tao, X.Y.; Wu, S.F.; Cheng, X.L.; Liu, F.F.; Li, D.J.; et al. Artificial heterogeneous interphase layer with boosted ion affinity and diffusion for Na/K-metal batteries. *Adv. Mater.* **2022**, *34*, 2109439. [CrossRef]
76. Xia, X.M.; Xu, S.T.; Tang, F.; Yao, Y.; Wang, L.F.; Liu, L.; He, S.N.; Yang, Y.X.; Sun, W.P.; Xu, C.; et al. A multifunctional interphase layer enabling superior sodium-metal batteries under ambient temperature and −40 °C. *Adv. Mater.* **2023**, *35*, 2209511. [CrossRef]
77. Xia, X.M.; Yang, Y.; Chen, K.Z.; Xu, S.T.; Tang, F.; Liu, L.; Xu, C.; Rui, X.H. Enhancing interfacial strength and wettability for wide-temperature sodium metal batteries. *Small* **2023**, *19*, 2300907. [CrossRef]
78. Li, D.J.; Sun, Y.J.; Li, M.H.; Cheng, X.L.; Yao, Y.; Huang, F.Y.; Jiao, S.H.; Gu, M.; Rui, X.H.; Ali, Z.; et al. Rational design of an artificial SEI: Alloy/solid electrolyte hybrid layer for a highly reversible Na and K metal anode. *ACS Nano* **2022**, *16*, 16966–16975. [CrossRef]
79. Chu, F.L.; Hu, J.L.; Tian, J.; Zhou, X.J.; Li, Z.; Li, C.L. In situ plating of porous Mg network layer to reinforce anode dendrite suppression in Li-metal batteries. *ACS Appl. Mater. Interfaces* **2018**, *10*, 12678–12689. [CrossRef] [PubMed]
80. Tu, Z.Y.; Choudhury, S.; Zachman, M.J.; Wei, S.Y.; Zhang, K.H.; Kourkoutis, L.F.; Archer, L.A. Fast ion transport at solid-solid interfaces in hybrid battery anodes. *Nature Energy* **2018**, *3*, 310–316. [CrossRef]
81. Ma, M.Y.; Lu, Y.; Yan, Z.H.; Chen, J. In situ synthesis of Bi layer on Na metal anode for fast nterfacial transport in $Na-O_2$ batteries. *Batter. Supercaps* **2019**, *2*, 663–667. [CrossRef]
82. Peng, Z.; Song, J.H.; Huai, L.Y.; Jia, H.P.; Xiao, B.W.; Zou, L.F.; Zhu, G.M.; Martinez, A.; Roy, S.; Murugesan, V. Enhanced stability of Li metal anodes by synergetic control of nucleation and the solid electrolyte interphase. *Adv. Energy Mater.* **2019**, *9*, 1901764. [CrossRef]
83. Liu, Y.; Li, Q.Z.; Lei, Y.Y.; Zhou, D.L.; Wu, W.W.; Wu, X.H. Stabilizing sodium metal anode by in-situ formed Ag metal layer. *J. Alloys Compd.* **2022**, *926*, 166850. [CrossRef]
84. Liu, C.Y.; Xie, Y.Y.; Li, H.X.; Xu, J.Y.; Zhang, Z.A. In situ construction of sodiophilic alloy interface enabled homogenous Na nucleation and deposition for sodium metal anode. *J. Electrochem. Soc.* **2022**, *169*, 080521. [CrossRef]
85. Lv, X.; Tang, F.; Yao, Y.; Xu, C.; Chen, D.; Liu, L.; Feng, Y.Z.; Rui, X.H.; Yu, Y. Sodium-gallium alloy layer for fast and reversible sodium deposition. *SusMat* **2022**, *2*, 699–707. [CrossRef]
86. Li, G.C.; Yang, Q.P.; Chao, J.; Zhang, B.; Wan, M.T.; Liu, X.X.; Mao, E.; Wang, L.; Yang, H.; Seh, Z.W.; et al. Enhanced processability and electrochemical cyclability of metallic sodium at elevated temperature using sodium alloy composite. *Energy Storage Mater.* **2021**, *35*, 310–316. [CrossRef]
87. Luo, W.; Lin, C.-F.; Zhao, O.; Noked, M.; Zhang, Y.; Rubloff, G.W.; Hu, L.B. Ultrathin surface coating enables the stable sodium metal anode. *Adv. Energy Mater.* **2017**, *7*, 1601526. [CrossRef]
88. Zhao, Y.; Goncharova, L.V.; Lushington, A.; Sun, Q.; Yadegari, H.; Wang, B.Q.; Xiao, W.; Li, R.Y.; Sun, X.L. Superior stable and long life sodium metal anodes achieved by atomic layer deposition. *Adv. Mater.* **2017**, *29*, 1606663. [CrossRef] [PubMed]
89. Jin, E.; Tantratian, K.; Zhao, C.; Codirenzi, A.; Goncharova, L.V.; Wang, C.H.; Yang, F.P.; Wang, Y.J.; Pirayesh, P.; Guo, J.H.; et al. Ionic conductive and highly-stable interface for alkali metal anodes. *Small* **2022**, *18*, 2203045. [CrossRef] [PubMed]
90. Sullivan, M.; Tang, P.; Meng, X.B. Atomic and molecular layer deposition as surface engineering techniques for emerging alkali metal rechargeable batteries. *Molecules* **2022**, *27*, 6170. [CrossRef]
91. Ye, L.; Liao, M.; Zhao, T.C.; Sun, H.; Zhao, Y.; Sun, X.M.; Wang, B.J.; Peng, H.S. A sodiophilic interphase-mediated, dendrite-free anode with ultrahigh specific capacity for sodium-metal batteries. *Angew. Chem. Int. Ed.* **2019**, *131*, 2–9. [CrossRef]
92. Liu, P.; Yi, H.T.; Zheng, S.Y.; Li, Z.P.; Zhu, K.J.; Sun, Z.Q.; Jin, T.; Jiao, L.F. Regulating deposition behavior of sodium ions for dendrite-free sodium-metal anode. *Adv. Energy Mater.* **2021**, *11*, 2101976. [CrossRef]

93. Zhang, S.J.; You, J.H.; He, Z.W.; Zhong, J.J.; Zhang, P.F.; Yin, Z.W.; Pan, F.; Ling, M.; Zhang, B.K.; Lin, Z. Scalable lithiophilic/sodiophilic porous buffer layer fabrication enables uniform nucleation and growth for lithium/sodium metal batteries. *Adv. Funct. Mater.* **2022**, *32*, 2200967. [CrossRef]
94. Wang, H.; Wang, C.L.; Matios, E.; Li, W.Y. Critical role of ultrathin graphene films with tunable thickness in enabling highly stable sodium metal anodes. *Nano Lett.* **2017**, *17*, 6808–6815. [CrossRef]
95. Tian, H.Z.; Seh, Z.W.; Yan, K.; Fu, Z.H.; Tang, P.; Lu, Y.Y.; Zhang, R.F.; Legut, D.; Cui, Y.; Zhang, Q.F. Theoretical investigation of 2D layered materials as protective films for lithium and sodium metal anodes. *Adv. Energy Mater.* **2017**, *7*, 1602528. [CrossRef]
96. He, X.; Jin, S.; Miao, L.C.; Cai, Y.C.; Hou, Y.P.; Li, H.X.; Zhang, K.; Yan, Z.H.; Chen, J. A 3D hydroxylated MXene/carbon nanotubes composite as a scaffold for dendrite-free sodium-metal Electrodes. *Angew Chem. Int. Ed.* **2020**, *59*, 16705–16711. [CrossRef]
97. Luo, J.M.; Wang, C.L.; Wang, H.; Hu, X.F.; Matios, E.; Lu, X.; Zhang, W.K.; Tao, X.Y.; Li, W.Y. Pillared MXene with ultralarge interlayer spacing as a stable matrix for high performance sodium metal anodes. *Adv. Funct. Mater.* **2019**, *29*, 1805946. [CrossRef]
98. Jiang, H.Y.; Lin, X.H.; Wei, C.L.; Zhang, Y.C.; Feng, J.K.; Tian, X.L. Sodiophilic Mg^{2+}-decorated Ti_3C_2 MXene for dendrite-free sodium metal batteries with carbonate-based electrolytes. *Small* **2022**, *18*, 2107637. [CrossRef]
99. Bao, C.Y.; Wang, J.H.; Wang, B.; Sun, J.G.; He, L.C.; Pan, Z.H.; Jiang, Y.P.; Wang, D.L.; Liu, X.M.; Dou, S.X.; et al. 3D sodiophilic Ti_3C_2 MXene@g-C_3N_4 hetero-interphase raises the stability of sodium metal anodes. *ACS Nano* **2022**, *16*, 17197–17209. [CrossRef]
100. Shi, R.J.; Shen, Z.; Yue, Q.Q.; Zhao, Y. Advances in functional organic material-based interfacial engineering on metal anodes for rechargeable secondary batteries. *Nanoscale* **2023**, *15*, 9256–9289. [CrossRef]
101. Li, N.W.; Shi, Y.; Yin, Y.X.; Zeng, X.X.; Li, J.Y.; Li, C.J.; Wan, L.J.; Wen, R.; Guo, Y.G. A flexible solid electrolyte interphase layer for long-life lithium metal anodes. *Angew. Chem. Int. Ed.* **2018**, *130*, 1521–1525. [CrossRef]
102. Ma, J.L.; Yin, Y.B.; Liu, T.; Zhang, X.B.; Yan, J.M.; Jiang, Q. Suppressing sodium dendrites by multifunctional polyvinylidene fluoride (PVDF) interlayers with nonthrough pores and high flux/affinity of sodium ions toward long cycle life sodium oxygen-batteries. *Adv. Funct. Mater.* **2018**, *28*, 1703931. [CrossRef]
103. Hou, Z.; Wang, W.H.; Yu, Y.K.; Zhao, X.X.; Chen, Q.W.; Zhao, L.F.; Di, Q.; Ju, H.X.; Quan, Z.W. Poly(vinylidene difluoride) coating on Cu current collector for high-performance Na metal anode. *Energy Storage Mater.* **2020**, *24*, 588–593. [CrossRef]
104. Shuai, Y.; Lou, J.; Pei, X.L.; Su, C.Q.; Ye, X.S.; Zhang, L.M.; Wang, Y.; Xu, Z.X.; Gao, P.P.; He, S.J.; et al. Constructing an in situ polymer electrolyte and a Na-rich artificial SEI layer toward practical solid-state Na metal batteries. *ACS Appl. Mater. Interfaces* **2022**, *14*, 45382–45391. [CrossRef]
105. Lu, Q.Q.; Omar, A.; Ding, L.; Oswald, S.; Hantusch, M.; Giebeler, L.; Nielsch, K.; Mikhailova, D. A facile method to stabilize sodium metal anodes towards high-performance sodium batteries. *J. Mater. Chem. A* **2021**, *9*, 9038–9047. [CrossRef]
106. Wei, S.Y.; Choudhury, S.; Xu, J.; Nath, P.; Tu, Z.Y.; Archer, L.A. Highly stable sodium batteries enabled by functional ionic polymer membranes. *Adv. Mater.* **2017**, *29*, 1605512. [CrossRef]
107. Zhu, M.; Wang, G.Y.; Liu, X.; Guo, B.K.; Xu, G.; Huang, Z.Y.; Wu, M.H.; Liu, H.K.; Dou, S.X.; Wu, C. Dendrite-free sodium metal anodes enabled by a sodium benzenedithiolate-rich protection layer. *Angew Chem. Int. Ed.* **2020**, *132*, 6658–6662. [CrossRef]
108. Wang, C.Z.; Zheng, Y.; Chen, Z.N.; Zhang, R.R.; He, W.; Li, K.X.; Yan, S.; Cui, J.Q.; Fang, X.L.; Yan, J.W.; et al. Robust anode-free sodium metal batteries enabled by artificial sodium formate interface. *Adv. Energy Mater.* **2023**, *13*, 2204125. [CrossRef]
109. Qian, J.; Li, Y.; Zhang, M.L.; Luo, R.; Wang, F.J.; Ye, Y.S.; Xing, Y.; Li, W.L.; Qu, W.J.; Wang, L.L.; et al. Protecting lithium/sodium metal anode with metal-organic framework based compact and robust shield. *Nano Energy* **2019**, *60*, 866–874. [CrossRef]
110. Zhu, M.Q.; Li, S.M.; Li, B.; Gong, Y.J.; Du, Z.G.; Yang, S.B. Homogeneous guiding deposition of sodium through main group II metals toward dendrite-free sodium anodes. *Sci. Adv.* **2019**, *5*, eaau6264. [CrossRef]
111. Miao, X.G.; Wang, P.; Sun, R.; Li, J.F.; Wang, Z.X.; Zhang, T.; Wang, R.T.; Li, Z.Q.; Bai, Y.J.; Hao, R.; et al. Liquid metal-organic frameworks in-situ derived interlayer for high-performance solid-state Na-metal batteries. *Adv. Energy Mater.* **2021**, *11*, 2102396. [CrossRef]
112. Hu, Y.Y.; Han, R.X.; Mei, L.; Liu, J.L.; Sun, J.C.; Yang, K.; Zhao, J.W. Design principles of MOF-related materials for highly stable metal anodes in secondary metal-based batteries. *Mater. Today Energy* **2021**, *19*, 100608. [CrossRef]
113. He, Y.B.; Qiao, Y.; Chang, Z.; Zhou, H.S. The potential of electrolyte filled MOF membranes as ionic sieves in rechargeable batteries. *Energy Environ. Sci.* **2019**, *12*, 2327–2344. [CrossRef]
114. Kim, Y.J.; Lee, H.; Noh, H.; Lee, J.; Kim, S.; Ryou, M.H.; Lee, Y.M.; Kim, H.T. Enhancing the cycling stability of sodium metal electrode by building an inorganic/organic composite protective layer. *ACS Appl. Mater. Interfaces* **2017**, *9*, 6000–6006. [CrossRef]
115. Wang, S.Y.; Jie, Y.L.; Sun, Z.H.; Cai, W.B.; Chen, Y.W.; Huang, F.Y.; Liu, Y.; Li, X.P.; Du, R.Q.; Cao, R.G.; et al. An implantable artificial protective layer enables stable sodium metal anodes. *ACS Appl. Energy Mater.* **2020**, *3*, 8688–8694. [CrossRef]
116. Hou, Z.; Wang, W.H.; Chen, Q.W.; Yu, Y.K.; Zhao, X.X.; Tang, M.; Zheng, Y.Y.; Quan, Z.W. Hybrid protective layer for stable sodium metal anodes at high utilization. *ACS Appl. Mater. Interfaces* **2019**, *11*, 37693–37700. [CrossRef]
117. Lv, X.; Tang, F.; Xu, S.T.; Yao, Y.; Yuan, Z.S.; Liu, L.; He, S.N.; Yang, Y.X.; Sun, W.P.; Pan, H.G.; et al. Construction of inorganic/organic hybrid layer for stable Na metal anode operated under wide temperatures. *Small* **2023**, *19*, 2300215. [CrossRef]
118. Xie, Y.Y.; Liu, C.Y.; Zheng, J.Q.; Li, H.X.; Zhang, L.Y.; Zhang, Z.A. NaF-rich protective layer on PTFE coating microcrystalline graphite for highly stable Na metal anodes. *Nano Res.* **2023**, *16*, 2436–2444. [CrossRef]
119. Xu, M.Y.; Li, Y.; Ihsan-Ul-Haq, M.; Mubarak, N.; Liu, Z.J.; Wu, J.X.; Luo, Z.T.; Kim, J.K. NaF-rich solid electrolyte interphase for dendrite-free sodium metal batteries. *Energy Storage Mater.* **2022**, *44*, 477–486. [CrossRef]

120. Tai, Z.X.; Liu, Y.J.; Yu, Z.P.; Lu, Z.Y.; Bondarchuk, O.; Peng, Z.J.; Liu, L.F. Non-collapsing 3D solid-electrolyte interphase for high-rate rechargeable sodium metal batteries. *Nano Energy* **2022**, *94*, 106947. [CrossRef]
121. Zhang, W.; Yang, X.K.; Wang, J.C.; Zheng, J.L.; Yue, K.; Liu, T.F.; Wang, Y.; Nai, J.W.; Liu, Y.J.; Tao, X.Y. Rapidly constructing sodium fluoride-rich interface by pressure and diglyme-induced defluorination reaction for stable sodium metal anode. *Small* **2023**, *19*, 2207540. [CrossRef]
122. Chen, Q.W.; Hou, Z.; Sun, Z.Z.; Pu, Y.Y.; Jiang, Y.B.; Zhao, Y.; He, H.; Zhang, T.X.; Huang, L.M. Polymer-inorganic composite protective layer for stable Na metal anodes. *ACS Appl. Energy Mater.* **2020**, *3*, 2900–2906. [CrossRef]
123. Xu, Y.; Wang, C.L.; Matios, E.; Luo, J.M.; Hu, X.F.; Yue, Q.; Kang, Y.J.; Li, W.Y. Sodium deposition with a controlled location and orientation for dendrite-free sodium metal batteries. *Adv. Energy Mater.* **2020**, *10*, 2002308. [CrossRef]
124. Zhang, J.L.; Wang, S.; Wang, W.H.; Li, B.H. Stabilizing sodium metal anode through facile construction of organic-metal interface. *J. Energy Chem.* **2022**, *66*, 133–139. [CrossRef]
125. Zhao, Y.; Goncharova, L.V.; Zhang, Q.; Kaghazchi, P.; Sun, Q.; Lushington, A.; Wang, B.Q.; Li, R.Y.; Sun, X.L. Inorganic-organic coating via molecular layer deposition enables long life sodium metal anode. *Nano Lett.* **2017**, *17*, 5653–5659. [CrossRef] [PubMed]
126. Zhang, S.M.; Zhao, Y.; Zhao, F.P.; Zhang, L.; Wang, C.H.; Li, X.N.; Liang, J.W.; Li, W.H.; Sun, Q.; Yu, C.; et al. Gradiently sodiated alucone as an interfacial stabilizing strategy for solid-state Na metal batteries. *Adv. Funct. Mater.* **2020**, *30*, 2001118. [CrossRef]
127. Pirayesh, P.; Tantratian, K.; Amirmaleki, M.; Yang, F.P.; Jin, E.Z.; Wang, Y.J.; Goncharova, L.V.; Guo, J.H.; Filleter, T.; Chen, L.; et al. From nano-alloy to nano-laminated interfaces for highly stable alkali metal anodes. *Adv. Mater.* **2023**, *35*, 2301414. [CrossRef]

Disclaimer/Publisher's Note: The statements, opinions and data contained in all publications are solely those of the individual author(s) and contributor(s) and not of MDPI and/or the editor(s). MDPI and/or the editor(s) disclaim responsibility for any injury to people or property resulting from any ideas, methods, instructions or products referred to in the content.

Article

Effect of Carrier Film Phase Conversion Time on Polyacrylate Polymer Electrolyte Properties in All-Solid-State LIBs

Shujian Zhang [1], Hongmo Zhu [2], Lanfang Que [3,*], Xuning Leng [4], Lei Zhao [1,*] and Zhenbo Wang [1,5,*]

[1] MIIT Key Laboratory of Critical Materials Technology for New Energy Conversion and Storage, State Key Lab of Urban Water Resources and Environment, Harbin Institute of Technology, School of Chemistry and Chemical Engineering, Harbin 150001, China; 652023783@qq.com

[2] Beijing Guodian Ruixin Technology Co., Ltd., Beijing 100085, China; zhuhongmo@goodrex.com.cn

[3] Institute of Materials Physical Chemistry, Engineering Research Center of Environment-Friendly Functional Materials, Ministry of Education, Huaqiao University, Xiamen 361021, China

[4] Shandong Yellow Sea Institute of Science and Technology Innovation, Headquarters Base No. 1, Qingdao Road, Rizhao 276800, China; lengxn14@mails.jlu.edu.cn

[5] Zhuhai Zhongli New Energy Sci-Tech Co., Ltd., Zhuhai 519000, China

* Correspondence: quelanfang@hqu.edu.cn (L.Q.); leizhao@hit.edu.cn (L.Z.); wangzhb@hit.edu.cn (Z.W.)

Abstract: To optimize the preparation process of polymer electrolytes by in situ UV curing and improve the performance of polymer electrolytes, we investigated the effect of carrier film phase conversion time on the properties of polymer electrolyte properties in all-solid-state LIBs. We compared several carrier films with phase conversion times of 24 h, 32 h, 40 h, and 48 h. Then, the physical properties of the polymer electrolytes were characterized and the properties of the polymer electrolytes were further explored. It was concluded that the carrier membrane with a phase transition time of 40 h and the prepared electrolyte had the best performance. The ionic conductivity of the sample was 1.02×10^{-3} S/cm at 25 °C and 3.42×10^{-3} S/cm at 60 °C. At its best cycle performance, it had the highest discharge-specific capacity of 155.6 mAh/g, and after 70 cycles, the discharge-specific capacity was 152.4 mAh/g, with a capacity retention rate of 98% and a discharge efficiency close to 100%. At the same time, the thermogravimetric curves showed that the samples prepared by this process had good thermal stability which can meet the various requirements of lithium-ion batteries.

Keywords: phase conversion time; UV curing; solid polymer electrolyte; porous carrier membrane; lithium-ion battery

Citation: Zhang, S.; Zhu, H.; Que, L.; Leng, X.; Zhao, L.; Wang, Z. Effect of Carrier Film Phase Conversion Time on Polyacrylate Polymer Electrolyte Properties in All-Solid-State LIBs. *Batteries* **2023**, *9*, 471. https://doi.org/10.3390/batteries9090471

Academic Editor: Douglas Ivey

Received: 4 August 2023
Revised: 2 September 2023
Accepted: 8 September 2023
Published: 19 September 2023

Copyright: © 2023 by the authors. Licensee MDPI, Basel, Switzerland. This article is an open access article distributed under the terms and conditions of the Creative Commons Attribution (CC BY) license (https://creativecommons.org/licenses/by/4.0/).

1. Introduction

Lithium-ion batteries with a high energy density and good cycle stability are widely used in portable devices, electric vehicles, and smart grids. Unfortunately, the liquid electrolytes used in the market for power batteries pose potential safety problems due to their low flash point [1]. As researchers pursue higher-energy power batteries, safety concerns have also emerged. As technology continues to evolve and application requirements continue to improve, lithium-ion batteries face several key challenges, such as the market demand for higher safety and energy density [2,3]. Traditional lithium-ion battery technology faces bottleneck problems in these aspects, so it is urgent to develop a new generation of lithium-ion battery technology [4,5].

All solid polymer electrolytes have the advantage of a wide electrochemical window, good thermal stability, low packaging requirements, and high production efficiency; they can significantly improve the operating conditions and safety of lithium-ion batteries in extreme environments [6,7]. The main challenge in achieving all-solid-state lithium batteries is to obtain solid electrolytes with considerable ionic conductivity [8,9]. The interface impedance between the solid electrolyte and the electrode also needs to be reduced. The plastic crystal solid-state polymer electrolyte prepared with succinonitrile (SN) not only overcomes the above shortcomings but also has good ionic conductivity at ambient temperature, showing a large electrochemical window, which greatly expands the application range of lithium-ion batteries [10,11]. At the same time, it has excellent electrochemical performance and no liquid exists, therefore, it shows excellent safety [12,13].

To further improve the performance of polymer electrolytes, the preparation process of the electrolyte carrier membrane is further optimized. The essence of the electrolyte carrier membrane is the PVDF-HFP phase conversion membrane [14,15]. In 1963, Leob and Sourirajan invented the phase conversion process for the first time and produced reverse osmosis membranes with asymmetric structures [16]. Thus, the polymer separation membrane was provided with an industrial value [17]. Since then, the phase conversion process has been widely studied and applied, gradually becoming the main flow method of polymerization separation membrane production [2,18].

The so-called phase conversion membrane is used to prepare a homogeneous polymer solution of a certain composition, which changes the thermodynamic state of the solution through certain physical methods, so that phase separation occurs from the homogeneous polymer solution and finally transforms it into a three-dimensional macromolecular network-type gel structure [19,20]. In this gel structure, the polymer is in a continuous phase and the dispersed phase is the pore structure left after the elution of the dilute polymer phase [21].

In general, when the casting liquid enters the gel bath and solidifies into a film, the major structure of the film is fixed [22,23]. However, the porosity and pore diameter can be adjusted by some post-treatment methods [24]. There are two kinds of post-treatment methods: heat treatment and solubilization treatment. Heat treatment is mainly used to reduce the porosity, while solvent treatment is mainly used to increase the porosity. Organic non-solvent treatment is when the newly formed wet film is soaked in some organic non-solvent, replacing the water in the wet film, and then dried [25]. These organic non-solvents are alcohols or hydrocarbons [1,26]. This is because the non-solvent for membrane formation by the phase separation method is generally water [19,27]. The main component of the dilute liquid remaining in the pore after film formation is water and its surface tension is as high as 72.3 dyn/cm. The pore size of the membrane is generally on the order of 10^{-8}–10^{-6}. According to the Laplace equation, in the drying process, the capillary tube stress is up to 1.45–145 bar, which easily causes capillary tube collapse, reduces the porosity, and damages the membrane property [28,29]. Wang compared the performance of membranes treated without organic non-solvent treatment and with methyl alcohol, ethyl alcohol, 1-propyl alcohol, and n-hexane in the drying condition [17,27]. It was found that the membrane permeability was increased by 3–4 times and the membrane pore diameter was slightly increased after organic non-solvent treatment. The increased flux is mainly provided by the increase in the effective porosity [8,30].

The processing time with organic non-solvent, that is, the phase conversion time of the carrier film, has a great influence on the porosity of the phase conversion film [31], thus greatly affecting the mechanical strength of the phase conversion membrane and the properties of the polymer electrolyte formed by the membrane [32]. In this paper, the effect of the phase transition time on the properties of the carrier film and polymer electrolytes is explored. The physical properties of carrier films prepared with various phase conversion times of 24 h, 32 h, 40 h, and 48 h are characterized. Further characterization of the polymer electrolyte is then carried out to obtain the best technology. The results showed that the carrier membrane prepared with a phase conversion time of 40 h has the best performance and that the corresponding polymer electrolyte performance is also the best.

2. Experiment

2.1. Materials

LiTFSI (99%, purchased from Aladdin Co., Ltd., Ontario, CA, USA), ethoxylated trimethylolpropane triacrylate (ETPTA, Mw ≈ 692, purchased from Aladdin Co., Ltd., Ontario, CA, USA), 2-hydroxy-2-methyl-1-phenyl-1-propanone (HMPP, a photoinitiator, purchased from Aldrich Industrial Inc., Wyoming, IL, USA), polyethersulfone (PSE) and polyvinylpyrrolidone (PVP) (purchased from Sinopharm Group Chemical Reagent and BASF, Shanghai, China), sulfoxane, butyrolactone, succinonitrile (SN), and adiponitrile (purchased from Aladdin Co., Ltd., Ontario, CA, USA), and PVDF-HFP (purchased from SOLVAY Co., Ltd., Shanghai, China) were obtained [14,33,34].

2.2. Preparation of Carrier Film and Polymer Electrolyte with Different Phase Transition Times

First, we prepared the carrier film according to the previous process. We dispersed PVDF-HFP, surface-modified alumina (wt15%), polyether sulfone, and polyvinylpyrrolidone (PVP) uniformly in a mixture of NMP and DMF. The specific preparation process is shown in Figure 1 [14,33,34].

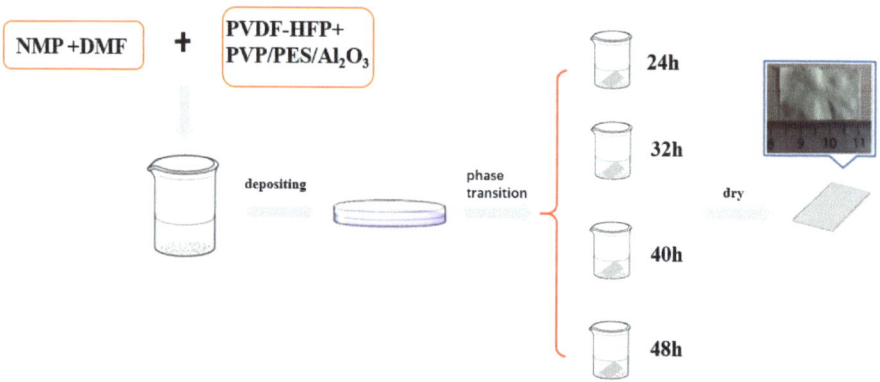

Figure 1. Flow chart of carrier film preparation with different phase transition times.

The membrane was then removed and placed in a mixture of ethanol–water (1:1). The soaking solution was replaced every 8 h, and the phase conversion time was set as 24 h, 32 h, 40 h, and 48 h for the experimental group. The phase conversion membrane was dried and the impurities in it dissolved and precipitated to obtain a carrier film with good performance. Then, the phase conversion film was cut into 16 cm diameter slices to prepare the polymer electrolyte for the next step. Next, the carrier membrane adsorbed the precursor liquid, and the polymer electrolyte was prepared by the UV curing method. Thus, we further characterized the electrochemical properties of electrolytes and batteries [10].

2.3. Preparation Method of Polymer Electrolyte

In this experiment, we used the UV curing method to prepare polymer electrolytes. First, the carrier films with different phase conversion times previously prepared were cut to the size of a button cell diaphragm for use. Then, the precursor liquid of the electrolyte was prepared. The specific method is to mix a certain proportion of 1.2 g succindinitrile (SN) with 0.8 g LiTFSI. Because the two will promote each other to lower the melting point, it will melt into a liquid in a few minutes. Then, 0.05 g of plasticizer was added to the mixed solution and stirred well. After that, 0.2 g of the polymeric monomer ETPTA of this electrolyte was added to the mixed solution and mixed thoroughly for a few minutes until well blended. Finally, the electrolyte precursor was prepared by adding a small amount of photoinitiator, about 0.01 g. The carrier film with the best surface effect was then selected to fully absorb the electrolyte precursor. When the quality of the carrier film did not change, it was covered on the electrode plate in the battery shell and irradiated with high-intensity ultraviolet light. After a few minutes, in situ UV polymerization curing was completed [14,33,34]. This method has the advantage of simple operation, high efficiency, and no pollution.

2.4. Material Characterization Method

The scanning electron microscope (SEM; Hitachi S-4800, Hitachi Co., Ltd., Tokyo, Japan) was used for characterization. To improve the conductivity of the sample, the sample was treated with gold spray, and the secondary electron mode was used to observe the electrolyte section. The structure of the sample was analyzed by X-ray diffraction (XRD). The role of Nicolet 6700 Fourier Transform (Thermo Electron Corporation, Waltham, MA, USA) infrared spectroscopy (FT-IR) is to analyze the functional groups present in the polymer electrolyte. The thermal stability of the polymer electrolyte prepared in this paper was analyzed by thermogravimetric analysis with a thermal analyzer (Q2000, TGA/DSC3+, Mettler, Switzerland) [14,33,34].

We usually use the AC impedance method to measure the ionic conductivity of polymer electrolytes. Ionic conductivity is calculated as $\sigma = L/(R \cdot S)$. First, the electrochemical stability of polymer electrolytes is generally analyzed using cyclic voltammetry curves. The electrolyte was assembled into a lithium metal/polymer electrolyte/stainless steel asymmetric battery which was analyzed with a scan rate of 0.1 mV/s [14,33,34].

3. Results and Discussion

Figure 2 shows the optical images of carrier films prepared at different phase conversion times. It can be seen that the carrier film was grayish-white, almost opaque, and it had good flexibility. From a macro point of view, there was no significant difference in the surface state of the carrier film prepared at the four different phase transition times, all of which were uniform and flat membrane structures.

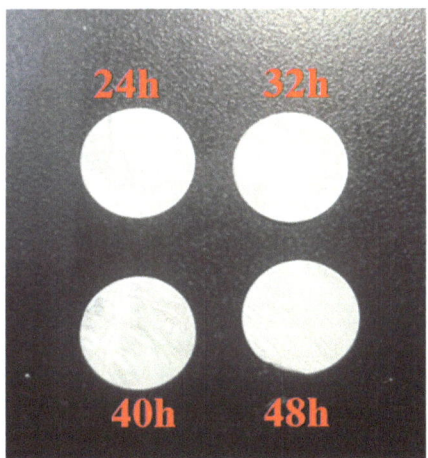

Figure 2. The optical images of the carrier films prepared with different phase transition times.

To further explore the effect of the phase transition time on the performance of the carrier membrane, the mechanical strength of the carrier membrane and the electrolyte prepared by the carrier membrane were tested. Figure 3a,b shows the mechanical property curves of the carrier membranes and electrolytes. It can be seen that the mechanical strength of the carrier film decreases gradually as the phase transition time increases. Interestingly, however, the mechanical strength of the electrolyte it forms is not monotonous. With the increase in the phase transition time from 24 h to 40 h, the mechanical strength of the electrolyte gradually increased. However, when the phase conversion time reached 48 h, the mechanical strength of the electrolyte began to decline to a value lower than that at 40 h. According to this phenomenon, we speculate that when the phase transition time is too long, it may affect the internal structure of the electrolyte and decrease the mechanical strength of the electrolyte. The most likely reason for this is that the phase transition time is too long so a large pore structure in the carrier membrane is formed. Because the pore structure is too sparse, it is difficult for the electrolyte to attach to the surface of the skeleton to form a uniform and stable structure.

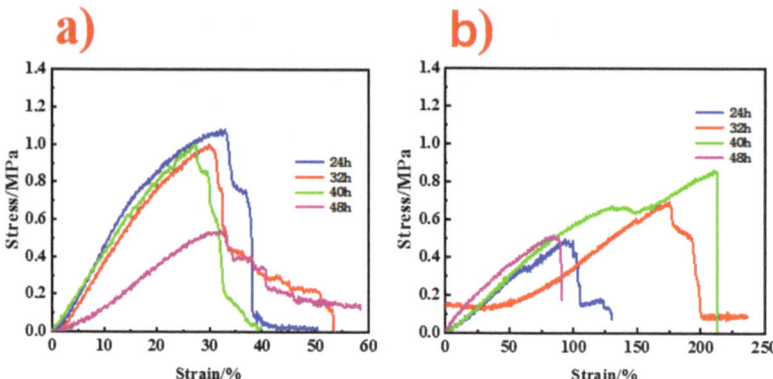

Figure 3. Mechanical properties of the carrier membrane and polymer electrolyte: (**a**) the carrier membrane and (**b**) the polymer electrolyte.

The research on the interface between electrodes and electrolytes in solid-state batteries has always been a hot topic. Li et al. combined solid polymer electrolytes with defect-rich Ga_2O_3 nanobricks to prepare high-performance lithium metal batteries and the effect was also very good [35]. Seongsoo et al. studied the interface contact between solid metal lithium battery electrodes and solid electrolytes and achieved very good results [36]. To verify our hypothesis, the micromorphologies of the carrier films prepared with different phase transition times and the corresponding electrolytes were analyzed. Figure 4a–h shows the SEM images of the carrier membranes and electrolytes. It can be seen from Figure 4a–d that with the increase in the phase transition time, the porosity and pore structure of the surface of the carrier film gradually increase, so the mechanical strength gradually decreases. This is also a good explanation for the phenomenon in Figure 3a. As can be seen from Figure 4e–h, the electrolyte formed after ultraviolet curing presents a folded shape on the surface of the carrier film, which is caused by the volatilization of the plasticizer. As displayed in Figure 4e–g, the fold structure gradually became dense, which is caused by the gradual increase in the porosity on the surface of the carrier film. However, in Figure 4h, the fold structure on the electrolyte surface suddenly appears less, and significant roughness appears on the electrolyte surface. This is caused by the excessive macropore structure in the carrier membrane which makes the pore structure too sparse so that the electrolyte cannot be well attached to the skeleton surface. Therefore, the phenomenon in Figure 3b can also be well explained.

Thermogravimetric analysis was performed on the carrier films prepared at different phase transition times and their corresponding electrolytes to explore their thermal stability. As seen in Figure 5a, the carrier film has good thermal stability, and thermal decomposition did not occur until temperatures in excess of 350 °C. This excellent thermal stability can fully meet the requirements of the working conditions of polymer electrolytes. The thermal stability of the carrier film with different phase transition times was not different. The results showed that the phase transition time only changed the pore size and porosity of the carrier film but did not change the composition and chemical properties of the carrier film. As shown in Figure 5b, the polymer electrolyte began to lose about 40% of its weight at 200 °C, mainly due to the thermal decomposition of butyronitrile. The thermal stability of polymer electrolytes was significantly better at 205–375 °C when the butyronitrile was almost completely decomposed. The weight loss above 375 °C was mainly due to the thermal decomposition of LiTFSI and the carrier film. Such high thermal stability can fully meet the requirements of lithium-ion battery working conditions. The thermal stability of electrolytes prepared by carrier membranes with different phase transition times was not different [11,24].

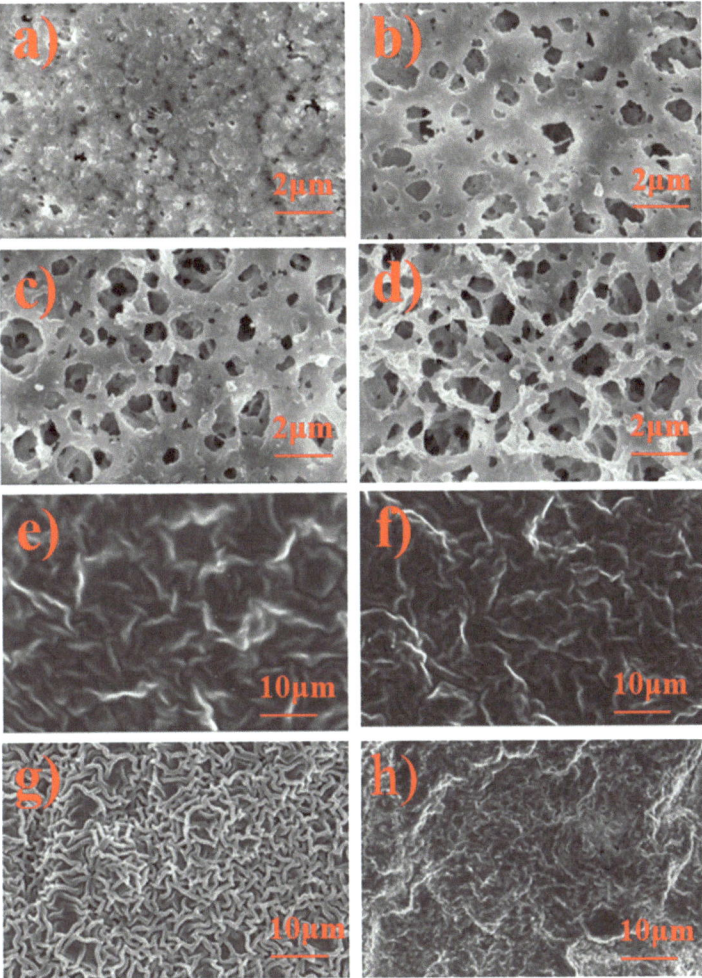

Figure 4. SEM images of the carrier membranes prepared at (**a**) 24 h, (**b**) 32 h, (**c**) 40 h, and (**d**) 48 h; SEM images of polymer electrolytes based on the carrier membranes prepared at (**e**) 24 h, (**f**) 32 h, (**g**) 40 h, and (**h**) 48 h.

At the same time, we also conducted an XRD analysis of the carrier films prepared at different phase conversion times and the corresponding electrolytes to explore the internal structure of the samples. Figure 6b shows that there was no significant difference in the lattice type of the electrolytes prepared by the carrier film with various phase transition times. However, it can be seen from Figure 6a that the absorption peak intensity increased with the increase in the phase transition time between 24 and 40 h. However, when the phase transition time reached 48 h, the absorption peak intensity decreased. On the surface of the 48 h sample, the internal structure of the carrier film changed suddenly, resulting in a decrease in the absorption peak strength. This phenomenon further confirmed our previous conjecture.

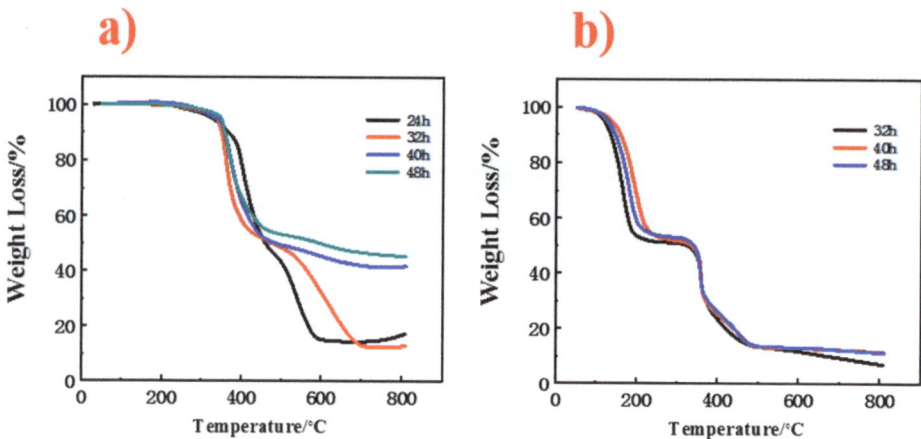

Figure 5. The TG curves of (**a**) the carrier membrane and (**b**) the polymer electrolyte.

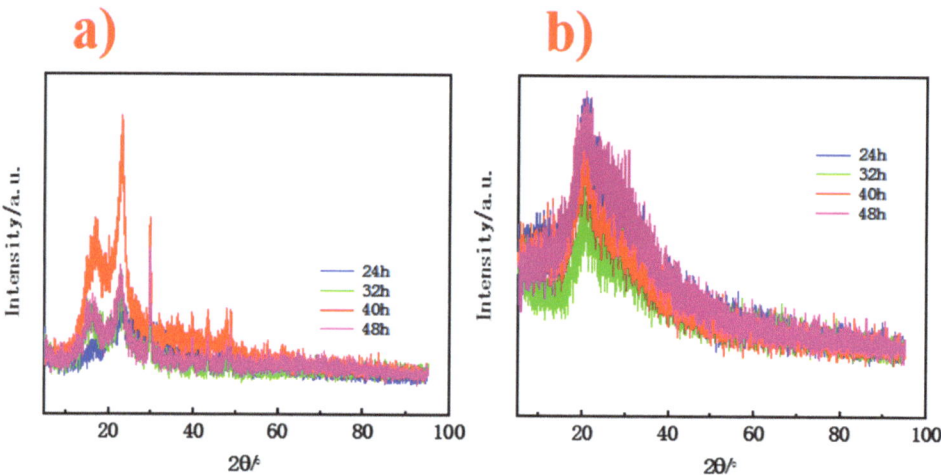

Figure 6. The XRD patterns of (**a**) the carrier membranes and (**b**) the polymer electrolytes.

Next, we explored the electrochemical properties of the polymer electrolytes made from carrier films with different phase transition times. First, we tested the cycling performance of the LiFePO$_4$/PCCE/Li battery at room temperature at 0.2 C. As seen from Figure 7a–d, the battery assembled by the polymer electrolyte with a phase conversion time of 40 h had the best cycle performance and the highest specific discharge capacity of 155.6 mAh/g after 70 cycles. On the 70th cycle, the discharge-specific capacity was 152.4 mAh/g with a capacity retention rate of 98% and discharge efficiency close to 100%. In contrast, in the other groups of batteries, the discharge capacity was reduced during the cycle. The specific discharge capacity of the sample with a phase conversion time of 48 h decreased the fastest, therefore, we concluded that the cycle performance is the best when the phase conversion time is 40 h.

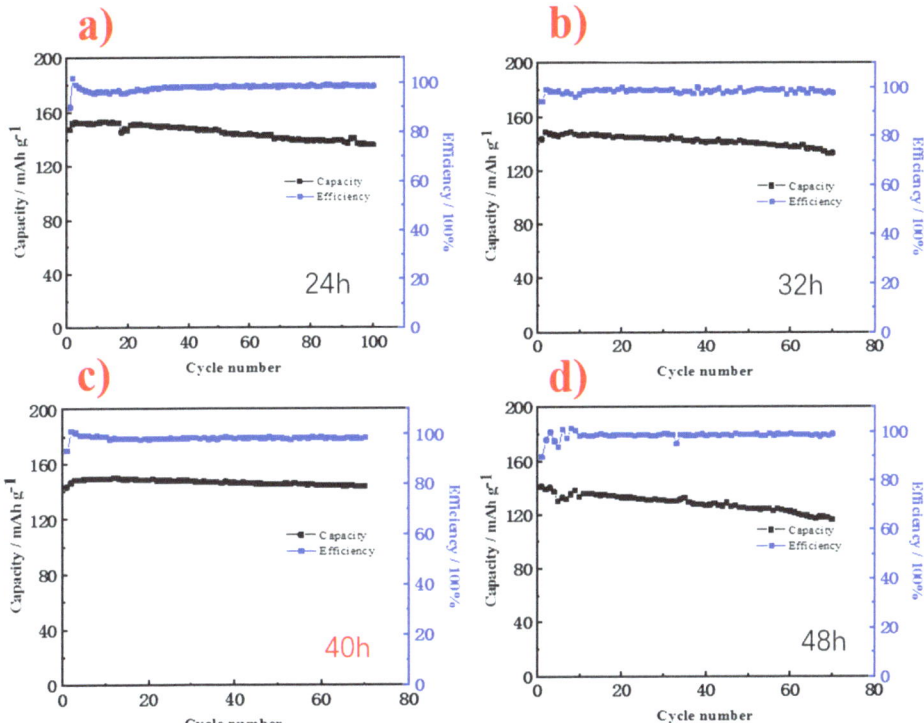

Figure 7. Cycle performance of the batteries composed of electrolytes with different phase transition durations: (**a**) 24 h, (**b**) 32 h, (**c**) 40 h, and (**d**) 48 h.

We next explored the ionic conductivity and the rate performance of the polymer electrolytes made from the carrier films with different phase transition times. Figure 8a shows the ionic conductivity of the polymer electrolytes at different temperatures. It can be seen that the sample with a phase conversion time of 40 h had the highest ionic conductivity. The ionic conductivity was 1.02×10^{-3} S/cm at 25 °C and 3.42×10^{-3} S/cm at 60 °C. The ionic conductivity of the sample with a phase conversion time of 48 h decreased significantly. This phenomenon further confirms our previous conjecture. The same conclusion can be drawn in Figure 8b. The battery operated similarly at low current densities, however, under the high-rate discharge, the sample with a phase conversion time of 40 h had the highest discharge-specific capacity while the sample with a phase conversion time of 48 h had the worst discharge effect. Based on the above results, we chose the sample with a phase transition time of 40 h as the key research object in the future, i.e., the optimum phase conversion time of carrier film is 40 h.

High-temperature performance is also an important index to evaluate lithium-ion batteries. For this purpose, we assembled the best process sample into LiFePO$_4$/S-PCCE/Li batteries and tested the high-temperature cycling performance at 55 °C. The results are shown in Figure 9. Figure 9a,b shows the charge-discharge cycle curves of the samples with a phase conversion time of 40 h at room temperature and high temperature. The sample still worked well at the high temperature of 55 °C, and the specific discharge capacity was still very high. This shows that the sample prepared by this process can meet the requirements of a lithium-ion battery working at high temperatures.

Finally, we studied the formation of the cathode-electrolyte-interface phase (CEI) membrane on the cathode surface of the battery composed of the optimal process samples and tested the TEM images of the LiFePO$_4$ cathode under different charging states.

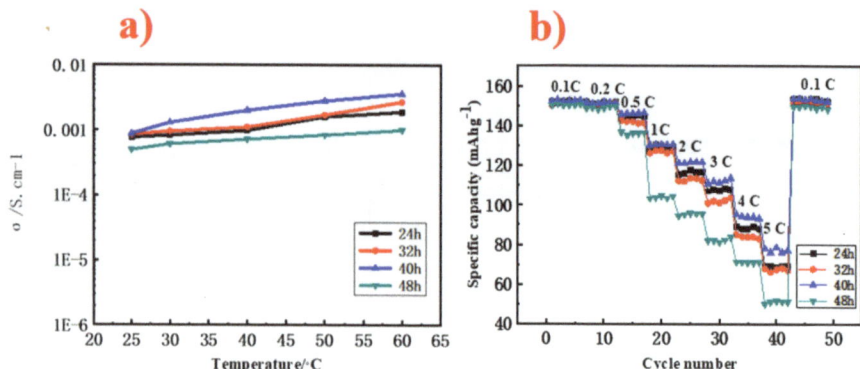

Figure 8. (**a**) Ionic conductivity of electrolytes with different phase transition times. (**b**) The rate performance of the batteries composed of electrolytes with different phase transition times.

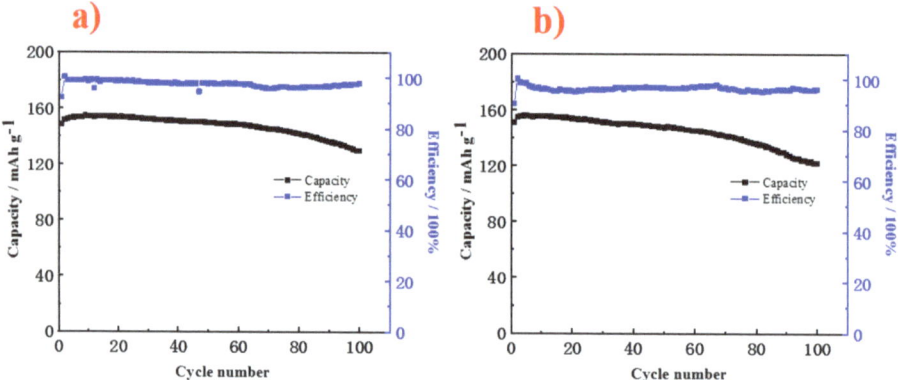

Figure 9. High-temperature performance of batteries with the optimum process electrolytes at (**a**) 25 °C and (**b**) 55 °C.

As shown in Figure 10a–c, a uniform CEI film with a thickness of about 8 nm is formed on the surface of the cathode when the cathode reaches the 100% charged state. When the battery power reaches 100%, the CEI film on the cathode surface is separated. This ensures that the polymer electrolyte is tightly bound to the cathode and the battery, and the battery is able to function normally and stably [14,33].

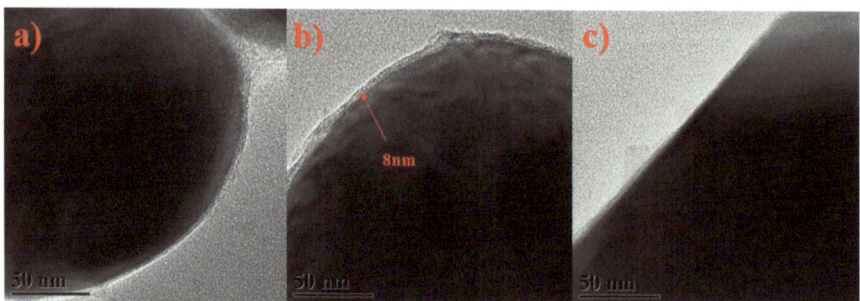

Figure 10. TEM images of the (**a**) LiFePO$_4$ cathode at the pristine state, (**b**) LiFePO$_4$ cathode with a charge state of 100%, and (**c**) LiFePO$_4$ cathode with a discharge state of 100%.

4. Conclusions

In conclusion, the phase conversion time of the carrier membrane has a great influence on the performance of polymer electrolytes. With the increase in the phase transition time, the pore structure and porosity of the carrier film increase. However, the excessive porosity of the skeleton structure makes it more difficult to attach the electrolyte, which leads to the deterioration in its performance. Through the analysis of carrier films prepared with multiple phase transition times, we saw that the sample with a phase transition time of 40 h had the best comprehensive performance. The ionic conductivity of the polymer electrolyte prepared with this sample was 1.02×10^{-3} S/cm at 25 °C. When the temperature reached 60 °C, it had an ionic conductivity of up to 3.42×10^{-3} S/cm. When assembled into a battery, it also had the best cycle performance, with the highest specific discharge capacity of 155.6 mAh/g. The carrier film also had good thermal stability; at the 70th cycle, the discharge-specific capacity was 152.4 mAh/g with a capacity retention rate of 98% and discharge efficiency close to 100%. Therefore, the 40 h phase conversion time will be selected as the optimal process in the follow-up study of this project. The method using in situ UV curing has the advantages of simple operation, energy saving, and high efficiency, and has a good practical application prospect.

Author Contributions: Regarding the contributions of the authors, they are as follows: investigation, H.Z.; writing—original draft preparation, S.Z.; writing—review and editing, L.Q.; supervision, L.Z.; project administration, X.L.; funding acquisition, Z.W. All authors have read and agreed to the published version of the manuscript.

Funding: National Natural Science Foundation of China (Grant No. 51902072 and 22075062), Heilongjiang Touyan Team (Grant No. HITTY-20190033), Heilongjiang Province "hundred million" project science and technology major special projects (2019ZX09A02), the State Key Laboratory of Urban Water Resource and Environment (Harbin Institute of Technology No. 2020DX11), and the Fundamental Research Funds for the Central Universities (Grant No. FRFCU5710051922).

Data Availability Statement: All data in this paper are original by the author.

Acknowledgments: We acknowledge the National Natural Science Foundation of China (Grant No. 51902072 and 22075062), Heilongjiang Touyan Team (Grant No. HITTY-20190033), Heilongjiang Province "hundred million" project science and technology major special projects (2019ZX09A02), the State Key Laboratory of Urban Water Resource and Environment (Harbin Institute of Technology No. 2020DX11), and the Fundamental Research Funds for the Central Universities (Grant No. FRFCU5710051922).

Conflicts of Interest: The authors declare no conflict of interest.

References

1. Zhang, Z.; Antonio, R.G.; Choy, K.L. Boron nitride enhanced polymer/salt hybrid electrolytes for all-solid-state lithium ion batteries. *J. Power Sources* **2019**, *435*, 226736. [CrossRef]
2. Zhang, Q.G.; Hu, W.W.; Liu, Q.L.; Zhu, A.M. Chitosan/polyvinylpyrrolidone-silica hybrid membranes for pervaporation separation of methanol/ethylene glycol azeotrope. *J. Appl. Polym. Sci.* **2013**, *129*, 3178–3184. [CrossRef]
3. Liu, L.; Mo, J.; Li, J.; Liu, J.; Yan, H.; Lyu, J.; Jiang, B.; Chu, L.; Li, M. Comprehensively-modified polymer electrolyte membranes with multifunctional PMIA for highly-stable all-solid-state lithium-ion batteries. *J. Energy Chem.* **2020**, *48*, 334–343. [CrossRef]
4. Pan, Q.; Barbash, D.; Smith, D.M.; Qi, H.; Gleeson, S.E.; Li, C.Y. Correlating Electrode-Electrolyte Interface and Battery Performance in Hybrid Solid Polymer Electrolyte-Based Lithium Metal Batteries. *Adv. Energy Mater.* **2017**, *7*, 1701231. [CrossRef]
5. Kongthong, T.; Tuantranont, S.; Primpray, V.; Sriprachuabwong, C.; Poochai, C. Rechargeable lithium-ion dual carbon batteries utilising a quasi-solid-state anion co-intercalation electrolyte and palm kernel shell-derived hard carbon. *Diam. Relat. Mater.* **2023**, *132*, 109680. [CrossRef]
6. Kim, S.-H.; Choi, K.-H.; Cho, S.-J.; Yoo, J.; Lee, S.-S.; Lee, S.-Y. Flexible/shape-versatile, bipolar all-solid-state lithium-ion batteries prepared by multistage printing. *Energy Environ. Sci.* **2018**, *11*, 321–330. [CrossRef]
7. Zhou, D.; Liu, R.; He, Y.-B.; Li, F.; Liu, M.; Li, B.; Yang, Q.-H.; Cai, Q.; Kang, F. SiO2 Hollow Nanosphere-Based Composite Solid Electrolyte for Lithium Metal Batteries to Suppress Lithium Dendrite Growth and Enhance Cycle Life. *Adv. Energy Mater.* **2016**, *6*, 1502214. [CrossRef]
8. Zhao, Q.; Stalin, S.; Zhao, C.-Z.; Archer, L.A. Designing solid-state electrolytes for safe, energy-dense batteries. *Nat. Rev. Mater.* **2020**, *5*, 229–252. [CrossRef]

9. Zhou, W.; Gao, H.; Goodenough, J.B. Low-Cost Hollow Mesoporous Polymer Spheres and All-Solid-State Lithium, Sodium Batteries. *Adv. Energy Mater.* **2016**, *6*, 1501802. [CrossRef]
10. Ha, H.-J.; Kil, E.-H.; Kwon, Y.H.; Kim, J.Y.; Lee, C.K.; Lee, S.-Y. UV-curable semi-interpenetrating polymer network-integrated, highly bendable plastic crystal composite electrolytes for shape-conformable all-solid-state lithium ion batteries. *Energy Environ. Sci.* **2012**, *5*, 6491. [CrossRef]
11. Wang, Q.J.; Zhang, P.; Zhu, W.Q.; Zhang, D.; Li, Z.J.; Wang, H.; Sun, H.; Wang, B.; Fan, L.-Z. A two-step strategy for constructing stable gel polymer electrolyte interfaces for long-life cycle lithium metal batteries. *J. Mater.* **2022**, *8*, 1048–1057. [CrossRef]
12. Li, J.; Zhu, L.; Xu, J.; Jing, M.; Yao, S.; Shen, X.; Li, S.; Tu, F. Boosting the performance of poly (ethylene oxide)-based solid polymer electrolytes by blending with poly (vinylidene fluoride-co-hexafluoropropylene) for solid-state lithium-ion batteries. *Int. J. Energy Res.* **2020**, *44*, 7831–7840. [CrossRef]
13. Sha, Y.; Yu, T.; Dong, T.; Wu, X.; Tao, H.; Zhang, H. In Situ Network Electrolyte Based on a Functional Polymerized Ionic Liquid with High Conductivity toward Lithium Metal Batteries. *ACS Appl. Energy Mater.* **2021**, *4*, 14755–14765. [CrossRef]
14. Zhang, S.; Lu, Y.; He, K.; Meng, X.; Que, L.; Wang, Z. Effect of UV light polymerization time on the properties of plastic crystal composite polyacrylate polymer electrolyte for all solid-state lithium-ion batteries. *J. Appl. Polym. Sci.* **2022**, *139*, e52001. [CrossRef]
15. Doyle, E.S.; Ford, H.O.; Webster, D.N.; Giannini, P.J.; Tighe, M.E.; Bartsch, R.; Peaslee, G.F.; Schaefer, J.L. Influence of Inorganic Glass Ceramic Particles on Ion States and Ion Transport in Composite Single-Ion Conducting Gel Polymer Electrolytes with Varying Chain Chemistry. *ACS Appl. Polym. Sci.* **2022**, *4*, 1095–1109. [CrossRef]
16. Deng, L.; Yu, F.-D.; Xia, Y.; Jiang, Y.-S.; Sui, X.-L.; Zhao, L.; Meng, X.-H.; Que, L.-F.; Wang, Z.-B. Stabilizing fluorine to achieve high-voltage and ultra-stable Na3V2(PO4)(2)F-3 cathode for sodium ion batteries. *Nano Energy* **2021**, *82*, 105659. [CrossRef]
17. Wang, D.L.; Li, K.; Teo, W.K. Porous PVDF asymmetric hollow fiber membranes prepared with the use of small molecular additives. *J. Membr. Sci.* **2000**, *178*, 13–23. [CrossRef]
18. Jeon, H.G.; Ito, Y.; Sunohara, Y.; Ichikawa, M. Dependence of photocurrent generation on the crystalline phase of titanyl phthalocyanine film in heterojunction photovoltaic cells. *Jpn. J. Appl. Phys.* **2015**, *54*, 091601. [CrossRef]
19. Mansourizadeh, A.; Ismail, A.F. A developed asymmetric PVDF hollow fiber membrane structure for CO2 absorption. *Int. J. Greenh. Gas Control* **2011**, *5*, 374–380. [CrossRef]
20. Xu, H.; Zhang, X.; Jiang, J.; Li, M.; Shen, Y. Ultrathin Li7La3Zr2O12@PAN composite polymer electrolyte with high conductivity for all-solid-state lithium-ion battery. *Solid State Ion.* **2020**, *347*, 115227. [CrossRef]
21. Wang, L.Y.; Yong, W.F.; Yu, L.E.; Chung, T.S. Design of high efficiency PVDF-PEG hollow fibers for air filtration of ultrafine particles. *J. Membr. Sci.* **2017**, *535*, 342–349. [CrossRef]
22. Liu, S.J.; Liu, M.H.; Xu, Q.; Zeng, G.F. Lithium Ion Conduction in Covalent Organic Frameworks. *Chin. J. Struct. Chem.* **2022**, *41*, 2211003-17.
23. Li, S.; Zhang, S.Q.; Shen, L.; Liu, Q.; Ma, J.B.; Lv, W.; He, Y.; Yang, Q. Progress and Perspective of Ceramic/Polymer Composite Solid Electrolytes for Lithium Batteries. *Adv. Sci.* **2020**, *7*, 1903088. [CrossRef]
24. Chen, Y.; Tian, Y.; Li, Z.; Zhang, N.; Zeng, D.; Xu, G.; Zhang, Y.; Sun, Y.; Ke, H.; Cheng, H. An AB alternating diblock single ion conducting polymer electrolyte membrane for all-solid-state lithium metal secondary batteries. *J. Membr. Sci.* **2018**, *566*, 181–189. [CrossRef]
25. Lu, Q.; Yang, J.; Lu, W.; Wang, J.; Nuli, Y. Advanced semi-interpenetrating polymer network gel electrolyte for rechargeable lithium batteries. *Electrochim. Acta* **2015**, *152*, 489–495. [CrossRef]
26. Yang, M.X.; Zhao, B.Q.; Li, J.Y.; Li, S.M.; Zhang, G.; Liu, S.Q.; Cui, Y.; Liu, H. Modified Poly(vinylidene fluoride-co-hexafluoropropylene) Polymer Electrolyte for Enhanced Stability and Polymer Degradation Inhibition toward the Li Metal Anode. *ACS Appl. Energy Mater.* **2022**, *5*, 9049–9057. [CrossRef]
27. Yao, W.; Zhang, Q.; Qi, F.; Zhang, J.; Liu, K.; Li, J.; Chen, W.; Du, Y.; Jin, Y.; Liang, Y.; et al. Epoxy containing solid polymer electrolyte for lithium ion battery. *Electrochim. Acta* **2019**, *318*, 302–313. [CrossRef]
28. Tao, C.; Gao, M.-H.; Yin, B.-H.; Li, B.; Huang, Y.-P.; Xu, G.; Bao, J.-J. A promising TPU/PEO blend polymer electrolyte for all-solid-state lithium ion batteries. *Electrochim. Acta* **2017**, *257*, 31–39. [CrossRef]
29. Kumar, A.; Sharma, R.; Suresh, M.; Das, M.K.; Kar, K.K. Structural and ion transport properties of lithium triflate/poly (vinylidene fluoride-co-hexafluoropropylene)-based polymer electrolytes: Effect of lithium salt concentration. *J. Elastom. Plast.* **2017**, *49*, 513–526. [CrossRef]
30. Zhou, D.; He, Y.-B.; Liu, R.; Liu, M.; Du, H.; Li, B.; Cai, Q.; Yang, Q.-H.; Kang, F. In Situ Synthesis of a Hierarchical All-Solid-State Electrolyte Based on Nitrile Materials for High-Performance Lithium-Ion Batteries. *Adv. Energy Mater.* **2015**, *5*, 1500353. [CrossRef]
31. Lv, P.; Yang, J.; Liu, G.; Liu, H.; Li, S.; Tang, C.; Mei, J.; Li, Y.; Hui, D. Flexible solid electrolyte based on UV cured polyurethane acrylate/succinonitrile-lithium salt composite compatibilized by tetrahydrofuran. *Compos. Part B Eng.* **2017**, *120*, 35–41. [CrossRef]
32. Wang, Y.; Fang, C.; Liu, W.; Pei, H.; Zhan, B.; Zou, W.; Yu, S.; Guo, R.; Xie, J. Constructing Self-Protected Li and Non-Li Candidates for Advanced Lithium Ion and Lithium Metal Batteries. *J. Phys. Chem. C* **2019**, *123*, 13318–13323. [CrossRef]
33. Zhang, S.J.; Lu, Y.; He, K.W.; Que, L.F.; Zhao, L.; Wang, Z.B. The plastic crystal composite polyacrylate polymer electrolyte with a semi-interpenetrating network structure for all-solid-state LIBs. *New J. Chem.* **2022**, *46*, 21640–21647. [CrossRef]
34. Lu, Y.; He, K.-W.; Zhang, S.-J.; Zhou, Y.-X.; Wang, Z.-B. UV-curable-based plastic crystal polymer electrolyte for high-performance all-solid-state Li-ion batteries. *Ionics* **2018**, *25*, 1607–1615. [CrossRef]

35. Li, H.; Xu, X.; Li, F.; Zhao, J.; Ji, S.; Liu, J.; Huo, Y. Defects-Abundant Ga2O3 Nanobricks Enabled Multifunctional Solid Polymer Electrolyte for Superior Lithium-Metal Batteries. *Chem. Eur. J.* **2023**, *29*, e202204035. [CrossRef]
36. Seongsoo, P.; Rashama, C.; Sang, A.H.; Hamzen, Q.; Moon, J.; Park, M.-S.; Kim, J.H. Ionic conductivity and mechanical properties of the solid electrolyte interphase in lithium metal batteries. *Energy Mater.* **2023**, *3*, 300005.

Disclaimer/Publisher's Note: The statements, opinions and data contained in all publications are solely those of the individual author(s) and contributor(s) and not of MDPI and/or the editor(s). MDPI and/or the editor(s) disclaim responsibility for any injury to people or property resulting from any ideas, methods, instructions or products referred to in the content.

Article

Optimizing Li Ion Transport in a Garnet-Type Solid Electrolyte via a Grain Boundary Design

Tao Sun [1], Xiaopeng Cheng [1,*], Tianci Cao [1], Mingming Wang [1], Jiao Tian [1], Tengfei Yan [1], Dechen Qin [1], Xianqiang Liu [1,*], Junxia Lu [1,*] and Yuefei Zhang [2]

[1] Faculty of Materials and Manufacturing, Beijing University of Technology, Beijing 100124, China; taosun@emails.bjut.edu.cn (T.S.); tccao@emails.bjut.edu.cn (T.C.); wmm@emails.bjut.edu.cn (M.W.); tianjiao@emails.bjut.edu.cn (J.T.); ytf1219@emails.bjut.edu.cn (T.Y.); qdc_4396@emails.bjut.edu.cn (D.Q.)

[2] Institute of Superalloys Science and Technology, School of Materials Science and Engineering, Zhejiang University, Hangzhou 310027, China; yfzhang76@zju.edu.cn

* Correspondence: xpcheng@bjut.edu.cn (X.C.); xqliu@bjut.edu.cn (X.L.); junxialv@bjut.edu.cn (J.L.)

Citation: Sun, T.; Cheng, X.; Cao, T.; Wang, M.; Tian, J.; Yan, T.; Qin, D.; Liu, X.; Lu, J.; Zhang, Y. Optimizing Li Ion Transport in a Garnet-Type Solid Electrolyte via a Grain Boundary Design. *Batteries* 2023, 9, 526. https://doi.org/10.3390/batteries9110526

Academic Editors: Chunwen Sun, Yongjie Zhao and Siqi Shi

Received: 27 June 2023
Revised: 25 August 2023
Accepted: 28 August 2023
Published: 24 October 2023

Copyright: © 2023 by the authors. Licensee MDPI, Basel, Switzerland. This article is an open access article distributed under the terms and conditions of the Creative Commons Attribution (CC BY) license (https://creativecommons.org/licenses/by/4.0/).

Abstract: Garnet-type solid electrolytes have gained considerable attention owing to their exceptional ionic conductivity and broad electrochemical stability window, making them highly promising for solid-state batteries (SSBs). However, this polycrystalline ceramic electrolyte contains an abundance of grain boundaries (GBs). During the repetitive electroplating and stripping of Li ions, uncontrolled growth and spreading of lithium dendrites often occur at GBs, posing safety concerns and resulting in a shortened cycle life. Reducing the formation and growth of lithium dendrites can be achieved by rational grain boundary design. Herein, the garnet-type solid electrolyte LLZTO was firstly coated with Al_2O_3 using the atomic layer deposition (ALD) technique. Subsequently, an annealing treatment was employed to introduce Al_2O_3 into grain boundaries, effectively modifying them. Compared with the Li/LLZTO/Li cells, the Li/LLZTO@Al_2O_3-annealed/Li symmetric batteries exhibit a more stable cycling performance with an extended period of 200 h at 1 mA cm^{-2}. After matching with the NMC811 cathode, the capacity retention rate of batteries can reach 96.8% after 50 cycles. The infusion of Al_2O_3 demonstrates its capability to react with LLZTO particles, creating an ion-conducting interfacial layer of Li-Al-O at the GBs. This interfacial layer effectively inhibits Li nucleation and filament growth within LLZTO, contributing to the suppression of lithium dendrites. Our work provides new suggestions for optimizing the synthesis of solid-state electrolytes, which can help facilitate the commercial application of solid-state batteries.

Keywords: lithium metal batteries; solid state electrolyte; ALD; grain boundaries

1. Introduction

As the demand for environmental protection increases, numerous countries have implemented policies aimed at reducing the emission of carbon dioxide. This drive has catalyzed the robust momentum towards the adoption of mobile electronic devices and electric vehicles, consequently engendering augmented requisites for lithium-ion batteries (LIBs) concerning both energy density and safety considerations [1–3]. Conventional LIBs traditionally employ liquid electrolytes typified by their inflammable nature, such as carbonates [4,5]. However, over prolonged cycling, the formation of lithium (Li) dendrites within the liquid electrolyte milieu can breach its integrity, precipitating short circuits and culminating in instances of combustion and detonation [6]. In this context, the integration of ceramic electrolytes and Li anodes within all-solid-state lithium metal batteries has garnered considerable scholarly and industrial interest owing to their inherent safety features and remarkable energy density attributes [7].

$Li_7La_3Zr_2O_{12}$ (LLZO), a garnet-type oxide, has emerged as a subject of extensive scholarly investigation as a prospective electrolyte material for all-solid-state lithium metal batteries (ASSLMBs) due to its remarkable attributes encompassing a broad electrochemical window, elevated ionic conductivity, and commendable mechanical robustness [8,9]. But, during cycling, the formidable concern of Li filament intrusion within the LLZO framework, particularly along the grain boundaries (GBs), has surfaced as a critical challenge [10,11]. Recent studies have shown that the generation of Li filaments is mainly attributed to the inhomogeneous interfacial ion transport and current leakage resulting from the heterogeneous electron transport at the GBs. Furthermore, pronounced incongruities in the physicochemical properties inherent to the LLZO grains vis-à-vis their adjacent GBs engender uneven ion concentrations along the LLZO–lithium metal interface [12–16]. The decomposition byproducts originating from LLZO, exemplified by constituents such as Zr_3O and La_2O_3, alongside the discernibly narrower bandgap (ranging approximately from 1 to 3 eV) characterizing the GBs within the LLZO matrix synergistically accentuate the magnitude of the electron-driven current leakage [17,18]. The presence of defects and Li_2CO_3 in the grain boundaries can cause uneven ion transport within the polycrystals. This intricate interplay can ultimately precipitate the selective reduction of Li ions at these GB sites, thereby instigating the localized genesis and accumulation of Li filaments, culminating in eventual short-circuiting events [19,20]. In the past, extensive efforts have been undertaken to mitigate the formation of dendrites and minimize detrimental side reactions, including the use of electrolyte additives [21–23] and nanostructured electrolytes [24,25]. Nevertheless, implementing these measures would inevitably introduce complexities and raise the cost of structural design. Another viable and effective strategy to alleviate irreversible lithium loss and hinder dendrite growth involves the application of a surface coating on the electrodes or the creation of a protective buffer layer on the surface of the lithium anode. This approach provides a viable solution without significantly altering the overall design and manufacturing process [26,27]. Most traditional GB modification methods using new substances are added to improve the physical and chemical characteristics at the GBs [27–29]. However, controlling the thickness of the introduced grain boundary modification layer is often challenging. Achieving a significant performance improvement requires precise control of the modification layer's thickness, which poses technical difficulties. Compared with other methods, an ALD-modified layer on the surface of LLZO permeates into the interior of the LLZO and can effectively modify the GBs, forming a GB modification layer with a controllable thickness. This approach also ensures a more uniform grain boundary modification layer. Several surface coating methods can be utilized for this purpose, including chemical vapor deposition, electrochemical deposition, and magnetron sputtering. However, among these options, ALD is considered an ideal strategy for improving a battery's lifetime. ALD offers a precise and uniform coating thickness, excellent conformality on complex electrode structures, and the ability to deposit a wide range of materials with a high degree of control, making it a promising technique for enhancing battery performance and longevity. This is because of its exceptional coating layer uniformity, high chemical stability, and good compactness [30–34]. Recently, there have been some studies on GB modification using a surface coating layer followed by annealing, including employing magnetron sputtering to deposit Indium (In) on the surface of LLZO, followed by annealing, resulting in an outstanding interface resistance [28]. However, due to the non-uniformity of the coating layer produced by magnetron sputtering, the modification layer that infiltrates into the grain boundary also varies. In addition to its function in inhibiting dendrite formation, the Al_2O_3 interlayer also plays a role in improving the Li^+ conductivity during the charging and discharging cycles. The Al_2O_3 interlayer reacts with Li^+, resulting in the formation of a Li-Al-O compound that exhibits favorable Li^+ conductivity. This phenomenon further improves the overall performance of the solid-state batteries by facilitating the efficient transport of Li^+ ions within the system [35]. The reported ion conductivity in Al-LLZO ceramics has been improved through the complex substitution of Al^{3+}-Li^+ and the segregation of Li_xAlO_y [36]. More importantly, by utilizing ALD technology, the controllable and uniform deposition of Al_2O_3 can be achieved on the LLZO surface to achieve a controllable and uniform GB modification layer.

In this work, we employed ALD to achieve the deposition of an Al_2O_3 coating layer with an approximate thickness of 10 nm onto the surface of $Li_{6.4}La_3Zr_{1.4}Ta_{0.6}O_{12}$ (LLZTO). Subsequent to a thermal treatment process conducted at 700 °C, the coated layer exhibited successful infiltration into the internal structure of the LLZTO, leading to a modification of the GBs. The electrochemical performance testing showed that the novel design structure effectively improved the ion conductivity of the electrolyte, and the capacity retention rate reached 96.8% after 50 cycles when matching with the NMC811 cathode. Our deduction posits that, over the course of battery charge and discharge cycles, the Al_2O_3 within the grain boundaries progressively undergoes a transformation into Li-Al-O. This chemical transformation serves to heighten the ion conduction capabilities at the grain boundaries, thereby significantly mitigating the onset and proliferation of Li dendritic formations. The implications of this study extend to the realm of interface engineering within solid-state electrolytes, providing practical and viable recommendations for the advancement of interfacial designs.

2. Experimental Section

2.1. Synthesis of LLZTO Solid Electrolyte

In the experimental setup, the materials used included high-purity LLZTO (with a purity of 99.99%) obtained from Kejing Zhida Co. Ltd (China). To act as a binder, polyethylene glycol (PEG) with a purity of 99% was added. The process began by weighing the PEG powders and subsequently dissolving them in deionized water at a ratio of 1:10. The dissolution process took place at a temperature of 40 °C. The cooled PEG solution was mixed with an appropriate amount of LLZTO powder. In order to prevent the block from being damaged during the demolding process, an open-flap tablet pressing mold was designed. After the PEG-containing mixture was prepared, it was pressed at a pressure of 20 MPa to form a compact block with a diameter of 12 mm. To obtain the LLZTO solid electrolyte bulks, a two-step sintering method was employed. The first step involved sintering the pressed block at a temperature of 800 °C for a duration of 120 min. Subsequently, the sintered block was further treated by sintering at a higher temperature of 1200 °C for a total duration of 600 min. This process resulted in the formation of the final LLZTO solid electrolyte bulks used for further experimentation and analysis. The entire sintering process was carried out within an alumina crucible, with the sintering atmosphere being selected as ambient air. The sintered LLZTO was ground by 460/1000/2000/3000 sandpaper in sequence to obtain a smooth surface. The sample without any additional coating is referred to as "bare LLZTO".

2.2. ALD Surface Coating Combined with Annealing to Realize the Grain Boundary Design

The deposition of Al_2O_3 was carried out within a reaction chamber using ALD equipment integrated into a glovebox. This setup ensured that the oxygen and water content remained below 0.1 ppm, creating a meticulously controlled environment for the deposition process. Trimethyl aluminum (TMA) and deionized water were chosen as the sources for aluminum and oxygen, respectively. The ALD process involved using argon as a cleaning gas with a flow rate of 50 standard cubic centimeters per minute (SCCM) to purge any excess precursor gas and maintain a stable and precise deposition environment. This meticulous control of the deposition process ensures the desired characteristics and performance of the Al_2O_3 interlayer for solid-state batteries. Operating at a temperature of 150 °C, the ALD process effectively produces a uniform and conformal Al_2O_3 layer on the surface of the sample. In the ALD process, each cycle comprises seven steps. Firstly, water vapor is introduced for 0.5 s, followed by the introduction of argon gas for 15 s to remove residual water vapor. Next, TMA is introduced for 0.3 s, and then argon gas is introduced for 15 s to remove any remaining TMA. Finally, a mechanical pump is used for 15 s to clear all residual gases. The inclusion of a 0.3 s static period after introducing the water vapor and TMA allows for a more thorough and efficient reaction process. Here, about 10 nm of Al_2O_3 was deposited on the LLZTO. This sample is called "LLZTO@Al_2O_3". The samples coated with Al_2O_3 were placed in a muffle furnace for annealing at 700 °C for about 2 h in an ambient air atmosphere to obtain the "LLZTO@Al_2O_3-annealed" sample.

2.3. Characterization of Materials

X-ray diffraction (XRD) using an Advance D8 instrument was utilized to analyze the phase structure of the material. The surface and cross-sectional morphology as well as the microstructure of the material were observed using a scanning electron microscope (SEM), specifically the TESCAN8000 model. To prepare specimens for transmission electron microscopy (TEM) and to analyze both the surface and cross-sectional morphology, a combination of a focused ion beam (FIB) and scanning electron microscopy (SEM), referred to as the FIB-SEM technique, was utilized. The specific FIB-SEM instrument used was the Helios Nanolab 600i manufactured by FEI. The transparent lamellae were obtained by Ga ion FIB milling (30 keV, 0.79 nA) and later fine polishing (2 keV, 9 pA). The microstructure characterization, energy-dispersive X-ray (EDX) analysis, and mapping were conducted using a Titan G2 TEM instrument, which operates at 300 kV and is manufactured by FEI.

2.4. Battery Sample Preparation and Electrochemical Testing

The investigation involved the use of Li||NCM811 full-cell and symmetric battery configurations. Three different SEs were utilized for the study, including bare LLZTO, LLZTO@Al_2O_3, and LLZTO@Al_2O_3-annealed. To prepare the $LiNi_{0.8}Mn_{0.1}Co_{0.1}O_2$ (NMC811) cathode, the following steps were followed. First, carbon black (CB) was mixed with NMC811 powders from Timcal, MTI Corporation, USA. The mixture was thoroughly ground to achieve a uniform distribution of components. Second, Poly (vinylidene fluoride) (PVDF) was added to the mixture in N-methyl-2-pyrrolidone (NMP). The mass fraction ratio employed was 8:1:1, corresponding to the NMC811, CB, and PVDF components, respectively. Third, the slurry obtained in the previous step was coated onto the surface of the solid electrolyte. Fourth, the coated solid electrolyte was subjected to drying in a vacuum drying oven at a temperature of 80 °C for a duration of 6 h under vacuum conditions. Fifth, the final cathode obtained after drying had an average active material loading mass of 1.8–2.0 mg cm^{-2}. Sixth, type-2032 coin cells were assembled or disassembled within an ultrapure argon-filled glovebox. This controlled environment was maintained to prevent exposure to moisture and ensure a controlled experimental setup. Seventh, the assembly of the coin cells was accomplished using a manual hydraulic coin cell sealing machine (MSK-110, MTI Corporation, USA). This procedure was employed to achieve appropriate sealing and to prevent any potential exposure to moisture or impurities.

A CT-2001A electrochemical test system was employed to assess the cycle and rate capability of the lithium symmetrical battery. Additionally, a CHI660E electrochemical workstation was utilized to measure the internal resistance of the lithium symmetrical battery. The testing frequency range spanned from 0.01 Hz to 1 MHz, while the voltage amplitude was set to 10 mV.

The experimental setup and testing procedures for the solid-state batteries were as follows. The working voltage window of the solid-state batteries was set to be 2.7–4.3 V. Two formation cycles were conducted at a low rate of 0.1 C (1 C = 180 mA g^{-1}) before initiating the cycling tests. After the formation cycles, the solid-state batteries were subjected to cycling tests at two different rates, namely 0.5 C and 1 C. The tests were carried out at a temperature of 25 °C. A liquid electrolyte with a composition of 1.0 M LiTFSI dissolved in a mixture of DME/DOL (1:1 Vol %) with 1.0% $LiNO_3$ was used. The LLZTO surface of the solid-state electrolyte disc, which had a diameter of 12 mm and a thickness of 0.5 mm, was impregnated with a liquid electrolyte. A total of 5 µL per side of the liquid electrolyte was applied. The solid-state batteries were assembled using a curling pressure of 50 MPa to ensure proper contact and adhesion between the components. The cycle and rate capability of the lithium symmetrical battery were assessed through the utilization of an electrochemical test system (CT-2001A). Furthermore, the internal resistance of the lithium symmetrical battery was gauged via an electrochemical workstation (CHI660E), which operated within a test frequency range spanning from 0.01 Hz to 1 MHz. Additionally, a voltage amplitude of 10 mV was employed in these measurements.

3. Results and Discussion

The synthetic route of LLZTO is illustrated in Figure 1. To fabricate LLZTO pellets, a mixture of LLZTO powder and PEG was compressed into circular plates using a specifically designed open-flap tablet pressing mold. The resulting plates had a diameter of 12 mm. Using traditional cold pressing molds can lead to an uneven stress distribution, causing edge cracking or fracturing during demolding. Cracks may be present on the green compacts formed by cold pressing, and these cracks do not heal during sintering, resulting in cracks in the final sintered products [37–39]. These internal cracks in the solid-state electrolyte significantly affect ion transport and electron transport while also adversely impacting the mechanical strength of the solid-state electrolyte. To address these issues, we developed an open-flap tablet mold to avoid structural damage during demolding. The use of this self-developed open-flap tablet mold preserves the integrity of the circular compacts and, after sintering, the compacts exhibit a smooth surface with good surface morphology. This approach helps eliminate the appearance of cracks and ensures better mechanical strength in the solid-state electrolyte. The smooth surface morphology and absence of cracks after sintering are crucial factors in enhancing the overall functionality and stability of the final solid-state electrolyte pellets.

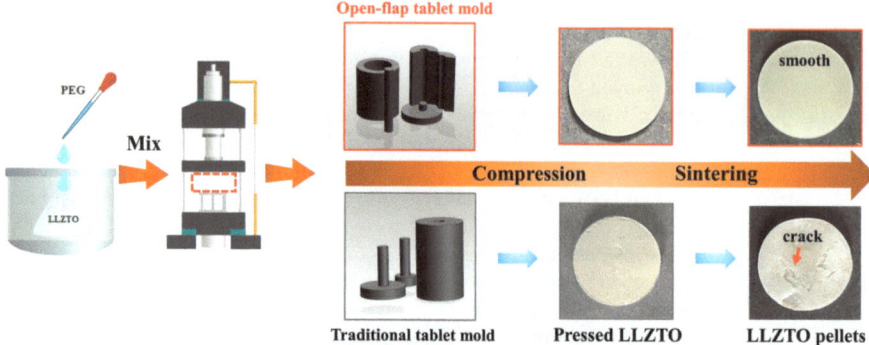

Figure 1. Schematic diagram of sintering LLZTO using a cold pressing process.

The X-ray diffraction (XRD) patterns of the LLZTO pellets are presented in Figure 2a. It is evident that the major peaks align well with the standard pattern of cubic-LLZO (ICSD#45-0109). This observation confirms the successful preparation of the cubic phase LLZTO material. The presence of characteristic peaks indicates that the crystal structure of the LLZTO pellets is consistent with the expected cubic phase, validating the synthesis process. To further illustrate the exceptional electrochemical performance of the prepared LLZTO, Electrochemical Impedance Spectroscopy (EIS) characterization was carried out. In Figure 2b, the inset showcases the equivalent circuit that was utilized to match the impedance spectra using Zview software. The impedance test was performed using a symmetrical battery structure, which has a simple design with stainless steel plates at both ends directly connected to LLZTO. At high frequencies, a capacitive arc related to the electrolyte grain boundary response was observed, while at low frequencies a linear response was seen due to lithium diffusion at the electrode. The real part impedance at the intersection of the capacitive arc and the linear response corresponds to the total impedance of LLZTO. Based on the impedance values [40], the Li$^+$ ionic conductivity at room temperature (approximately 25 °C) was calculated to be about 4.6×10^{-4} S cm^{-1}. The SEM image of the cleaved cross-section of a sintered LLZTO pellet is depicted in Figure 2c. The outcomes presented above indicate the successful synthesis of high-quality

LLZTO pellets. In solid oxide electrolytes, the infiltration of Li is closely linked to the local electronic band structure. Indeed, at the GBs of solid electrolytes like LLZTO, the electronic conductivity is typically higher compared with the bulk material. This increased electronic conductivity can lead to the accumulation of electrons at the GBs, creating localized regions with a high electron concentration. The presence of these accumulated electrons at the GBs can facilitate the premature reduction of Li$^+$ ions during the charge and discharge cycles in solid-state batteries, which can cause safety issues and lead to short circuiting or capacity loss in the battery. At the GBs, the electronic conductivity increases, causing many electrons to be enriched at the grain boundary, resulting in the premature reduction of Li$^+$ and the formation of Li filaments. After 50 cycles, the LLZTO from the disassembled battery was soaked in ethanol, which dissolved the lithium filaments produced by local deposition in LLZTO during cycling. Several cracks appeared in the LLZTO pellets after cycling, resulting in a significant reduction in contact between particles compared with the original sample (Figure 2d). This phenomenon can be detected more clearly at a lower magnification, as shown in Figure S2, and this crack is quite different from the mechanical crack caused during disassembly (Figure S2c). It may be due to the accumulation and growth of "dead Li" at the GBs of the original LLZTO, leading to the formation of cracks.

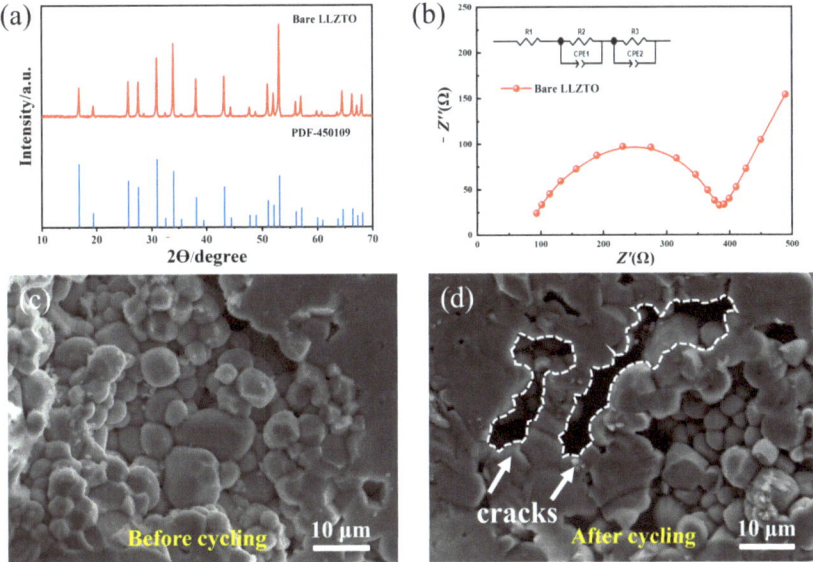

Figure 2. Characterization of the as-prepared LLZTO garnet electrolyte pellet. (**a**) XRD comparison of LLZTO. (**b**) EIS spectra of the LLZTO electrolyte at a temperature of 25 °C. (**c**) Cross-section SEM image of LLZTO pellets. (**d**) Cross−section SEM image of LLZTO after 50 cycles.

Herein, we attempted to prevent the formation of Li filaments at GBs by modifying them through annealing after coating. Figure 3a depicts the cross-sectional SEM morphology of LLZTO coated with Al$_2$O$_3$ through ALD, while Figure 3b provides a TEM image of the enclosed region within Figure 3a. The imagery reveals that the LLZTO surface is consistently enveloped by a coating layer of approximately 10 nm thickness. Figure 3c–e shows the EDS mapping of an LLZTO garnet pellet that has been coated with Al$_2$O$_3$. In the EDS mapping, it was observed that Zr and Al elements are present in the LLZTO garnet, with the presence of Al being attributed to the deposited Al$_2$O$_3$. The distribution of Al demonstrates that the Al$_2$O$_3$ coating is exclusively present on the surface of LLZTO, while there is no presence of Al within its interior. Furthermore, we studied the microstructure change of the GBs during the annealing process. Figure 3f shows the cross-sectional SEM image of the LLZTO@Al$_2$O$_3$-annealed sample, and Figure 3g presents a magnified image of the boxed area in Figure 3f, where the position with a deeper contrast is the GBs in the LLZTO. Impressively, after annealing, it was found that Al infiltrated into the GBs as shown in Figure 3h,i. In LLZTO, most of the grain boundaries (GBs) are typically narrow and may slightly widen near triple junctions where three grains meet. These GBs exhibit structural asymmetry and are primarily composed of grains that are randomly oriented. Hence, the presence of Al_2O_3 entering the interior of LLZTO is more prominently characterized at the triple junction's location. A schematic diagram of the Al_2O_3 infusion is shown in Figure 3k. The high temperature cannot lead to the LLZTO grains' decomposition and promotes surface Al_2O_3 diffusion. Therefore, Al is enriched at the GBs and does not enter the interior of the grains, forming a uniform GB modification layer.

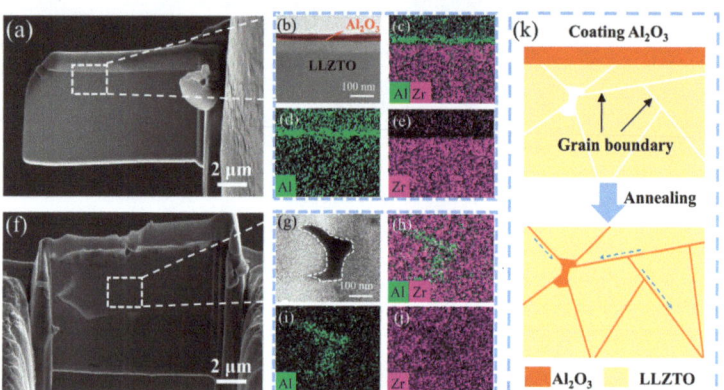

Figure 3. (**a**) Cross-sectional SEM images. (**b**–**e**) HAADF-STEM images and the corresponding EDS mapping of the LLZTO@Al_2O_3 garnet pellet. (**f**) Cross-sectional SEM images. (**g**–**j**) STEM-HAADF images and the corresponding EDS mapping of the LLZTO@Al_2O_3-annealed pellet. (**k**) Schematic diagram of the LLZTO surface coating enriched and modified at grain boundaries through annealing.

Electrochemical testing was conducted to validate the role of GB engineering in the study. Impedance diagrams of the coating and annealing samples (Figure 4a) showed that the internal resistance of the annealing sample was attenuated compared with the unmodified sample. The addition of aluminum oxides to the grain boundaries and surface coating of solid-state batteries improves their stability against electronic transmission; thus, the samples with Al_2O_3 exhibited higher impedance. The annealed sample exhibited an ionic conductivity of 4.6×10^{-4} s cm^{-2} (as shown in Figure 4b). Despite detecting a relatively higher internal resistance in the LLZTO sample coated with Al_2O_3, the ion conductivity of the LLZTO sample demonstrated an improvement after undergoing the annealing treatment with the Al_2O_3 coating layer. The symmetric cells are commonly used to test the cycling performance of electrolytes [41]. The structural diagram is shown in Figure 4d. The Li/LLZTO@Al_2O_3-annealed/Li symmetric cell was able to cycle stably for 200 h at 1 mA cm^{-2}, while the Li/LLZTO/Li symmetric cell cycled for only 60 h at 1 mA cm^{-2} (Figure 4c). Figure S5 illustrates a magnified view of the region highlighted in Figure 4c. The solid electrolyte was then paired with an NMC811 cathode to study the effectiveness of LLZTO@Al_2O_3-annealed on the full-cell cycling performance. The electrochemical performance of Li/Bare LLZTO/NMC811 and Li/LLZTO@Al_2O_3-annealed/NMC811 cells is presented in Figure 4e. The Li/Bare LLZTO/NMC811 and Li/LLZTO@Al_2O_3-annealed/NMC811 batteries initially exhibited a similar specific capacity, around 160 mAh g^{-1}, during the early cycles. After 50 cycles, the Li/Bare LLZTO/NMC811 battery experienced continuous capacity decay, with only 80.3 mAh g^{-1} of capacity remaining. On the contrary, the Li/LLZTO@Al_2O_3-annealed/NMC811 battery exhibited a significantly higher discharge capacity, retaining 161.6 mAh g^{-1} after undergoing 50 cycles. The capacity retention rates were calculated to be 46.7% for the bare sample and a notable 96.8% for the annealed sample battery. These results further confirmed that, compared with unmodified samples, LLZTO with a grain boundary modification layer exhibits higher ion conductivity and excellent cycling performance.

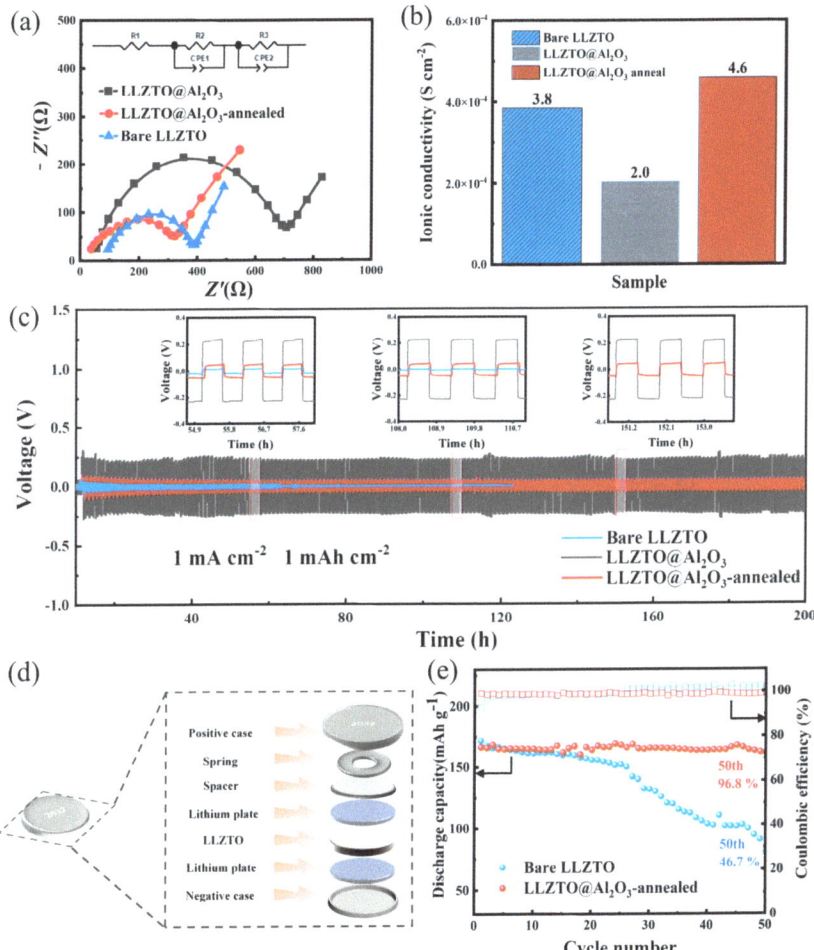

Figure 4. (**a,b**) The impedance distribution and ion conductivity maps of three types of samples, bare, coated, and annealed after coating, are presented in this figure. (**c**) Constant current cycle curves of Li/LLZTO/Li, Li/LLZTO@Al$_2$O$_3$/Li, and Li/LLZTO@Al$_2$O$_3$−anealed/Li symmetrical batteries at a current density of 1 mA cm^{-2} are presented in this figure. (**d**) The schematic diagram of the Li/LLZTO/Li symmetric battery assembly. (**e**) Long cycle curves of the entire battery assembled with bare and annealed samples and an NMC811 positive electrode.

The LLZTO with a uniform GB modification layer exhibited excellent cycling performance, and our research endeavored to unravel the underlying mechanisms behind the functioning of the modifying layer in this context. As shown in Figure 5a, after 50 cycles, the bare sample exhibited numerous gaps in the SEM image and less contact between particles, which may have resulted from localized lithium deposition during cycling. During the cycling process's progression, Li filaments grew, causing the battery to short circuit. However, the annealed sample did not show obvious gaps or cracks after 50 cycles (Figure 5b). Observed in a larger field of view, the annealed sample showed fewer traces of lithium deposition compared with the pristine sample (Figure S3). Building upon the foundation of the experimental analysis, we formulated the following hypotheses. In the initial unmodified sample depicted in Figure 5c, intrinsic defects and pores inherent to LLZTO serve as initiation sites for the nucleation of lithium dendrites, which subsequently undergo progressive growth. This process results in the generation of a substantial number of residual cracks. In samples modified by grain boundaries, the initial nucleation of lithium dendrites was suppressed, resulting in the absence of a significant number of cracks observed during 50 cycles. During battery cycling, Al_2O_3 at the GBs gradually becomes a Li-Al-O mixture, ensuring efficient ion conduction at the GBs. The uniform and dense coating of Al_2O_3 by ALD ensures a consistent and stable interface, reducing the likelihood of localized flaws. This indicates that the modification layer at the grain boundary improves the ion conduction, reduces the initial deposition in the intrinsic void pores, makes the ion transport channel more uniform, and reduces the growth of dendrites. Overall, the use of ALD to coat an ultrathin layer of Al_2O_3 and anneal it is a promising method for modifying the grain boundaries of LLZTO.

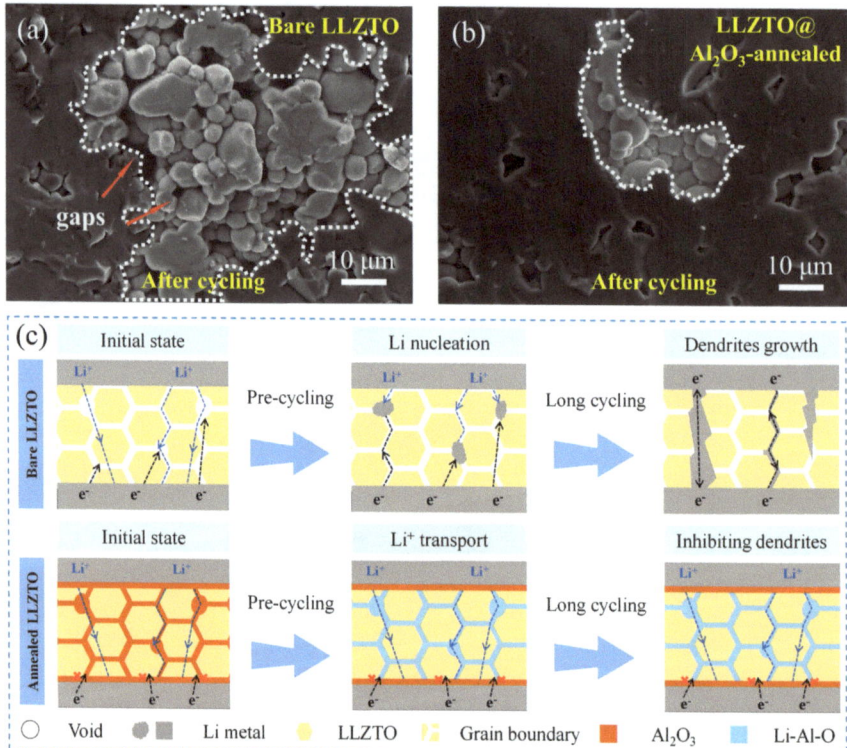

Figure 5. (**a**,**b**) SEM image of pristine and annealed samples disassembled after 50 cycles; (**c**) Schematic diagram of electron and ion transport and grain boundary modification using an annealing process in LLZTO.

4. Conclusions

The key to prolonging the cycling lifespan of garnet-type solid-state electrolytes is inhibiting the nucleation and growth of internal dendrites. In this report, a solid-state electrolyte (SSE) with low interfacial resistance was achieved by constructing a grain boundary structure. An ultrathin layer of Al_2O_3 was deposited on LLZTO using ALD equipment and, after annealing, the Al_2O_3 layer infiltrated into the inner grain boundaries. It was demonstrated that the meaningful design exhibited a high Li^+ conductivity of about 4.6×10^{-4} s cm^{-2}. Based on this, the Li/LLZTO@Al$_2$O$_3$-annealed/Li symmetric cells showed low resistance and great interface stability, which achieved excellent cycling performances. This design emphasizes the significance of tailoring the SSE interfacial component at the grain boundary to echo the Li dendrite penetration challenges. This study presents an effective approach for the development of high-performance solid-state batteries, including those with alkali metal anodes.

Supplementary Materials: The following supporting information can be downloaded at: https://www.mdpi.com/article/10.3390/batteries9110526/s1. Figure S1: (a,b) SEM image of pristine and coated samples disassembled after 50 cycles; Figure S2: (a) Cross-section SEM image of LLZTO pellets. (b) Cross-section SEM image and (c) the edges of bulk image of LLZTO after 50 cycles; Figure S3: (a,b) SEM image of pristine and annealed samples disassembled after 50 cycles; Figure S4: Voltage-capacity curves of Li plating in coated and annealed sample in the overpotential for Li nucleation at 1.0 mA cm^{-2}, respectively; Figure S5: Constant current cycle curves of Li/LLZTO/Li, Li/LLZTO@Al$_2$O$_3$/Li, and Li/LLZTO@Al$_2$O$_3$-anealed/Li symmetrical batteries at a current density of 0.5 mA cm^{-2} are presented in this figure; Figure S6: Voltage-versus-capacity plot of NMC811 cells using bare LLZTO and LLZTO@Al$_2$O$_3$-annealed as electrolyte and Li metal as anodes, respectively.

Author Contributions: Conceptualization, X.C. and X.L.; methodology, T.C. and T.S.; validation, T.S., M.W. and T.Y.; formal analysis, T.S., T.C. and X.C.; investigation, T.S.; resources, D.Q. and J.T.; data curation, T.S.; writing—original draft preparation, T.S.; writing—review and editing, X.C.; visualization, T.S.; supervision, X.L. and J.L.; project administration, X.C.; funding acquisition, X.L. and Y.Z. All authors have read and agreed to the published version of the manuscript.

Funding: We acknowledge the funding support from the National Natural Science Foundation of China (22075006), the China Postdoctoral Science Foundation (2021M700296), the Beijing Postdoctoral Research Foundation (2022-ZZ-051), the R&D Program of the Beijing Municipal Education Commission (KM20231005004), and the National Natural Science Foundation of China for Youth Science Fund (12204025).

Data Availability Statement: The data presented in this study are available upon request from the corresponding author.

Conflicts of Interest: The authors declare no conflict of interest.

References

1. Gao, J.; Zhu, J.; Li, X.; Li, J.; Guo, X.; Li, H.; Zhou, W. Rational Design of Mixed Electronic-Ionic Conducting Ti-Doping Li$_7$La$_3$Zr$_2$O$_{12}$ for Lithium Dendrites Suppression. *Adv. Funct. Mater.* **2021**, *31*, 2001918. [CrossRef]
2. Liu, F.; Xu, R.; Wu, Y.; Boyle, D.T.; Yang, A.; Xu, J.; Zhu, Y.; Ye, Y.; Yu, Z.; Zhang, Z.; et al. Dynamic spatial progression of isolated lithium during battery operations. *Nature* **2021**, *600*, 659–663. [CrossRef]
3. Rehnlund, D.; Wang, Z.; Nyholm, L. Lithium-Diffusion Induced Capacity Losses in Lithium-Based Batteries. *Adv. Mater.* **2022**, *34*, 2108827. [CrossRef]
4. Wu, R.; Cao, T.; Liu, H.; Cheng, X.; Liu, X.; Zhang, Y. Ultralong Lifespan for High-Voltage LiCoO$_2$ Enabled by In Situ Reconstruction of an Atomic Layer Deposition Coating Layer. *ACS Appl. Mater. Interfaces* **2022**, *14*, 25524–25533. [CrossRef]
5. Krause, C.H.; Butzelaar, A.J.; Diddens, D.; Dong, D.; Théato, P.; Bedrov, D.; Hwang, B.J.; Winter, M.; Brunklaus, G. Quasi-solid single ion conducting polymer electrolyte membrane containing novel fluorinated poly (arylene ether sulfonimide) for lithium metal batteries. *J. Power Sources* **2021**, *484*, 229267. [CrossRef]
6. Jiang, J.-H.; Wang, A.-B.; Wang, W.-K.; Jin, Z.-Q.; Fan, L.-Z. P (VDF-HFP)-poly (sulfur-1, 3-diisopropenylbenzene) functional polymer electrolyte for lithium–sulfur batteries. *J. Energy Chem.* **2020**, *46*, 114–122. [CrossRef]
7. Banerjee, A.; Wang, X.; Fang, C.; Wu, E.A.; Meng, Y.S. Interfaces and interphases in all-solid-state batteries with inorganic solid electrolytes. *Chem. Rev.* **2020**, *120*, 6878–6933. [CrossRef]
8. Samson, A.J.; Hofstetter, K.; Bag, S.; Thangadurai, V. A bird's-eye view of Li-stuffed garnet-type Li$_7$La$_3$Zr$_2$O$_{12}$ ceramic electrolytes for advanced all-solid-state Li batteries. *Energy Environ. Sci.* **2019**, *12*, 2957–2975. [CrossRef]
9. Zhao, N.; Khokhar, W.; Bi, Z.; Shi, C.; Guo, X.; Fan, L.-Z.; Nan, C.-W. Solid garnet batteries. *Joule* **2019**, *3*, 1190–1199. [CrossRef]
10. Vishnugopi, B.S.; Dixit, M.B.; Hao, F.; Shyam, B.; Cook, J.B.; Hatzell, K.B.; Mukherjee, P.P. Mesoscale interrogation reveals mechanistic origins of lithium filaments along grain boundaries in inorganic solid electrolytes. *Adv. Energy Mater.* **2022**, *12*, 2102825. [CrossRef]
11. Wang, C.; Wang, H.; Tao, L.; Wang, X.; Cao, P.; Lin, F.; Xin, H.L. Direct Observation of Nucleation and Growth Behaviors of Lithium by In Situ Electron Microscopy. *ACS Energy Lett.* **2023**, *8*, 1929–1935. [CrossRef]

12. Liu, X.; Garcia-Mendez, R.; Lupini, A.R.; Cheng, Y.; Hood, Z.D.; Han, F.; Sharafi, A.; Idrobo, J.C.; Dudney, N.J.; Wang, C. Local electronic structure variation resulting in Li 'filament' formation within solid electrolytes. *Nat. Mater.* **2021**, *20*, 1485–1490. [CrossRef]
13. Tantratian, K.; Yan, H.; Ellwood, K.; Harrison, E.T.; Chen, L. Unraveling the Li penetration mechanism in polycrystalline solid electrolytes. *Adv. Energy Mater.* **2021**, *11*, 2003417. [CrossRef]
14. Huo, H.; Gao, J.; Zhao, N.; Zhang, D.; Holmes, N.G.; Li, X.; Sun, Y.; Fu, J.; Li, R.; Guo, X. A flexible electron-blocking interfacial shield for dendrite-free solid lithium metal batteries. *Nat. Commun.* **2021**, *12*, 176. [CrossRef]
15. Li, Z.; Fu, J.; Zheng, S.; Li, D.; Guo, X. Self-Healing Polymer Electrolyte for Dendrite-Free Li Metal Batteries with Ultra-High-Voltage Ni-Rich Layered Cathodes. *Small* **2022**, *18*, 2200891. [CrossRef]
16. Cojocaru-Mirédin, O.; Schmieg, J.; Müller, M.; Weber, A.; Ivers-Tiffée, E.; Gerthsen, D. Quantifying lithium enrichment at grain boundaries in $Li_7La_3Zr_2O_{12}$ solid electrolyte by correlative microscopy. *J. Power Sources* **2022**, *539*, 231417. [CrossRef]
17. Huo, H.; Li, X.; Sun, Y.; Lin, X.; Doyle-Davis, K.; Liang, J.; Gao, X.; Li, R.; Huang, H.; Guo, X. Li_2CO_3 effects: New insights into polymer/garnet electrolytes for dendrite-free solid lithium batteries. *Nano Energy* **2020**, *73*, 104836. [CrossRef]
18. Delluva, A.A.; Kulberg-Savercool, J.; Holewinski, A. Decomposition of trace Li_2CO_3 during charging leads to cathode interface degradation with the solid electrolyte LLZO. *Adv. Funct. Mater.* **2021**, *31*, 2103716. [CrossRef]
19. Tian, H.-K.; Xu, B.; Qi, Y. Computational study of lithium nucleation tendency in $Li_7La_3Zr_2O_{12}$ (LLZO) and rational design of interlayer materials to prevent lithium dendrites. *J. Power Sources* **2018**, *392*, 79–86. [CrossRef]
20. Aguesse, F.; Manalastas, W.; Buannic, L.; Lopez del Amo, J.M.; Singh, G.; Llordés, A.; Kilner, J. Investigating the dendritic growth during full cell cycling of garnet electrolyte in direct contact with Li metal. *ACS Appl. Mater. Interfaces* **2017**, *9*, 3808–3816. [CrossRef]
21. Biao, J.; Han, B.; Cao, Y.; Li, Q.; Zhong, G.; Ma, J.; Chen, L.; Yang, K.; Mi, J.; Deng, Y. Inhibiting formation and reduction of Li_2CO_3 to LiC_x at grain boundaries in garnet electrolytes to prevent Li penetration. *Adv. Mater.* **2023**, *35*, 2208951. [CrossRef] [PubMed]
22. Zhai, L.; Yang, K.; Jiang, F.; Liu, W.; Yan, Z.; Sun, J. High-performance solid-state lithium metal batteries achieved by interface modification. *J. Energy Chem.* **2023**, *79*, 357–364. [CrossRef]
23. Shi, C.; Hamann, T.; Takeuchi, S.; Alexander, G.V.; Nolan, A.M.; Limpert, M.; Fu, Z.; O'Neill, J.; Godbey, G.; Dura, J.A. 3D Asymmetric Bilayer Garnet-Hybridized High-Energy-Density Lithium–Sulfur Batteries. *ACS Appl. Mater. Interfaces* **2022**, *15*, 751–760. [CrossRef]
24. Liu, B.; Zhang, L.; Xu, S.; McOwen, D.W.; Gong, Y.; Yang, C.; Pastel, G.R.; Xie, H.; Fu, K.; Dai, J. 3D lithium metal anodes hosted in asymmetric garnet frameworks toward high energy density batteries. *Energy Storage Mater.* **2018**, *14*, 376–382. [CrossRef]
25. Fu, K.K.; Gong, Y.; Hitz, G.T.; McOwen, D.W.; Li, Y.; Xu, S.; Wen, Y.; Zhang, L.; Wang, C.; Pastel, G. Three-dimensional bilayer garnet solid electrolyte based high energy density lithium metal–sulfur batteries. *Energy Environ. Sci.* **2017**, *10*, 1568–1575. [CrossRef]
26. Wang, J.-G.; Liu, H.; Zhou, R.; Liu, X.; Wei, B. Onion-like nanospheres organized by carbon encapsulated few-layer MoS_2 nanosheets with enhanced lithium storage performance. *J. Power Sources* **2019**, *413*, 327–333. [CrossRef]
27. Zhang, Y.; Wang, G.; Tang, L.; Wu, J.; Guo, B.; Zhu, M.; Wu, C.; Dou, S.X.; Wu, M. Stable lithium metal anodes enabled by inorganic/organic double-layered alloy and polymer coating. *J. Mater. Chem. A* **2019**, *7*, 25369–25376. [CrossRef]
28. Müller, M.; Schmieg, J.; Dierickx, S.; Joos, J.; Weber, A.; Gerthsen, D.; Ivers-Tiffée, E. Reducing impedance at a Li-metal anode/garnet-type electrolyte interface implementing chemically resolvable in layers. *ACS Appl. Mater. Interfaces* **2022**, *14*, 14739–14752. [CrossRef]
29. Qin, Z.; Xie, Y.; Meng, X.; Qian, D.; Mao, D.; Zheng, Z.; Wan, L.; Huang, Y. Grain boundary engineering in Ta-doped garnet-type electrolyte for lithium dendrite suppression. *ACS Appl. Mater. Interfaces* **2022**, *14*, 40959–40966. [CrossRef]
30. Zhao, Y.; Zheng, K.; Sun, X. Addressing interfacial issues in liquid-based and solid-state batteries by atomic and molecular layer deposition. *Joule* **2018**, *2*, 2583–2604. [CrossRef]
31. Xie, J.; Sendek, A.D.; Cubuk, E.D.; Zhang, X.; Lu, Z.; Gong, Y.; Wu, T.; Shi, F.; Liu, W.; Reed, E.J. Atomic layer deposition of stable $LiAlF_4$ lithium ion conductive interfacial layer for stable cathode cycling. *ACS Nano* **2017**, *11*, 7019–7027. [CrossRef]
32. Yu, F.; Du, L.; Zhang, G.; Su, F.; Wang, W.; Sun, S. Electrode engineering by atomic layer deposition for sodium-ion batteries: From traditional to advanced batteries. *Adv. Funct. Mater.* **2020**, *30*, 1906890. [CrossRef]
33. Liu, Y.; Sun, Q.; Zhao, Y.; Wang, B.; Kaghazchi, P.; Adair, K.R.; Li, R.; Zhang, C.; Liu, J.; Kuo, L.-Y. Stabilizing the interface of NASICON solid electrolyte against Li metal with atomic layer deposition. *ACS Appl. Mater. Interfaces* **2018**, *10*, 31240–31248. [CrossRef]
34. Cheng, X.; Zheng, J.; Lu, J.; Li, Y.; Yan, P.; Zhang, Y. Realizing superior cycling stability of Ni-Rich layered cathode by combination of grain boundary engineering and surface coating. *Nano Energy* **2019**, *62*, 30–37. [CrossRef]
35. Zhang, Y.; Gao, X.; Tang, Z.; Mei, Y.; Xiang, X.; Deng, J. Study on the suppression of lithium dendrites based on solid electrolyte $Li_7La_3Zr_2O_{12}$ with annealing treatment of Al interlayer. *J. Alloys Compd.* **2022**, *904*, 163908. [CrossRef]
36. Zhuang, L.; Huang, X.; Lu, Y.; Tang, J.; Zhou, J.; Ao, X.; Yang, Y.; Tian, B. Phase transformation and grain-boundary segregation in Al-Doped $Li_7La_3Zr_2O_{12}$ ceramics. *Ceram. Int.* **2021**, *47*, 22768–22775. [CrossRef]
37. Park, Y.S.; Lee, J.M.; Yi, E.J.; Moon, J.-W.; Hwang, H. All-Solid-State Lithium-Ion Batteries with Oxide/Sulfide Composite Electrolytes. *Materials* **2021**, *14*, 1998. [CrossRef] [PubMed]

38. Seo, J.-H.; Nakaya, H.; Takeuchi, Y.; Fan, Z.; Hikosaka, H.; Rajagopalan, R.; Gomez, E.D.; Iwasaki, M.; Randall, C.A. Broad temperature dependence, high conductivity, and structure-property relations of cold sintering of LLZO-based composite electrolytes. *J. Eur. Ceram. Soc.* **2020**, *40*, 6241–6248. [CrossRef]
39. Yang, L.; Huang, X.; Zou, C.; Tao, X.; Liu, L.; Luo, K.; Zeng, P.; Dai, Q.; Li, Y.; Yi, L.; et al. Rapid preparation and performances of garnet electrolyte with sintering aids for solid-state Li–S battery. *Ceram. Int.* **2021**, *47*, 18196–18204. [CrossRef]
40. Xie, H.; Li, C.; Kan, W.H.; Avdeev, M.; Zhu, C.; Zhao, Z.; Chu, X.; Mu, D.; Wu, F. Consolidating the grain boundary of the garnet electrolyte LLZTO with Li_3BO_3 for high-performance $LiNi_{0.8}Co_{0.1}Mn_{0.1}O_2$/$LiFePO_4$ hybrid solid batteries. *J. Mater. Chem. A* **2019**, *7*, 20633–20639. [CrossRef]
41. Zamudio-García, J.; Caizán-Juanarena, L.; Porras-Vázquez, J.M.; Losilla, E.R.; Marrero-López, D. A review on recent advances and trends in symmetrical electrodes for solid oxide cells. *J. Power Sources* **2022**, *520*, 230852. [CrossRef]

Disclaimer/Publisher's Note: The statements, opinions and data contained in all publications are solely those of the individual author(s) and contributor(s) and not of MDPI and/or the editor(s). MDPI and/or the editor(s) disclaim responsibility for any injury to people or property resulting from any ideas, methods, instructions or products referred to in the content.

Article

Micro-Sized MoS$_6$@15%Li$_7$P$_3$S$_{11}$ Composite Enables Stable All-Solid-State Battery with High Capacity

Mingyuan Chang [1,2,3], Mengli Yang [1], Wenrui Xie [1], Fuli Tian [1], Gaozhan Liu [1], Ping Cui [1,4], Tao Wu [2,3,5,*] and Xiayin Yao [1,4,*]

1. Ningbo Institute of Materials Technology and Engineering, Chinese Academy of Sciences, Ningbo 315201, China
2. Municipal Key Laboratory of Clean Energy Technologies of Ningbo, University of Nottingham Ningbo China, Ningbo 315100, China
3. Department of Chemical and Environmental Engineering, University of Nottingham Ningbo China, Ningbo 315100, China
4. Center of Materials Science and Optoelectronics Engineering, University of Chinese Academy of Sciences, Beijing 100049, China
5. Key Laboratory of Carbonaceous Wastes Processing and Process Intensification of Zhejiang, Ningbo 315100, China
* Correspondence: authors: tao.wu@nottingham.edu.cn (T.W.); yaoxy@nimte.ac.cn (X.Y.)

Abstract: All-solid-state lithium batteries without any liquid organic electrolytes can realize high energy density while eliminating flammability issues. Active materials with high specific capacity and favorable interfacial contact within the cathode layer are crucial to the realization of good electrochemical performance. Herein, we report a high-capacity polysulfide cathode material, MoS$_6$@15%Li$_7$P$_3$S$_{11}$, with a particle size of 1–4 μm. The MoS$_6$ exhibited an impressive initial specific capacity of 913.9 mAh g^{-1} at 0.1 A g^{-1}. When coupled with the Li$_7$P$_3$S$_{11}$ electrolyte coating layer, the resultant MoS$_6$@15%Li$_7$P$_3$S$_{11}$ composite showed improved interfacial contact and an optimized ionic diffusivity range from 10^{-12}–10^{-11} cm^2 s^{-1} to 10^{-11}–10^{-10} cm^2 s^{-1}. The Li/Li$_6$PS$_5$Cl/MoS$_6$@15%Li$_7$P$_3$S$_{11}$ all-solid-state lithium battery delivered ultra-high initial and reversible specific capacities of 1083.8 mAh g^{-1} and 851.5 mAh g^{-1}, respectively, at a current density of 0.1 A g^{-1} within 1.0–3.0 V. Even under 1 A g^{-1}, the battery maintained a reversible specific capacity of 400 mAh g^{-1} after 1000 cycles. This work outlines a promising cathode material with intimate interfacial contact and superior ionic transport kinetics within the cathode layer as well as high specific capacity for use in all-solid-state lithium batteries.

Keywords: MoS$_6$@15%Li$_7$P$_3$S$_{11}$ composite; high capacity; interfacial contact; ionic transport kinetics; all-solid-state lithium battery

Citation: Chang, M.; Yang, M.; Xie, W.; Tian, F.; Liu, G.; Cui, P.; Wu, T.; Yao, X. Micro-Sized MoS$_6$@15%Li$_7$P$_3$S$_{11}$ Composite Enables Stable All-Solid-State Battery with High Capacity. *Batteries* **2023**, *9*, 560. https://doi.org/10.3390/batteries9110560

Academic Editor: Yong-Joon Park

Received: 19 October 2023
Revised: 11 November 2023
Accepted: 15 November 2023
Published: 17 November 2023

Copyright: © 2023 by the authors. Licensee MDPI, Basel, Switzerland. This article is an open access article distributed under the terms and conditions of the Creative Commons Attribution (CC BY) license (https://creativecommons.org/licenses/by/4.0/).

1. Introduction

Commercial lithium-ion batteries have been utilized broadly in the fields of plug-in hybrid/electric vehicles, smartphones, laptop computers, and other portable electronic devices for the past several decades. They significantly reduce environmental pollution and greenhouse gas emissions. The energy densities of lithium-ion batteries are close to their limitation because of the low theoretical capacities of oxide cathodes (i.e., LiFePO$_4$, LiCoO$_2$, LiMn$_2$O$_4$, and LiNi$_x$Mn$_y$Co$_z$O$_2$) and graphite anode materials [1]. Lithium–sulfur batteries that use lithium metal anodes instead of graphite can theoretically reach high specific capacity values [2]. However, there are safety concerns, such as leakage and risk of explosion, associated with organic liquid electrolyte-based lithium–sulfur batteries, and they exhibit poor rate capability and cycling stability due to the huge changes in the volume of sulfur that occur during the charge/discharge process. All-solid-state lithium batteries with nonflammable inorganic solid-state electrolytes have attracted considerable attention in the field of energy storage [3–5]. Among the various solid electrolytes, sulfide solid electrolytes have broad application potential because of their high ionic conductivity and low elastic modulus. For example, the $Li_{10}GeP_2S_{12}$ sulfide solid electrolyte [6] possesses a high ionic conductivity of around 10^{-2} S cm^{-1} and surpasses the liquid electrolytes used in traditional lithium-ion batteries. The electrochemical stability of the Li_6PS_5Cl electrolyte was also investigated, and it displayed excellent stability when in contact with lithium metal [7].

Not only is the selection of the sulfide solid electrolyte essential, but the selection of an appropriate cathode material is also key to achieving high performance all-solid-state batteries. The classical oxide cathode materials and transition metal sulfides display different reaction mechanisms during the charge/discharge process. For classical lithium intercalation process in oxide cathode materials, because the 3d metal cationic band is much higher than the p band of oxygen, the ion–electron transfer reactions occur from lithium to the lowest unoccupied energy level of the transition metal d band in ionic oxides. The discharge/charge (lithium ion intercalation/de-intercalation) process relies on the d metal level to host/release the associated electrons; this is called the cation-driven redox process. For transition metal sulfide materials, the p band of sulfur is located in a higher position and is therefore closer or even penetrate the d band of transition metal. Due to the charge transfer of these two bands, the anionic redox process is triggered, and this can improve the reversible capacity of the battery [8]. The use of transition metal sulfides such as a-TiS_4 [9], FeS_2 [10], NiS_2 [11,12], and TiS_3 [12] in all-solid-state batteries has been widely studied due to the low cost of raw materials and their abundant yields. Meanwhile, their relatively high transport kinetics can promote conductivity and reduce charge transfer resistance. Furthermore, sulfide electrolytes show superior compatibility with transition metal sulfide cathodes due to their similar chemical potential, thus realizing high energy density [13,14]. However, all-solid-state lithium batteries with sulfide electrolytes and transition metal sulfide cathodes still suffer from inferior interfacial contact within the cathode layer.

Traditional sulfide cathode materials based on insertion reactions can accommodate lithium ions in their lattices without serious structural changes occuring during the intercalation/de-intercalation reaction of $xLi^+ + MS_y + xe^- \leftrightarrow Li_xMS_y$, in which M represents the transition metal [15]. Although insertion reaction-based cathode materials exhibit stability and long life cycles when used in all-solid-state lithium batteries, they still have intrinsic problems related to their limited space for lithium ions result in to a low specific capacity [8,16]. Nevertheless, in transition metal sulfide cathode materials with anionic redox driven chemistry (a-TiS_4 [9], MoS_3 [17], and FeS_2 [10]), sulfur fully or partially exists in the state of S_2^{2-} pairs, and this has a strong influence on specific capacity. For example, MoS_2 only has S^{2-} pairs and has a theoretical specific capacity of 670 mAh g^{-1} [18,19]. MoS_3 has both S_2^{2-} and S^{2-} pairs and has a higher theoretical specific capacity of 837 mAh g^{-1} [17]. The electronic structure of the S_2^{2-} group allows them to donate or receive electrons. Their ability as donors is attributed to the π^*_g orbital, which can release electrons that convert S_2^{2-} to S^{2-}. Because of the reversible reaction of $S_2^{2-} + 2e^- = 2S^{2-}$, multi-electron reactions proceed during the charge/discharge process, resulting in a high specific capacity [8,19]. In addition, transition metal sulfides with anionic redox driven chemistry display a high voltage plateau of about 2 V, which is close to that of a lithium–sulfur battery.

Even though MoS_2 and MoS_3 demonstrate high theoretical specific capacities, the actual reversible specific capacities of MoS_2 and MoS_3 in all-solid-state lithium batteries are only 592 mAh g^{-1} (0.1 C, 0.1–3.0 V) and 747 mAh g^{-1} (0.05 A g^{-1}, 0.5–3.0 V), respectively [19,20]. Ionic transport kinetics and interfacial contact within the cathode layer become crucial challenges. Coating a thin electrolyte layer onto the surfaces of active materials is an effectively strategy to enhance ionic transport kinetics and the contact between the solid electrolyte and the active material [17,21]. Compared with MoS_2 and MoS_3, MoS_6 is considered one of the most promising cathode materials for all-solid-state lithium batteries because of its ultra-high theoretical specific capacity of 1117 mAh g^{-1}, which is a result of its high S_2^{2-} content as well as its amorphous nature, according to which possess open and random transmission paths to achieve cycling stability [22]. It therefore has potential applications in high energy all-solid-state batteries.

In this work, a micro-sized cathode material composite, $MoS_6@15\%Li_7P_3S_{11}$, is designed for use in all-solid-state lithium batteries. First, MoS_6 was synthesized using a wet chemical method. After the in situ coating of the $Li_7P_3S_{11}$ solid electrolyte onto the MoS_6, the ionic diffusivity enhanced from 10^{-12}–10^{-11} cm^2 s^{-1} to 10^{-11}–10^{-10} cm^2 s^{-1}. The $Li/Li_6PS_5Cl/MoS_6@15\%Li_7P_3S_{11}$ all-solid-state lithium batteries exhibited high initial and reversible discharge capacities of 1083.8 mAh g^{-1} and 851.5 mAh g^{-1}, respectively, at 0.1 A g^{-1} and 25 °C within 1.0–3.0 V. Because of the remarkable intimate interfacial contact between the active material MoS_6 and the solid electrolyte, these all-solid-state batteries can realize a long cycle life of 1000 cycles with a high reversible specific capacity of 400 mAh g^{-1} at 1 A g^{-1} and 25 °C.

2. Materials and Methods

2.1. Synthesis of $(NH_4)_2Mo_2S_{12}$

$(NH_4)_2Mo_2S_{12}$, which was used as a precursor, was synthesized using wet chemical methods. Briefly, 4.5 g of $(NH_4)_6Mo_7O_{24}\cdot 4H_2O$ and 3 g of $NH_2OH\cdot HCl$ were dissolved in 60 mL of water (50 °C). Next, 12 g of sulfur powder and 60 mL of $(NH_4)_2S$ (6−20% aqueous solution) were mixed and stirred for 1 h. The two abovementioned solutions were mixed together and placed in an oven at 50 °C and 90 °C for 1 h and 4 h, respectively. After filtering, the filtrate was mixed with 20 mL of $(NH4)_2S$ and allowed to stand for 36 h. It was then filtered and washed with iced water, ethanol carbon disulfide, and iced diethyl ether. The obtained $(NH_4)_2Mo_2S_{12}$ displayed a crystal structure which matched the standard peaks in the PDF card (JCPDS: F73-0900) (Figure S1) [23].

2.2. Synthesis of Micro-Sized MoS_6

The micro-sized MoS_6 was prepared via an aqueous solution reaction corresponding to the reaction of $(NH_4)_2Mo_2S_{12} + I_2 = 2NH_4I + 2MoS_6$. Accordingly, 0.28 g of $(NH_4)_2Mo_2S_{12}$ and 0.2 g of iodine were separately dissolved in N,N-dimethylformamide. Afterward, the two solutions were thoroughly mixed together and stirred continuously for 30 min. The obtained MoS_6 was filtered, washed with N,N-dimethylformamide, CS_2, and acetone, and finally dried and stored in an argon atmosphere.

2.3. Synthesis of $MoS_6@15\%Li_7P_3S_{11}$ Composite

A in situ liquid phase reaction and an annealing process were conducted in order to prepare the $MoS_6@15\%Li_7P_3S_{11}$ composite [17]. The prepared MoS_6 was mixed with P_2S_5 and Li_2S powders and then underwent magnetic stirring in anhydrous acetonitrile at 60 °C for 12 h. After removing the residual solvent, the $Li_7P_3S_{11}$ precursor was in situ coated with MoS_6, and the precursor was then collected and heated at 260 °C for about 1 h to obtain the $MoS_6@15\%Li_7P_3S_{11}$ composite.

Details concerning the synthesis of $Li_{10}GeP_2S_{12}$, Li_6PS_5Cl, and $75\%Li_2S$-$24\%P_2S_5$-$1\%P_2O_5$ can be found elsewhere. The ionic conductivities of these solid electrolytes were 8.27×10^{-3} S cm^{-1} ($Li_{10}GeP_2S_{12}$), 6.11×10^{-3} S cm^{-1} (Li_6PS_5Cl), and 1.54×10^{-3} S cm^{-1} ($75\%Li_2S$-$24\%P_2S_5$-$1\%P_2O_5$).

2.4. Material Characterization

X-ray diffraction (XRD) measurements were conducted using a Bruker D8 Advance Davinci (Karlsruhe, Germany) with Cu $K\alpha$ radiation of λ = 1.54178 Å in a 2θ range of 10–60° to determine the crystal structure of the samples. Raman spectra results were recorded in a range of 250 to 580 cm^{-1} using a Raman spectrophotometer (Renishaw inVia Reflex, Gloucestershire, UK) with a 532 nm laser. The atom ratio of the MoS_6 was analyzed via an inductively coupled plasma emission spectrometer analysis (ICP-OES, Spectro Arcos, Spectro, Dusseldorf, Germany). Field emission scanning electron microscopy (SEM, S-4800, Hitachi, Tokyo, Japan) and field emission scanning electron microscopy energy dispersive spectroscopy (SEM-EDS, S-4800, Hitachi, Tokyo, Japan) were conducted using an accelerated voltage of 15 kV to confirm the morphology, particle size, and elemental distribution. High-resolution transmission electron microscopy (HRTEM) was performed using an FEI Tecnai F20 (Hillsboro, OR, USA) with an accelerated voltage of 200 kV to confirm the existence of the $Li_7P_3S_{11}$ solid sulfide electrolyte thin layer on the surface of the MoS_6. An electrochemical workstation (Solartron 1470E, Bognor Regis, UK) was employed to conduct cyclic voltammetry (CV) measurements and electrochemical impedance spectroscopy (EIS). All of the samples used for the various measurements were prepared in glove boxes, and all of the measurements were obtained at room temperature.

2.5. Assemby and Evaluation of All-Solid-State Lithium Batteries

To analyze the electrochemical performances of the synthesized samples described above, the obtained MoS_6, $MoS_6@15\%Li_7P_3S_{11}$ composite, $Li_{10}GeP_2S_{12}$, and Super P were homogeneously mixed with a weight ratio of 40:50:10. For the preparation of the $Li/Li_6PS_5Cl/MoS_6$ all-solid-state batteries, Li_6PS_5Cl (150 mg) solid electrolyte pellets (ϕ = 10 mm) were fabricated via cold pressing at 240 MPa. The previously synthesized composite cathodes (~1 mg/cm^2) were then homogeneously spread onto the electrolyte surface and cold pressing was applied again at 240 MPa. Finally, metallic lithium foil with a 10 mm diameter was attached to the other side of the Li_6PS_5Cl layer at 360 MPa. For the purpose of comparison, $Li/75\%Li_2S-24\%P_2S_5-1\%P_2O_5/Li_{10}GeP_2S_{12}/MoS_6$ all-solid-state batteries were fabricated as well. Bilayer pellets (ϕ = 10 mm) consisting of $Li_{10}GeP_2S_{12}$ and orthorhombic phase $75\%Li_2S-24\%P_2S_5-1\%P_2O_5$ were constructed at 240 MPa as well. The abovementioned cathodes were spread onto the sides of the $Li_{10}GeP_2S_{12}$ solid electrolytes homogeneously, and this was followed by cold pressing at 240 MPa. Pieces of metallic lithium foil (ϕ = 10 mm) were placed onto the sides of the $75\%Li_2S-24\%P_2S_5-1\%P_2O_5$ solid electrolytes by pressing them together at 360 MPa. The bilayer solid electrolyte pellets were necessary due to the instability caused by the side-reaction of the $Li_{10}GeP_2S_{12}$ and the lithium metal. All of these processes were carried out in a glove box with a dry argon atmosphere.

Galvanostatic charge and discharge tests were conducted at room temperature on a multi-channel battery test system under various current densities with voltages ranging from 1.0 V to 3.0 V. The galvanostatic intermittent titration technique (GITT) was used at 1 A g^{-1} for 1 min followed by a 120 min rest. The Li ion diffusion coefficient (D) was determined using Equation (1), which is based on the Fick's second law [24]:

$$D = \frac{4}{\pi\tau}\left(\frac{n_m v_m}{S}\right)^2 \left(\frac{\Delta E_s}{\Delta E_t}\right)^2 \quad (1)$$

where τ is the duration of the pulse, n_m and v_m are the molar mass (mol) and volume (cm^3/mol) of the active material, respectively, S is the cell interfacial area, and ΔE_s and ΔE_t are the voltage drops of the pulse and discharge processes [25].

3. Results and Discussions

To better demonstrate the properties of the prepared sample, the Mo/S atom ratio was measured using an inductively coupled plasma emission spectrometer (ICP) (Table S1). The actual remaining weight ratios of the Mo and S were 31.5 wt% and 63.5 wt%, respectively, indicating an Mo/S ratio of 6.036, which is in agreement with the theoretical value of 6. The procedure for synthesizing the $MoS_6@15\%Li_7P_3S_{11}$ composite is schematically illustrated in Figure 1. A $Li_7P_3S_{11}$ solid electrolyte precursor was in situ coated onto the MoS_6 surface during facile liquid phase deposition. After a 260 °C annealing treatment, the $MoS_6@15\%Li_7P_3S_{11}$ composite was successfully prepared.

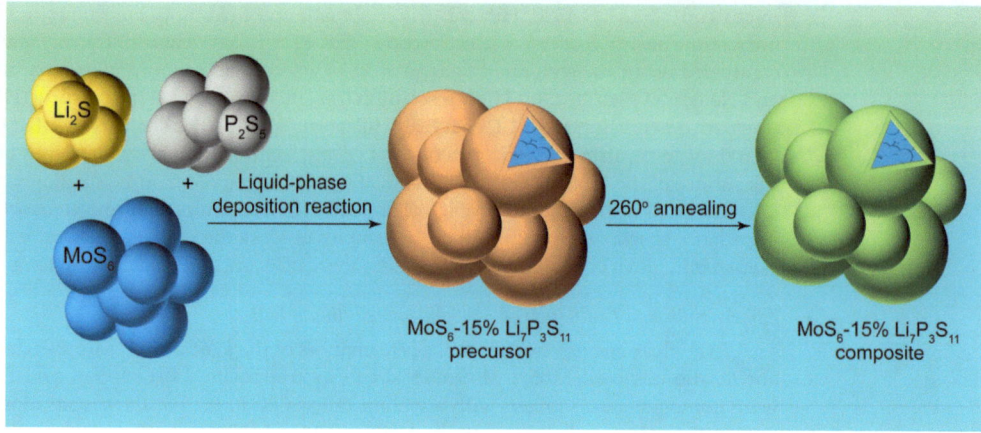

Figure 1. Schematic diagram of procedure for preparing MoS$_6$@15%Li$_7$P$_3$S$_{11}$ composite.

The XRD patterns of the MoS$_6$ and MoS$_6$@15%Li$_7$P$_3$S$_{11}$ composite are shown in Figure 2a, and they confirm the amorphous nature of MoS$_6$. No characteristic peaks were detected for the Li$_7$P$_3$S$_{11}$ sulfide electrolytes, indicating the low amount of Li$_7$P$_3$S$_{11}$ sulfide electrolytes in the coating layer. In order to confirm the existence of the Li$_7$P$_3$S$_{11}$ sulfide solid electrolytes in the MoS$_6$@15%Li$_7$P$_3$S$_{11}$ composite, Raman spectroscopy was performed (Figure 2b). The peaks located in the ranges of 286–385 cm^{-1} and 518–550 cm^{-1} correspond to vibrations of molybdenum sulfide bonds and bridging disulfide/terminal disulfide [26,27]. Furthermore, the peak located at 421 cm^{-1} can be attributed to the PS$_4^{3-}$ in the Li$_7$P$_3$S$_{11}$ sulfide solid electrolyte [21].

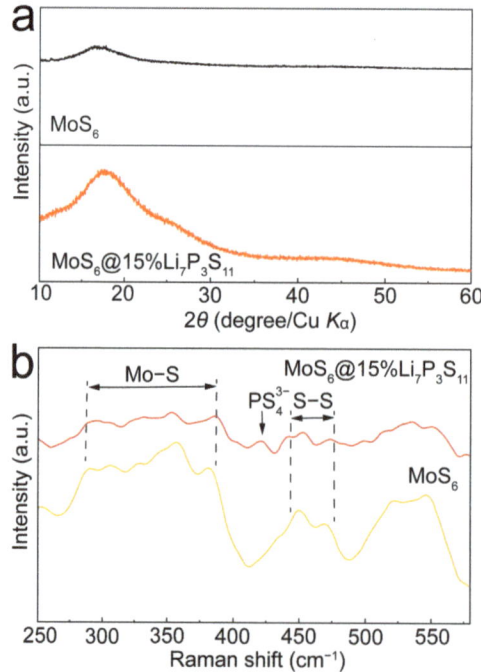

Figure 2. (**a**) XRD patterns of MoS$_6$ and MoS$_6$@15%Li$_7$P$_3$S$_{11}$ composite. (**b**) Raman spectra of MoS$_6$ and MoS$_6$@15%Li$_7$P$_3$S$_{11}$ composite.

The morphology and microstructure of the MoS$_6$ and MoS$_6$@15%Li$_7$P$_3$S$_{11}$ composite were observed using SEM and HRTEM (Figure 3). The particle sizes of the MoS$_6$ and MoS$_6$@15%Li$_7$P$_3$S$_{11}$ composite were in the range of 1–4 μm (Figure 3a,b). EDS mapping of the MoS$_6$@15%Li$_7$P$_3$S$_{11}$ composite clearly illustrates the good distribution of molybdenum (blue), phosphorus (purple), and sulfur (yellow) (Figure 3c). The HRTEM images further confirmed that Li$_7$P$_3$S$_{11}$ solid electrolytes were uniformly growing on the surface of the MoS$_6$ (Figure 3d,e), and this facilitates improvements in ionic transport kinetics and the formation of intimate interfacial contact. The d-spacings of 0.38, 0.35, and 0.309 nm correspond to the d_{030}, d_{202}, and d_{-211} spacing of the Li$_7$P$_3$S$_{11}$, respectively (Figure 3f) [21].

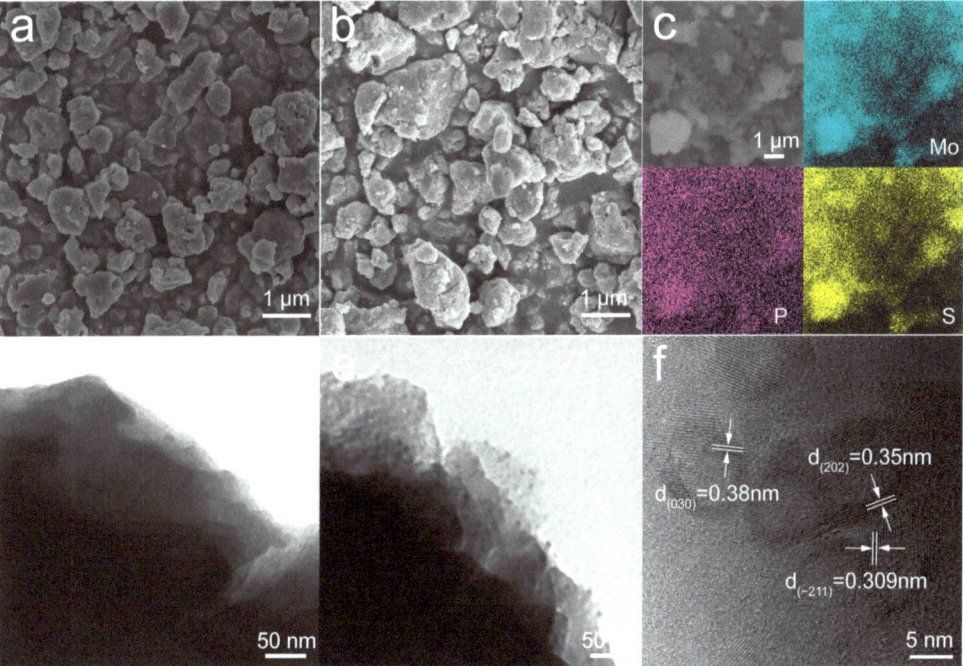

Figure 3. SEM images of (**a**) MoS$_6$ and (**b**) MoS$_6$@15% Li$_7$P$_3$S$_{11}$, EDS mapping of (**c**) MoS$_6$@15%Li$_7$P$_3$S$_{11}$ composite, TEM images of (**d**) MoS$_6$ and (**e**) MoS$_6$@15% Li$_7$P$_3$S$_{11}$, and HRTEM images of (**f**) MoS$_6$@15% Li$_7$P$_3$S$_{11}$.

The electrochemical performance of the MoS$_6$ was tested in all-solid-state lithium batteries. Two types of battery configuration were employed at 25 °C and 0.2 A g^{-1}, i.e., Li/Li$_6$PS$_5$Cl/MoS$_6$ and Li/75%Li$_2$S-24%P$_2$S$_5$-1%P$_2$O$_5$/Li$_{10}$GeP$_2$S$_{12}$/MoS$_6$ all-solid-state batteries. As Figure S2 shows, the Li/Li$_6$PS$_5$Cl/MoS$_6$ all-solid-state battery exhibited a reversible specific capacity of 361.8 mAh g^{-1} after 200 cycles, which was higher than the value exhibited by the Li/Li$_{10}$GeP$_2$S$_{12}$/75%Li$_2$S-24%P$_2$S$_5$-1%P$_2$O$_5$/MoS$_6$ all-solid-state battery (78.3 mAh g^{-1}), indicating a better capacity retention. For the purpose of comparison, cycles of 50MoS$_6$:50Li$_{10}$GeP$_2$S$_{12}$, 40MoS$_6$:50Li$_{10}$GeP$_2$S$_{12}$:10Super P, and 40MoS$_6$:40Li$_{10}$GeP$_2$S$_{12}$:20Super P were performed at 0.1 A g^{-1} and 25 °C to determine the appropriate mixing ratio, as is shown in Figure S3. The capacity retention of the 40MoS$_6$:50Li$_{10}$GeP$_2$S$_{12}$:10Super P was 544.8 mAh g^{-1} after 10 cycles; this was higher than the 394.69 mAh g^{-1} and 438.29 mAh g^{-1} values observed for the 50MoS$_6$:50Li$_{10}$GeP$_2$S$_{12}$ and 40MoS$_6$:40Li$_{10}$GeP$_2$S$_{12}$:20Super P, respectively, and indicated a high reversible discharge capacity.

To reveal the electrochemical reaction mechanisms of the MoS$_6$ and MoS$_6$@15% Li$_7$P$_3$S$_{11}$ composite, CV curves were generated for the Li/Li$_6$PS$_5$Cl/MoS$_6$ and Li/Li$_6$PS$_5$Cl/

MoS$_6$@15%Li$_7$P$_3$S$_{11}$ all-solid-state batteries for the first three cycles, and these are shown in Figure 4a,b. During the first cathodic scan, there was a reduction peak at around 1.75 V which can be attributed to the lithiation process of the MoS$_6$ and a further conversion process. During the initial anodic scanning period, the oxidation peak which occurred at 2.25 V corresponds to the de-lithiation processes and the forming of molybdenum sulfides. The CV curves of the MoS$_6$@15%Li$_7$P$_3$S$_{11}$ composite are similar to those of the MoS$_6$; they show the same electrochemical reaction process [21,28]. Even so, the redox peaks of the MoS$_6$@15%Li$_7$P$_3$S$_{11}$ composite are narrower and stronger than those of the MoS$_6$, illustrating enhancements in the reversibility and electrochemical reaction kinetics resulting from favorable ionic diffusion. In the second cycle, the CV curves display an apparent tendency to overlap, which indicates that the MoS$_6$@15%Li$_7$P$_3$S$_{11}$ composite has excellent cycling stability. Clearly, the redox reactions of the MoS$_6$ and MoS$_6$@15%Li$_7$P$_3$S$_{11}$ composite occur within the potential window of 1.0–3.0 V, which was selected to further evaluate electrochemical performances.

Figure 4c,d present the galvanostatic discharge–charge curves of the first three cycles of the MoS$_6$ and MoS$_6$@15% Li$_7$P$_3$S$_{11}$ composite in the all-solid-state lithium batteries at 0.1 A g^{-1} and 25 °C. The all-solid-sate lithium battery with the MoS$_6$ cathodes delivered a high initial discharge capacity of 913.9 mAh g^{-1}. After coupling with the Li$_7$P$_3$S$_{11}$ solid-electrolyte thin layer, the all-solid-state lithium battery with the MoS$_6$@15%Li$_7$P$_3$S$_{11}$ composite cathodes delivered initial and reversible discharge capacities of 1083.8 mAh g^{-1} and 851.5 mAh g^{-1}, respectively (Figure 4d), which were the one of highest values observed among the many sulfide-based cathode materials, i.e., rGO-MoS$_3$, cubic FeS$_2$, Co$_9$S$_8$@Li$_7$P$_3$S$_{11}$, and MoS$_2$@Li$_7$P$_3$S$_{11}$ [10,17,19,21]. As is shown in Table S2, the initial discharge capacities of the rGO-MoS$_3$, FeS$_2$, Co$_9$S$_8$@Li$_7$P$_3$S$_{11}$, and MoS$_2$@Li$_7$P$_3$S$_{11}$ were 1241.4 mAh g^{-1}, 750 mAh g^{-1}, 633 mAh g^{-1}, and 868.4 mAh g^{-1}, respectively, while their respective reversible capacities were around 760 mAh g^{-1}, 730 mAh g^{-1}, 574 mAh g^{-1}, and 669.2 mAh g^{-1}. The increased reversible discharge capacity compared with that of the MoS$_6$ could be attributed to the better interface compatibility between the active material and the solid electrolyte. In addition, the initial Coulombic efficiency of the MoS$_6$@15%Li$_7$P$_3$S$_{11}$ composite was 73.2%, which was higher than that of the MoS$_6$ (65.6%). These values were calculated using the equation $Coulombic\ efficiency = \frac{C_{charge}}{C_{discharge}}$, where C is the specific capacity of the charge or discharge process. As is shown in Figure 4e, the MoS$_6$@15%Li$_7$P$_3$S$_{11}$ composite exhibited a remarkably stable cyclic performance with an impressive reversible specific capacity of 693.2 mAh g^{-1} after 20 cycles, while the MoS$_6$ only showed a reversible specific capacity of 517.7 mAh g^{-1}. The excellent electrochemical performance of the MoS$_6$@15%Li$_7$P$_3$S$_{11}$ composite can be attributed to the Li$_7$P$_3$S$_{11}$ thin layer, which can improve the cathode–electrolyte compatibility. In fact, the coating ratio of the Li$_7$P$_3$S$_{11}$ solid electrolyte also affected the electrochemical performance of the MoS$_6$@Li$_7$P$_3$S$_{11}$ composite (Figure S4). The Li/Li$_6$PS$_5$Cl/MoS$_6$@10%Li$_7$P$_3$S$_{11}$ and Li/Li$_6$PS$_5$Cl/MoS$_6$@20%Li$_7$P$_3$S$_{11}$ delivered initial discharges of 967.8 mAh g^{-1} and 991.4 mAh g^{-1}, respectively. After 20 cycles, the capacity retention values of the 10%Li$_7$P$_3$S$_{11}$- and 20%Li$_7$P$_3$S$_{11}$-coated electrodes were 476.1 mAh g^{-1} and 496.3 mAh g^{-1}, respectively. Obviously, the Li/Li$_6$PS$_5$Cl/MoS$_6$@15%Li$_7$P$_3$S$_{11}$ displayed the highest reversible capacity. To explore the principles of cathode kinetics and capacity variation for the Li/Li$_6$PS$_5$Cl/MoS$_6$ and Li/Li$_6$PS$_5$Cl/MoS$_6$@15%Li$_7$P$_3$S$_{11}$ all-solid-state lithium batteries, EIS was conducted with an amplitude of 15 mV from 10^6 to 0.1 Hz. The corresponding equivalent circuit model is shown in Figure 4f, in which R_e is the resistance of the electrolyte, the semicircle shows the emergence of the interfacial charge transfer resistance (R_{ct}), the constant phase angle element (CPE) is used to indicate the behavior of the non-ideal capacitance of the double-layer, and Z_w represents the Warburg resistance, which indicates that the lithium ions diffuse into the bulk electrodes [21,28]. The fitted R_e and R_{ct} results are listed in Table S3. After the first cycle, the EIS plots of the MoS$_6$ and MoS$_6$@15%Li$_7$P$_3$S$_{11}$ composite-based all-solid-state lithium batteries were straight lines. The MoS$_6$@15%Li$_7$P$_3$S$_{11}$ composite exhibited an R_e value of 75.89 Ω, which was lower than that of the MoS$_6$ (105.42 Ω). After 20 cycles, the MoS$_6$

exhibited higher R_e (384.11 Ω) and R_{ct} (52.96 Ω) values. In contrast, the MoS$_6$@15%Li$_7$P$_3$S$_{11}$ composite exhibited respective R_e and R_{ct} values of 124.79 Ω and 10.78 Ω due to the intimate interfacial contact.

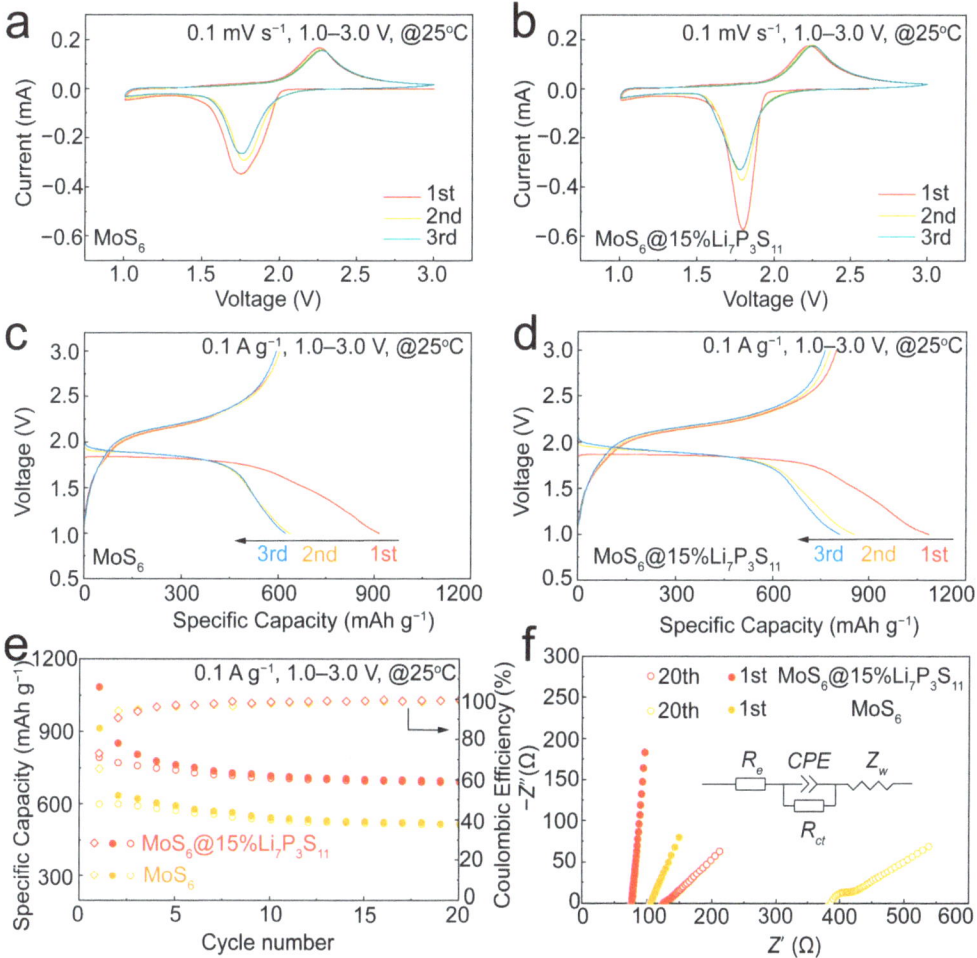

Figure 4. CV curves of (**a**) MoS$_6$ and (**b**) MoS$_6$@15%Li$_7$P$_3$S$_{11}$ composite; galvanostatic discharge/charge profiles of (**c**) MoS$_6$ and (**d**) MoS$_6$@15%Li$_7$P$_3$S$_{11}$ composite cathodes at 0.1 A g^{-1}; (**e**) cyclic performances of MoS$_6$ and MoS$_6$@15%Li$_7$P$_3$S$_{11}$ composite at 0.1 A g^{-1} within 1.0–3.0 V (the solid circles represent discharge capacities); (**f**) Nyquist plots and equivalent circuit diagram of MoS$_6$ and MoS$_6$@15%Li$_7$P$_3$S$_{11}$ composite cathodes after 1st and 20th cycles at 0.1 A g^{-1} within 1.0–3.0 V.

The rate capabilities of the MoS$_6$ and MoS$_6$@15%Li$_7$P$_3$S$_{11}$ composite cathodes were measured under various current densities ranging from 0.1 to 2 A g^{-1} (Figure 5a). The MoS$_6$@15%Li$_7$P$_3$S$_{11}$ composite cathodes exhibited superior reversible discharge capacities of 801.5, 648.1, 536.3, 454.4, and 370.8 mAh g^{-1} under current densities of 0.1, 0.2, 0.5, 1, and 2 A g^{-1}, respectively, while the MoS$_6$ only exhibited reversible discharge capacities 683.9, 516.8, 407.7, 326.2, and 255.8 mAh g^{-1}. The excellent rate capability of the MoS$_6$@15%Li$_7$P$_3$S$_{11}$ composite can be attributed to its enhanced ionic diffusivity, which led to improved electrochemical reaction kinetics. The Ragone plot shown in Figure 5b gives the relationship between the average power density and energy density. The power density of the MoS$_6$ and MoS$_6$@15%Li$_7$P$_3$S$_{11}$ composite were calculated using the equation

$$\text{power density} = \frac{ED \times \frac{m_{active\ materials}}{m_{cathode\ layer}}}{t},$$ where ED is the energy density of the active materials, m is the loading weight of the active materials or the cathode layer, and t is the duration of the discharge process. The MoS_6 and $MoS_6@15\%Li_7P_3S_{11}$ composite numbers that were used in the energy density and power density calculations are listed in Tables S4 and S5. At current densities of 0.1 and 2.0 A g^{-1}, the $MoS_6@15\%Li_7P_3S_{11}$ composite cathodes delivered energy and power densities of 588 Wh kg^{-1} and 1358 W kg^{-1}, respectively, based on the total cathode layer which is composed of the $MoS_6@15\%Li_7P_3S_{11}$ composite, $Li_{10}GeP_2S_{12}$, and super P. These values were significantly higher than the energy and power densities of 495.8 Wh kg^{-1} and 1332.2 W kg^{-1} exhibited by the MoS_6 at the same respective current densities. The long-term cycling stability of the $MoS_6@15\%Li_7P_3S_{11}$ composite cathodes at 1 A g^{-1} is further shown in Figure 5c, which shows that the cathodes exhibited a high reversible capacity of 400 mAh g^{-1} after 1000 cycles.

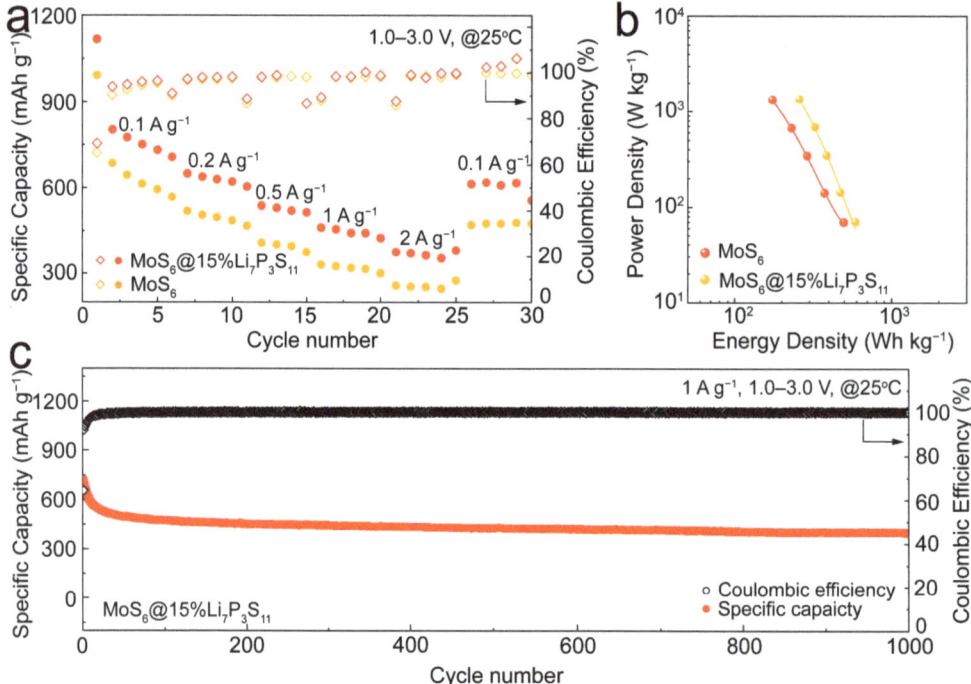

Figure 5. (**a**) Rate performances of MoS_6 and $MoS_6@15\%Li_7P_3S_{11}$ composite under different current densities. (**b**) Ragone plot deduced from the rate performances shown in (**a**). (**c**) Long-term cyclic performance of $MoS_6@15\%Li_7P_3S_{11}$ composite at 1 A g^{-1}.

CV measurements were conducted to illustrate the electrochemical reaction kinetics. The relationship between the peak current (i) and the scan rate obeys the power law: $i = av^b$. The b-value is fitted using a log(v)-log(i) plot. A b-value of 1.0 indicates a surface-mediated mode, while a b-value of 0.5 indicates a diffusion-controlled mode. The CV curves shown in Figure 6a,c show similar shapes and gradually broadened redox peaks. At the same scan rate, the curve intensity of the $MoS_6@15\%Li_7P_3S_{11}$ composite was higher than that of the MoS_6, indicating that the $MoS_6@15\%Li_7P_3S_{11}$ composite can maintain fast electrochemical kinetics with an increased scan rate. As shown in Figure 6b, the fitted b-values of the reduction peak and the oxidation peak were 0.50 and 0.67, respectively, for the $MoS_6@15\%Li_7P_3S_{11}$ composite; these values are lower than the fitted b-values of 0.62 and 0.76 recorded for the MoS_6 (Figure 6d), indicating that the electrochemical reaction kinetics are dominated by diffusion-controlled processes. This condition allows lithium ions to become fully intercalated and thereby realize high reversible capacity [17,29].

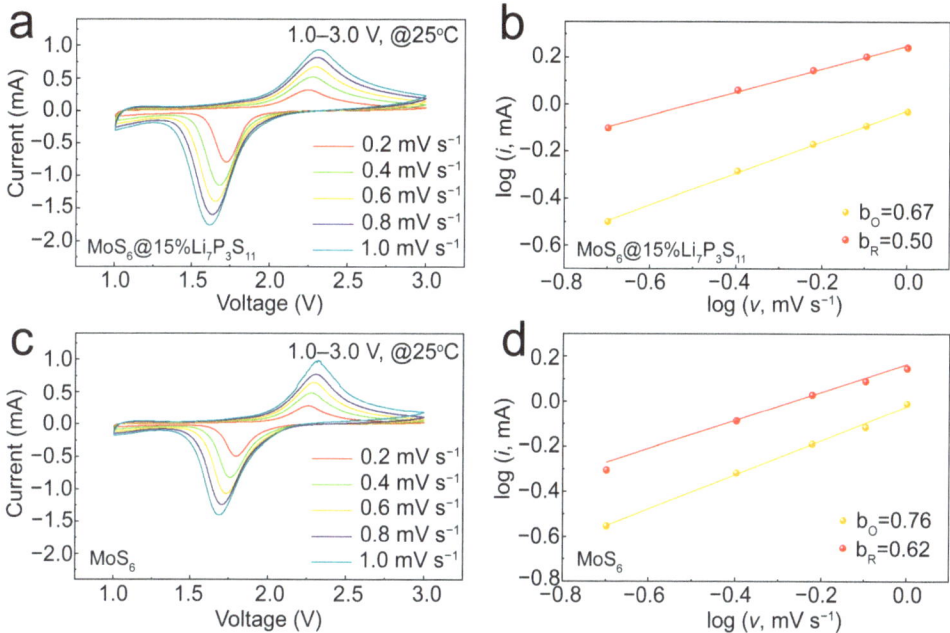

Figure 6. CV curves at different scan rates for (**a**) MoS$_6$@15%Li$_7$P$_3$S$_{11}$ composite and (**c**) MoS$_6$. The log (peak current) vs. log (scan rate) fitted plots at reduction and oxidation peaks of (**b**) MoS$_6$@15%Li$_7$P$_3$S$_{11}$ composite and (**d**) MoS$_6$.

To further quantify the ionic transport kinetics of the Li/Li$_6$PS$_5$Cl/MoS$_6$@15%Li$_7$P$_3$S$_{11}$ and Li/Li$_6$PS$_5$Cl/MoS$_6$ all-solid-state batteries, the GITT was used and calculated were made to determine the lithium ion diffusion coefficients at 1 A g^{-1} and 1.0–3.0 V. As is shown in Figure 7, the ion diffusion coefficient range of the MoS$_6$@15%Li$_7$P$_3$S$_{11}$ composite was calculated to be 10^{-11}–10^{-10} cm^2 s^{-1}, which is higher than that of the MoS$_6$ (10^{-12}–10^{-11} cm^2 s^{-1}). Obviously, the ionic diffusivity of the MoS$_6$@15%Li$_7$P$_3$S$_{11}$ composite was enhanced significantly; this is beneficial for the rapid transportation of lithium ions and thus significantly improved the electrochemical performances related to high discharge capacity and rate capability.

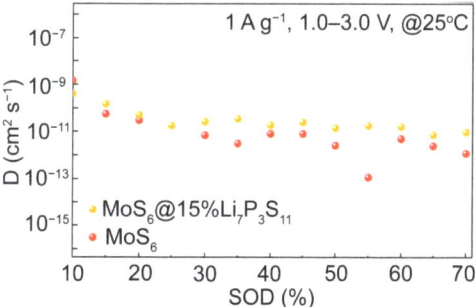

Figure 7. GITT plot of Li/Li$_6$PS$_5$Cl/MoS$_6$@15%Li$_7$P$_3$S$_{11}$ composite and Li/Li$_6$PS$_5$Cl/MoS$_6$ all-solid-state lithium batteries.

4. Conclusions

In summary, the micro-sized MoS$_6$ with high S$_2^{2-}$ content demonstrated a high initial theoretical capacity of 1117 mAh g^{-1}. After coating the MoS$_6$ with Li$_7$P$_3$S$_{11}$ solid electrolytes, the resulting MoS$_6$@15%Li$_7$P$_3$S$_{11}$ composite was able to achieve an improved ion diffusion coefficient range of 10^{-11}–10^{-10} cm^2 s^{-1}, which was higher than the value of the MoS$_6$ (10^{-12}–10^{-11} cm^2 s^{-1}). The Li/Li$_6$PS$_5$Cl/MoS$_6$@15%Li$_7$P$_3$S$_{11}$ all-solid-sate lithium batteries showed a high initial discharge capacity of 1083.8 mAh g^{-1} at 0.1 A g^{-1} and a long cycle life of 1000 cycles at 1A g^{-1} and 25 °C. In addition, the MoS$_6$@15%Li$_7$P$_3$S$_{11}$ composite displayed a high energy density of 588 Wh kg^{-1} and a power density of 1358 W kg^{-1} based on the total cathode layer. This contribution provides a new sulfide-based cathode material with high specific capacity and a superior ion diffusion coefficient which has promising application potential for use in all-solid-state lithium batteries.

Supplementary Materials: The follow supporting information can be downloaded at: https://www.mdpi.com/article/10.3390/batteries9110560/s1, Figure S1: XRD pattern of (NH$_4$)$_2$Mo$_2$S$_{12}$; Figure S2: Cyclic performances of Li/Li$_6$PS$_5$Cl/MoS$_6$ and Li/75%Li$_2$S-24%P$_2$S$_5$-1%P$_2$O$_5$/Li$_{10}$GeP$_2$S$_{12}$/MoS$_6$ all-solid-state batteries under 0.2 A g^{-1}; Figure S3: Cyclic performances of all-solid-state batteries with 50MoS$_6$:50Li$_{10}$GeP$_2$S$_{12}$, 40MoS$_6$:50Li$_{10}$GeP$_2$S$_{12}$:10Super P, and 40MoS$_6$:40Li$_{10}$GeP$_2$S$_{12}$:20Super P cathodes under 0.1 A g^{-1}; Figure S4: Cyclic performances of MoS$_6$@10%Li$_7$P$_3$S$_{11}$, MoS$_6$@15%Li$_7$P$_3$S$_{11}$, and MoS$_6$@20%Li$_7$P$_3$S$_{11}$ cathodes under 0.1 A g^{-1}; Table S1: Inductively coupled plasma emission spectrometer analysis of MoS$_6$; Table S2: Electrochemical performances comparisons of various active materials in all-solid-state-batteries; Table S3: EIS fitting results of MoS$_6$ and MoS$_6$@15%Li$_7$P$_3$S$_{11}$ under 0.1 A g^{-1} after 20 cycles; Table S4: MoS$_6$ values used in energy density and power density calculations; Table S5: MoS$_6$@15%Li$_7$P$_3$S$_{11}$ composite values used in energy density and power density calculations.

Author Contributions: Conceptualization: X.Y.; methodology: M.C. and X.Y.; general experimental investigation and electrochemical analysis: M.C., M.Y., W.X., F.T. and G.L.; characterization: M.C.; supervision and research guidance: P.C., T.W. and X.Y.; funding acquisition: X.Y. All authors have read and agreed to the published version of the manuscript.

Funding: This work was supported by National Natural Science Foundation of China (Grant No. U1964205), the Jiangsu Provincial S&T Innovation Special Programme for Carbon Peak and Carbon Neutrality (Grant No. BE2022007), and the Youth Innovation Promotion Association CAS (Y2021080).

Data Availability Statement: The data presented in this study are available on request from the corresponding author. The data are not publicly available due to privacy policy.

Conflicts of Interest: The authors declare no conflict of interest.

References

1. Zhao, W.G.; Zou, L.F.; Jia, H.P.; Zheng, J.M.; Wang, D.H.; Song, J.H.; Hong, C.Y.; Liu, R.; Xu, W.; Yang, Y.; et al. Optimized Al Doping Improves Both Interphase Stability and Bulk Structural Integrity of Ni-Rich NMC Cathode Materials. *ACS Appl. Energy Mater.* **2020**, *3*, 3369–3377. [CrossRef]
2. Bruce, P.G.; Freunberger, S.A.; Hardwick, L.J.; Tarascon, J.M. Li-O$_2$ and Li-S batteries with high energy storage. *Nat. Mater.* **2012**, *11*, 19–29. [CrossRef]
3. Kato, Y.; Hori, S.; Saito, T.; Suzuki, K.; Hirayama, M.; Mitsui, A.; Yonemura, M.; Iba, H.; Kanno, R. High-power all-solid-state batteries using sulfide superionic conductors. *Nat. Energy* **2016**, *1*, 16030. [CrossRef]
4. Yi, Y.K.; Hai, F.; Guo, J.Y.; Tian, X.L.; Zheng, S.T.; Wu, Z.D.; Wang, T.; Li, M.T. Progress and Prospect of Practical Lithium-Sulfur Batteries Based on Solid-Phase Conversion. *Batteries* **2023**, *9*, 27. [CrossRef]
5. Shi, C.M.; Alexander, G.V.; O'Neill, J.; Duncan, K.; Godbey, G.; Wachsman, E.D. All-Solid-State Garnet Type Sulfurized Polyacrylonitrile/Lithium-Metal Battery Enabled by an Inorganic Lithium Conductive Salt and a Bilayer Electrolyte Architecture. *ACS Energy Lett.* **2023**, *8*, 1803–1810. [CrossRef]
6. Kamaya, N.; Homma, K.; Yamakawa, Y.; Hirayama, M.; Kanno, R.; Yonemura, M.; Kamiyama, T.; Kato, Y.; Hama, S.; Kawamoto, K.; et al. A lithium superionic conductor. *Nat. Mater.* **2011**, *10*, 682–686. [CrossRef] [PubMed]
7. Boulineau, S.; Courty, M.; Tarascon, J.M.; Viallet, V. Mechanochemical synthesis of Li-argyrodite Li$_6$PS$_5$X (X = Cl, Br, I) as sulfur-based solid electrolytes for all solid state batteries application. *Solid State Ion.* **2012**, *221*, 1–5. [CrossRef]
8. Grayfer, E.D.; Pazhetnov, E.M.; Kozlova, M.N.; Artemkina, S.B.; Fedorov, V.E. Anionic Redox Chemistry in Polysulfide Electrode Materials for Rechargeable Batteries. *ChemSusChem* **2017**, *10*, 4805–4811. [CrossRef]

9. Sakuda, A.; Ohara, K.; Fukuda, K.; Nakanishi, K.; Kawaguchi, T.; Arai, H.; Uchimoto, Y.; Ohta, T.; Matsubara, E.; Ogumi, Z.; et al. Amorphous Metal Polysulfides: Electrode Materials with Unique Insertion/Extraction Reactions. *J. Am. Chem. Soc.* **2017**, *139*, 8796–8799. [CrossRef]
10. Yersak, T.A.; Macpherson, H.A.; Kim, S.C.; Le, V.D.; Kang, C.S.; Son, S.B.; Kim, Y.H.; Trevey, J.E.; Oh, K.H.; Stoldt, C.; et al. Solid State Enabled Reversible Four Electron Storage. *Adv. Energy Mater.* **2013**, *3*, 120–127. [CrossRef]
11. Wang, T.S.; Hu, P.; Zhang, C.J.; Du, H.P.; Zhang, Z.H.; Wang, X.G.; Chen, S.G.; Xiong, J.W.; Cui, G.L. Nickel Disulfide-Graphene Nanosheets Composites with Improved Electrochemical Performance for Sodium Ion Battery. *ACS Appl. Mater Interfaces* **2016**, *8*, 7811–7817. [CrossRef]
12. Arsentev, M.; Missyul, A.; Petrov, A.V.; Hammouri, M. TiS$_3$ Magnesium Battery Material: Atomic-Scale Study of Maximum Capacity and Structural Behavior. *J. Phys. Chem. C* **2017**, *121*, 15509–15515. [CrossRef]
13. Dong, S.W.; Sheng, L.; Wang, L.; Liang, J.; Zhang, H.; Chen, Z.H.; Xu, H.; He, X.M. Challenges and Prospects of All-Solid-State Electrodes for Solid-State Lithium Batteries. *Adv. Funct. Mater.* **2023**, 2304371. [CrossRef]
14. Chong, P.D.; Zhou, Z.W.; Wang, K.H.; Zhai, W.H.; Li, Y.F.; Wang, J.B.; Wei, M.D. The Stabilizing of 1T-MoS$_2$ for All-Solid-State Lithium-Ion Batteries. *Batteries* **2023**, *9*, 26. [CrossRef]
15. Wang, Y.F.; Goikolea, E.; de Larramendi, I.R.; Lanceros-Méndez, S.; Zhang, Q. Recycling methods for different cathode chemistries—A critical review. *J. Energy Storage* **2022**, *56*, 106053. [CrossRef]
16. Shchegolkov, A.V.; Komarov, F.F.; Lipkin, M.S.; Milchanin, O.V.; Parfimovich, I.D.; Shchegolkov, A.V.; Semenkova, A.V.; Velichko, A.V.; Chebotov, K.D.; Nokhaeva, V.A. Synthesis and Study of Cathode Materials Based on Carbon Nanotubes for Lithium-Ion Batteries. *Inorg. Mater. Appl. Res.* **2021**, *12*, 1281–1287. [CrossRef]
17. Zhang, Q.; Ding, Z.G.; Liu, G.Z.; Wan, H.L.; Mwizerwa, J.P.; Wu, J.H.; Yao, X.Y. Molybdenum trisulfide based anionic redox driven chemistry enabling high-performance all-solid-state lithium metal batteries. *Energy Storage Mater.* **2019**, *23*, 168–180. [CrossRef]
18. Wu, F.X.; Yushin, G. Conversion cathodes for rechargeable lithium and lithium-ion batteries. *Energy Environ. Sci.* **2017**, *10*, 435–459. [CrossRef]
19. Xu, R.C.; Wang, X.L.; Zhang, S.Z.; Xia, Y.; Xia, X.H.; Wu, J.B.; Tu, J.P. Rational coating of Li$_7$P$_3$S$_{11}$ solid electrolyte on MoS$_2$ electrode for all-solid-state lithium ion batteries. *J. Power Sources* **2018**, *374*, 107–112. [CrossRef]
20. Ye, H.; Ma, L.; Zhou, Y.; Wang, L.; Han, N.; Zhao, F.; Deng, J.; Wu, T.; Li, Y.; Lu, J. Amorphous MoS$_3$ as the sulfur-equivalent cathode material for room-temperature Li-S and Na-S batteries. *Proc. Natl. Acad. Sci. USA* **2017**, *114*, 13091–13096. [CrossRef]
21. Yao, X.; Liu, D.; Wang, C.; Long, P.; Peng, G.; Hu, Y.S.; Li, H.; Chen, L.; Xu, X. High-Energy All-Solid-State Lithium Batteries with Ultralong Cycle Life. *Nano Lett.* **2016**, *16*, 7148–7154. [CrossRef] [PubMed]
22. Ye, H.L.; Wang, L.; Deng, S.; Zeng, X.Q.; Nie, K.Q.; Duchesne, P.N.; Wang, B.; Liu, S.; Zhou, J.H.; Zhao, F.P.; et al. Amorphous MoS$_3$ Infiltrated with Carbon Nanotubes as an Advanced Anode Material of Sodium-Ion Batteries with Large Gravimetric, Areal, and Volumetric Capacities. *Adv. Energy Mater.* **2017**, *7*, 160102. [CrossRef]
23. Wang, X.; Du, K.; Wang, C.; Ma, L.; Zhao, B.; Yang, J.; Li, M.; Zhang, X.X.; Xue, M.; Chen, J. Unique Reversible Conversion-Type Mechanism Enhanced Cathode Performance in Amorphous Molybdenum Polysulfide. *ACS Appl. Mater. Interfaces* **2017**, *9*, 38606–38611. [CrossRef] [PubMed]
24. Horner, J.S.; Whang, G.; Ashby, D.S.; Kolesnichenko, I.V.; Lambert, T.N.; Dunn, B.S.; Talin, A.A.; Roberts, S.A. Electrochemical Modeling of GITT Measurements for Improved Solid-State Diffusion Coefficient Evaluation. *ACS Appl. Energy Mater.* **2021**, *4*, 11460–11469. [CrossRef]
25. Shen, Z.; Cao, L.; Rahn, C.D.; Wang, C.Y. Least Squares Galvanostatic Intermittent Titration Technique (LS-GITT) for Accurate Solid Phase Diffusivity Measurement. *J. Electrochem. Soc.* **2013**, *160*, A1842–A1846. [CrossRef]
26. Mabayoje, O.; Wygant, B.R.; Wang, M.; Liu, Y.; Mullins, C.B. Sulfur-Rich MoS$_6$ as an Electrocatalyst for the Hydrogen Evolution Reaction. *ACS Appl. Energy Mater.* **2018**, *1*, 4453–4458. [CrossRef]
27. Tran, P.D.; Tran, T.V.; Orio, M.; Torelli, S.; Truong, Q.D.; Nayuki, K.; Sasaki, Y.; Chiam, S.Y.; Yi, R.; Honma, I.; et al. Coordination polymer structure and revisited hydrogen evolution catalytic mechanism for amorphous molybdenum sulfide. *Nat. Mater.* **2016**, *15*, 640–646. [CrossRef]
28. Huang, J. Diffusion impedance of electroactive materials, electrolytic solutions and porous electrodes: Warburg impedance and beyond. *Electrochim. Acta* **2018**, *281*, 170–188. [CrossRef]
29. Li, W.B.; Huang, J.F.; Cao, L.Y.; Feng, L.L.; Yao, C.Y. Controlled construction of 3D self-assembled VS$_4$ nanoarchitectures as high-performance anodes for sodium-ion batteries. *Electrochim. Acta* **2018**, *274*, 334–342. [CrossRef]

Disclaimer/Publisher's Note: The statements, opinions and data contained in all publications are solely those of the individual author(s) and contributor(s) and not of MDPI and/or the editor(s). MDPI and/or the editor(s) disclaim responsibility for any injury to people or property resulting from any ideas, methods, instructions or products referred to in the content.

MDPI AG
Grosspeteranlage 5
4052 Basel
Switzerland
Tel.: +41 61 683 77 34

Batteries Editorial Office
E-mail: batteries@mdpi.com
www.mdpi.com/journal/batteries

Disclaimer/Publisher's Note: The statements, opinions and data contained in all publications are solely those of the individual author(s) and contributor(s) and not of MDPI and/or the editor(s). MDPI and/or the editor(s) disclaim responsibility for any injury to people or property resulting from any ideas, methods, instructions or products referred to in the content.